AF347518

Coordination Polymers and Metal-Organic Frameworks: Structures and Applications—A Themed Issue in Honor of Professor Christoph Janiak on the Occasion of His 60th Birthday

Coordination Polymers and Metal-Organic Frameworks: Structures and Applications—A Themed Issue in Honor of Professor Christoph Janiak on the Occasion of His 60th Birthday

Editor

Catherine E. Housecroft

MDPI • Basel • Beijing • Wuhan • Barcelona • Belgrade • Manchester • Tokyo • Cluj • Tianjin

Editor
Catherine E. Housecroft
University of Basel
Switzerland

Editorial Office
MDPI
St. Alban-Anlage 66
4052 Basel, Switzerland

This is a reprint of articles from the Special Issue published online in the open access journal *Chemistry* (ISSN 2624-8549) (available at: https://www.mdpi.com/journal/chemistry/special_issues/for_Christoph_Janiak).

For citation purposes, cite each article independently as indicated on the article page online and as indicated below:

LastName, A.A.; LastName, B.B.; LastName, C.C. Article Title. *Journal Name* **Year**, *Volume Number*, Page Range.

ISBN 978-3-0365-1958-6 (Hbk)
ISBN 978-3-0365-1959-3 (PDF)

Contents

About the Editor

Catherine E. Housecroft is a Professor of Chemistry in the Department of Chemistry at the University of Basel, Switzerland. Her research group, run jointly with Professor Edwin Constable, is engaged in aspects of materials, coordination, and structural and supramolecular chemistries. Much of the emphasis is upon the design and development of materials which will contribute to mankind achieving the 17 Sustainable Development Goals designated in 2015 by the United Nations General Assembly. Compounds of interest range from coordination compounds which can absorb or emit light to coordination networks which can act as hosts for small molecules.

Preface to "Coordination Polymers and Metal-Organic Frameworks: Structures and Applications—A Themed Issue in Honor of Professor Christoph Janiak on the Occasion of His 60th Birthday"

I consider the invitation by Professor Catherine Housecroft to write this preface as one of the greatest moments in my professional life. The reason is simple: The book at hand is devoted to Professor Christoph Janiak on the occasion of his 60th birthday! Professor Janiak is one of the great scientific names in inorganic chemistry with excellent contributions to the fields of metallosupramolecular chemistry, metal–organic frameworks (MOFs), inorganic materials from ionic liquids and hybrid (inorganic/organic) materials. I have attended some of his lectures in international conferences, but I never worked or discussed with him. However, I feel that I know him quite well through his innovative research, which has influenced many inorganic chemistry groups around the world.

The1950s were considered the renaissance of inorganic chemistry. Since then, the field has grown rapidly in many new directions. It reaches into solid-state chemistry, supramolecular chemistry, polymer chemistry, biochemistry, organic synthesis via organometallics, homogeneous and heterogeneous catalysis, energy storage and energy-related technology, analytical, physical and theoretical chemistry. The challenges that the world presents to inorganic chemists have never been greater; nor have the contributions that inorganic chemists make been greater than at this time. Professor Janiak has been one of the leaders responsible for this "transition".

Following BSc and MSc Chemistry studies at the Technical University (TU) of Berlin and the University of Oklahoma (USA), PhD research at the TU of Berlin, prestigious post-doctoral positions at Cornell University (working with Nobel Laureate Roald Hoffman) and in industry, and habilitation at TU, Professor Janiak was appointed to the Associate Professor level at the University of Freiburg in 1996. Since 2010, he has been Full Professor in Bioinorganic Chemistry and Catalysis at the Heinrich-Heine University of Düsseldorf. He has received many awards, including a Heisenberg fellowship (1997), and he was visiting Professor at the Universities of Angers, France, during 2006 and Wuhan (China) in the period 2015–2018. He is currently a Guest Professor at the Hoffmann Institute of Advanced Materials at Shenzhen Polytechnic in China. He is member of the Editorial and Advisory Boards of several journals, including *CrystEngComm* (Royal Society of Chemistry) and *Inorganica Chimica Acta*. Professor Janiak is co-author of three inorganic chemistry textbooks which are very popular in German Universities. He is Fellow of the Royal Society of Chemistry(FRSC).

The research merits of Professor Janiak are dizzying to normal scientists! He has authored and co-authored more than 550 research papers in peer-reviewed journals ($\sim$140 in the period 2019 up to now), and his h-index is about 80. His research has received more than 27,000 heterocitations! The research interests of Professor Janiak involve the synthesis and properties of MOFs and porous organic polymers, mixed-matrix membranes, metal nanoparticles, ionic liquids and catalysis. Three of his seminal reviews have influenced thousands of inorganic chemists working in the fields of metallosupramolecular chemistry, crystal engineering and porous coordination networks. These reviews have been cited close to 6700 times. The first review (*J. Chem. Soc., Dalton Trans.*, **2000**, 3885–3896) is the "Bible" of most inorganic chemists who use single-crystal X-ray crystallography as a research tool. It gives a full geometrical analysis on π-π stacking in metal complexes with aromatic nitrogen-containing ligands. The review was based on a Cambridge Structural Database search and reveals that a face-to-face π-π alignment, where most of the ring-plane area overlaps, is a rare phenomenon. The usual π interaction is an offset or slipped stacking, i.e., the rings are parallel displaced. Such parallel structures also have contributions from π-σ attraction, which increases with increasing offset. This work also clarifies the difference between π-π interactions and C-H$\cdots\pi$ attraction forces. The second review-perspective (*Dalton Trans.*, **2003**, 2781–2804) is the main literature source of scientists who prepare and characterize coordination polymers. The development in the field was assessed in terms of properties in the areas of catalysis, chirality, conductivity, magnetism, spin-crossover behavior, luminescence, non-linear optics (NLO) and porosity or zeolitic behavior; such properties are the basis for potential applications. An important conclusion was the prediction that applications of coordination polymers do not necessarily include structurally authenticated compounds but metal–ligand combinations that "work", irrespective of a known three-dimensional structure obtained through single-crystal X-ray crystallography. The third review-perspective summarizes MOFs, Materials Institute Lavoisier (MILs), iso-reticular MOFs (IR-MOFs), porous coordination networks, zeolitic MOFs and porous coordination polymers with selected examples of concepts for linkers, syntheses, post-synthesis modifications, metal nanoparticle formations in MOFs, porosity and zeolitic behavior. Most importantly, the potential applications of such materials in gas storage (hydrogen, carbon dioxide, methane, etc.), catalysis, conductivity and luminescence are outlined. I anticipate that another review of Professor Janiak, published almost one year ago (*Chem. Soc. Rev.*, **2020**, *49*, 2751–2798), will become another guide for scientists working in MOF chemistry; it has already received more than 80 citations. This review describes MOFs which contain open metal sites, coordinatively unsaturated sites or open coordination sites. It combines for the first time all aspects of open metal sites in MOFs. The presence of open metal sites in MOFs causes enhanced interactions with guest molecules because they are the primary adsorption sites for hydrogen, carbon monoxide, carbon dioxide, nitrogen monoxide and acetylene (but not for methane and possibly sulfur dioxide).

The contents of the above-mentioned reviews reflect many of the research activities in Janiak's group. However, he also works in other projects. I would need a whole chapter in this book to briefly describe his additional activities. I confine myself to two exciting projects. The first (see, for example, the paper in *Pharmaceutics*, **2021**, *13*, 295) describes the augmented therapeutic potential of the glutaminase inhibitor CB839 in glioblastoma stem cells using gold nanoparticle delivery. This is the first attempt to combine CB839 with gold nanotechnology, and scientists hope to overcome delivery hurdles and increase bioavailability in target sites. The second (see, *Inorg. Chem.*, **2020**, *59*, 17028–17037) exploits the fluorescent-dye properties of uranyl complexes with 8-hydroxyquinoline and its derivatives as ligands to fabricate blue light-emitting diodes employing solution processing methods.

I am sure that Professor Janiak will produce new exciting results and make innovations in various aspects of inorganic and materials chemistry in the future. Since he has been a constant inspiration to us, this book is a small tribute to him as a person and to his excellent science. Together with his former and current students, his colleagues and the members of the whole chemical community, I wish Professor Christoph Janiak good health, happiness and new scientific achievements in the coming years!

Spyros P. Perlepes
University of Patras, Patras, Greece

 *chemistry*

Editorial

Coordination Polymers and Metal-Organic Frameworks: Structures and Applications—A Themed Issue in Honor of Professor Christoph Janiak on the Occasion of His 60th Birthday

Catherine E. Housecroft

Department of Chemistry, University of Basel, Mattenstrasse 24a, BPR 1096, 4058 Basel, Switzerland; catherine.housecroft@unibas.ch

Citation: Housecroft, C.E. Coordination Polymers and Metal-Organic Frameworks: Structures and Applications—A Themed Issue in Honor of Professor Christoph Janiak on the Occasion of His 60th Birthday. *Chemistry* **2021**, *3*, 831–833. https://doi.org/10.3390/chemistry3030060

Received: 14 July 2021
Accepted: 27 July 2021
Published: 28 July 2021

Publisher's Note: MDPI stays neutral with regard to jurisdictional claims in published maps and institutional affiliations.

This themed issue of *Chemistry* is in honor of Professor Christoph Janiak on the occasion of his 60th birthday, and celebrates his innovative contributions to the fields of supramolecular chemistry, coordination polymers, networks and metal-organic frameworks, inorganic/organic hybrid materials and inorganic materials from ionic liquids.

Professor Janiak currently holds a chair in Bioinorganic Chemistry and Catalysis at the Heinrich Heine University of Düsseldorf and his research interests span metal-organic frameworks (MOFs), coordination polymers, chirality, supramolecular interactions and metal nanoparticles in ionic liquids. Two of his seminal reviews have influenced many of us working in the fields of heterocyclic chemistry and coordination networks: "A critical account on π–π stacking in metal complexes with aromatic nitrogen-containing ligands" (*Dalton Trans.* 2000, 3885) and "Engineering coordination polymers towards applications" (*Dalton Trans.* 2003, 2781), which have been cited close to 6000 times.

The issue begins with a preface written by Spyros Perlepes, which provides a personalized account of the achievements of Christoph Janiak. This sets the scene for many of the topics covered in the eleven contributions described in this editorial.

Baca and Kögerler have contributed a review dealing with cluster-based coordination polymers in which the clusters act as nodes in the assemblies. The focus is on recent developments in architectures containing Mn/Fe-oxo pivalate and isobutyrate building blocks, and on the synthetic strategies and the magnetic behavior of these coordination polymers [1].

Carboxylate linkers are popular choices in the assembly of coordination polymers and networks, and therefore it is not surprising that these donors feature in several of the papers in this themed issue. The metal-organic framework UiO-66 is formed from $ZrCl_4$ and terephthalic acid, and is a highly stable assembly. Yang, Schröder and coworkers have demonstrated the ability of a defective form of UiO-66 to take up I_2, and this investigation is relevant to the uptake and storage of radioactive I_2 [2]. Additionally related to adsorption of toxic small molecules (in this case CS_2), Li and coworkers have described the conjugate bases of 4′,4‴,4‴′′,4‴′′′-(ethene-1,1,2,2-tetrayl)tetrakis(([1,1′-biphenyl]-4-carboxylic acid)), H_4tcbpe, and a tetrafluoro-derivative, H_4tcbpe-f, which act as tetratopic ligands to assemble $[Zr_6(\mu_3\text{-}O)_4(OH)_8(tcbpe)_2(H_2O)_4]$ and $[Zr_6(\mu_3\text{-}O)_4(OH)_8(tcbpe\text{-}f)_2(H_2O)_4]$; these assemblies are strongly luminescent and have been applied as fluorescent sensors for the detection of CS_2 [3]. Chhetri, Chen and coworkers have detailed a series of coordination polymers constructed from copper(II) or cadmium(II) combined with semi-rigid N,N'-bis(3-pyridyl)terephthalamide and dicarboxylic acids; the assemblies encompass 1D, 2D and 3D architectures [4]. The chelidonate ligand has the potential to offer several different metal-binding modes. In their paper, Carballo et al. have combined chelidonic acid and either 4,4′-bipyridine or 1,2-bis(4-pyridyl)ethane with zinc(II) and copper(II) to give a series of coordination polymers in which the chelidonate dianion acts either as a non-coordinating counterion, or as a bridging ligand [5].

Linkers incorporating pyridine donors are also popular among coordination polymer chemists. The work by Carballo et al. [5] introduced these donors, and this is complemented by a report from Andruh, Avarvari and coworkers concerning the coordination behavior of three bis(pyridine) ligands containing benzothiadiazole spacers. Both discrete and polymeric assemblies are described with zinc(II) and silver(I) [6]. Two contributions focus on terpyridines. Harrowfield and coworkers provide a detailed analysis of the weak interactions in the single crystal structures of 33 metal(II) complexes incorporating twelve 4'-functionalized 2,2':6',2''-terpyridine (tpy) ligands. The structural and packing details that are revealed in this investigation should be valuable in the future design of functional materials based on tpy [7]. While 2,2':6',2''-terpyridine is the ubiquitous isomer of terpyridine, 3,2':6',3''-terpyridine has been far less studied. We have contributed to this issue of *Chemistry* with an investigation of the effects of the conformational flexibility of 3,2':6',3''-tpy (functionalized in the 4'-position with a biphenyl domain) on the structures and crystal packing of 1D coordination polymers bearing $Cu_2(\mu\text{-OAc})_4$ paddle-wheel units [8].

The final three contributions are focused either on the magnetic or electrochemical properties of arrays with different nuclearities or on organic donor-acceptor Stenhouse adducts, respectively. Cimpoesu, Ferbinteanu and coworkers describe a series of pyrazolato-bridged, multinuclear copper(II) coordination compounds; the magnetic properties of these Cu_3, Cu_6 and Cu_7-containing species are dominated by the strong antiferromagnetism across the pyrazolate bridges [9]. Neville and coworkers have investigated the use of the *N*-(pyridin-4-yl)benzamide ligand in {Fe(II)Pd(II)} and {Fe(II)Pt(II)}-containing 2D Hofmann-type frameworks to facilitate spin crossover cooperativity through hydrogen bonding involving the amide moiety [10]. A departure from the main theme of the issue comes with an investigation of donor–acceptor Stenhouse adducts. These represent a class of photoswitches that undergo light-initiated triene cyclization. However, D'Alessandro and coworkers have demonstrated that electrochemical oxidation is an effective alternative stimulus [11].

I would like to take this opportuniy to thank all the teams of researchers who have contributed to this issue, which is dedicated to Christoph Janiak. Together, we express our sincere good wishes on the occasion of Christoph's 60th birthday.

References

1. Baca, S.; Kögerler, P. Cluster-Based Coordination Polymers of Mn/Fe-Oxo Pivalates and Isobutyrates. *Chemistry* **2021**, *3*, 314–326. [CrossRef]
2. Maddock, J.; Kang, X.; Liu, L.; Han, B.; Yang, S.; Schröder, M. The Impact of STructural Defects on Iodine Adsorption in UiO-66. *Chemistry* **2021**, *3*, 525–531. [CrossRef]
3. Velasco, E.; Osumi, Y.; Teat, S.J.; Jensen, S.; Tan, K.; Thonhauser, T.; Li, J. Fluorescent Detection of Carbon Disulfide by a Highly Emissive and Robust Isoreticular Series of Zr-Based Luminescent Metal Organic Frameworks (LMOFs). *Chemistry* **2021**, *3*, 327–337. [CrossRef]
4. Chen, C.-J.; Chen, C.-L.; Liu, Y.-H.; Lee, W.-T.; Hu, J.-H.; Chhetri, P.M.; Chen, J.-D. Coordination Polymers Constructed from Semi-Rigid *N,N'*-Bis(3-pyridyl)terephthalamide and Dicarboxylic Acids: Effect of Ligand Isomerism, Flexibility, and Identity. *Chemistry* **2021**, *3*, 1–12. [CrossRef]
5. Carballo, R.; Lago, A.B.; Pino-Cuevas, A.; Gómez-Paz, O.; Fernández-Hermida, N.; Vázquez-López, E.M. Neutral and Cationic Chelidonate Coordination Polymers with *N,N'*-Bridging Ligands. *Chemistry* **2021**, *3*, 256–268. [CrossRef]
6. Mocanu, T.; Plyuta, N.; Cauchy, T.; Andruh, M.; Avarvari, N. Dimensionality Control in Crystalline Zinc(II) and Silver(I) Complexes with Ditopic Benzothiadiazole-Dipyridine Ligands. *Chemistry* **2021**, *3*, 269–287. [CrossRef]
7. Lee, Y.H.; Kim, J.Y.; Kusumoto, S.; Ohmagari, H.; Hasegawa, M.; Thuéry, P.; Harrowfield, J.; Hayami, S.; Kim, Y. Functionalised Terpyridines and Their Metal Complexes—Solid-State Interactions. *Chemistry* **2021**, *3*, 199–227. [CrossRef]
8. Rocco, D.; Novak, S.; Prescimone, A.; Constable, E.C.; Housecroft, C.E. Manipulating the Conformation of 3,2':6',3''-Terpyridine in [Cu₂(μ-OAc)₄(3,2':6',3''-tpy)]ₙ 1D-Polymers. *Chemistry* **2021**, *3*, 182–198. [CrossRef]
9. Toader, A.M.; Buta, M.C.; Cimpoesu, F.; Toma, A.-I.; Zalaru, C.M.; Cinteza, L.O.; Ferbinteanu, M. New Syntheses, Analytic Spin Hamiltonians, Structural and Computational Characterization for a Series of Tri-, Hexa- and Hepta-Nuclear Copper(II) Complexes with Prototypic Patterns. *Chemistry* **2021**, *3*, 411–439. [CrossRef]

10. Ong, X.; Ahmed, M.; Xu, L.; Brennan, A.T.; Hua, C.; Zenere, K.A.; Xie, Z.; Kepert, C.J.; Powell, B.J.; Neville, S.M. Spin-Crossover 2-D Hofmann Frameworks Incorporating an Amide-Functionalized Ligand: *N*-(pyridin-4-yl)benzamide. *Chemistry* **2021**, *3*, 360–372. [CrossRef]
11. Shepherd, N.D.; Moore, H.S.; Beves, J.E.; D'Alessandro, D.M. Electrochemical Switching of First-Generation Donor-Acceptor Stenhouse Adducts (DASAs): An Alternative Stimulus for Triene Cyclisation. *Chemistry* **2021**, *3*, 728–733. [CrossRef]

Review

Cluster-Based Coordination Polymers of Mn/Fe-Oxo Pivalates and Isobutyrates [†]

Svetlana Baca [1,*] and Paul Kögerler [2]

[1] Institute of Applied Physics, Academiei 5, 2028 Chisinau, Moldova
[2] Institute of Inorganic Chemistry, RWTH Aachen University, Landoltweg 1, 52074 Aachen, Germany; paul.koegerler@ac.rwth-aachen.de
* Correspondence: sbaca_md@yahoo.com; Tel.: +373-2-273-9661
† Dedicated to Professor Christoph Janiak on the occasion of his 60th birthday.

Abstract: Polynuclear coordination clusters can be readily arranged in cluster-based coordination polymers (CCPs) by appropriate bridging linkers. This review provides an overview of our recent developments in exploring structurally well-defined Mn/Fe-oxo pivalate and isobutyrate building blocks in the formation of CCPs assemblies with an emphasis on synthetic strategies and magnetic properties.

Keywords: cluster-based coordination polymer; carboxylate; iron(III); manganese(II,III)

1. Introduction

Cluster-based coordination polymers (CCPs) have received significant attention over the past decade due to their promising potential as multifunctional materials for technological and industrial applications [1–6]. Typically, CCPs are constructed from metal coordination clusters that are bridged by polytopic organic ligands forming multidimensional systems such as one-dimensional (1D) chains, two-dimensional (2D) layers and three-dimensional (3D) metal–organic frameworks (MOFs) [7]. Two synthetic pathways have been explored by synthetic chemists for the fabrication of the CCPs. In formation reactions, one can utilize simple salts or presynthesized well-defined polynuclear metal clusters, following Robson's classical node (typically consisting of a single metal ion) and spacer (polytopic coordination ligand) approach [8] or Yaghi and O'Keeffe's "secondary building units" (SBUs, representing a rigid metal carboxylate cluster with external connectivity that mimics triangle, square, tetrahedral, hexagonal or octahedral patterns) strategy [9,10]. Later, Zaworotko et al. developed a design strategy that exploits metal–organic polyhedra as "supermolecular building blocks" (SBBs), which combine a greater range of scale (nanometer scale) and high symmetry and, thus, can afford improved control over the topology of the resulting coordination polymers [11].

To date, considerable efforts have been devoted to developing a range of CCPs by using rigid carboxylate clusters of paramagnetic transition metals as building units that contain multiple metal ions linked by multiple coordination ligands, especially for producing molecular magnetic arrays [12]. In comparison to other typically employed structural building blocks, polynuclear carboxylate-based clusters offer distinct advantages for engineering CCPs: (i) they can afford specific control over a CCPs' topology through precise adjusting the coordination environment of metal ions and thus providing the easy accessibility of spacers to a vacant coordination site at the periphery of the cluster, to which a separate cluster can be attached; (ii) ease in managing and fine-tuning their shape and size via increasing or decreasing the nuclearity; (iii) and feasibility to vary their physical properties since their physical characteristics (magnetic, spectroscopic and redox behavior) can be determined and modified prior to network formation [4,5]. Furthermore, the carboxylate ligands can be partially substituted, e.g., by redox-active inorganic ligands such as

Citation: Baca, S.; Kögerler, P. Cluster-Based Coordination Polymers of Mn/Fe-Oxo Pivalates and Isobutyrates . *Chemistry* **2021**, *3*, 314–326. https://doi.org/10.3390/chemistry3010023

Received: 5 February 2021
Accepted: 22 February 2021
Published: 24 February 2021

Publisher's Note: MDPI stays neutral with regard to jurisdictional claims in published maps and institutional affiliations.

magnetically functionalized polyoxoanions [13] or paramagnetic organic ligands [14–17]; thus, the final assemblies can reveal some additional properties and functions such as charge-state switching of magnetic ground states and anisotropy. Such tailor-made multidimensional CCPs can be applied in numerous important fields such as catalysis, gas storage, pharmaceuticals, etc. [18].

Mn/Fe-oxo carboxylate clusters incorporating multiple spin centers exhibit fascinating magnetic phenomena such as single-molecule magnet behavior (SMM) [19–21] and represent useful building blocks for the creation of Mn and Fe-based CCPs. In particular, Christou, Tasiopoulos et al. [22] prepared 3D CCPs with nanometer-sized channels that exhibit SMM characteristics and are based on large polynuclear $\{Mn_{19}O_{10}\}$ acetate building units, $[Mn_{19}NaO_{10}(OH)_4(ac)_9(pd)_9(H_2O)_3]_n$ and $[Mn_{19}NaO_{10}(OH)_4(ac)_9(mpd)_9(H_2O)_3]_n$ (where Hac = acetic acid), using 1,3-propanediol (H_2pd) and 2-methyl-1,3-propanediol (H_2mpd). Later, fascinating extended networks based on polynuclear $\{Mn_{17}O_8\}$ acetate building units connected by N_3^- or OCN^- anions into 1D and 2D CCPs, namely $[Mn_{17}O_8(ac)_4(pd)_{10}(py)_6(N_3)_5]_n$, $[Mn_{17}O_8(ac)_4(mpd)_{10}(py)_8(N_3)_5]_n$ and $[Mn_{17}O_8(ac)_2(pd)_{10}(py)_4(OCN)_7]_n$ have also been reported [23,24]. The use of such large polynuclear carboxylate building blocks offers additional possibilities for crystal engineering of interesting new materials with high porosity and could lead to attractive materials properties. This review provides an overview of our recent developments in exploring $Mn^{II,III}$-oxo and Fe^{III}-oxo pivalate and isobutyrate building blocks with stable $\{M_3(\mu_3\text{-}O)\}$, $\{M_4(\mu_3\text{-}O)_2\}$ (M = Mn, Fe), $\{Mn_6(\mu_4\text{-}O)_2\}$ or $\{Fe_6(\mu_3\text{-}O)_2\}$ cores in the formation of CCPs assemblies with an emphasis on the synthetic strategies and magnetic properties.

2. Oxo-Trinuclear Mn/Fe-Based CCPs

Trinuclear oxo-centered carboxylate coordination clusters of general composition $[M_3(\mu_3\text{-}O)(O_2CR)_6(L)_3]^{+/0}$ (where L = a neutral terminal ligand) are the most frequently used building blocks in the construction of CCPs. For assembling 1D oxo-trinuclear Mn/Fe CCPs, several synthetic strategies have been explored, e.g., using simple soluble metal salts or well-known pre-designed "basic carboxylate", e.g., μ_3-oxo trinuclear Mn/Fe carboxylate clusters. The combination of these starting materials with organic ligands, usually in "one-pot" syntheses at temperatures starting from room temperature and up to solvothermal conditions in different solvents, gave the expected CCPs. The first 1D CCP, $[Mn_3O(ac)_7(Hac)]_n$, which is composed of $[Mn_3(\mu_3\text{-}O)(ac)_6(Hac)]$ acetate clusters interlinked by acetate bridges, was reported by Hessel and Romers in 1969 [25], whereas Rentschler and Albores synthesized the first 1D CCP composed of $[Fe_3(\mu_3\text{-}O)(piv)_6(H_2O)]$pivalate clusters interlinked by dicyanamide (dca) bridges in 2008, $[Fe_3O(piv)_6(H_2O)(dca)]_n$ (where Hpiv = pivalic acid) [26].

Cronin, Kögerler et al. [27] suggested an effective route to assembling 1D CCPs through metal building block linkers in 2006. A helical $\{[(Fe_3O(aa)_6(H_2O))(MoO_4)(Fe_3O(aa)_6(H_2O)_2)]\cdot2(MeCN)\cdot H_2O\}_n$ CCP (where Haa = acrylic acid) has been synthesized by linking the $[Fe_3O(aa)_6(H_2O)_3]^+$ cations with $[MoO_4]^{2-}$ dianions derived from $(Bu_4N)_2[Mo_6O_{19}]$ in MeCN. As an extension, Bu et al. [28] in 2015, introduced $[M(H_2O)_2(fa)_4]^{2-}$ formate building block linkers for connecting μ_3-oxo trinuclear neutral $[Fe_3O(fa)_7]$ formate clusters into $\{(NH_4)_2[Fe_3O(fa)_7]_2[M(H_2O)_2(fa)_4]\}_n$ CCPs (where Hfa = formic acid; M^{II} = Fe; Mn; Mg) with the formation of anionic double-strained chains.

The family of 1D Mn and Fe pivalate or isobutyrate CCPs (Table 1) has been generated by using simple metal carboxylates or employing pre-designed oxo-trinuclear building blocks in reactions [29,30]. Thus, mixing a hot ethanol solution of hexamethylenetetramine (hmta) with Mn^{II} isobutyrate in tetrahydrofuran yields the chain coordination polymer $\{[Mn_3O(is)_6(hmta)_2]\cdot EtOH\}_n$ (where His = isobutyric acid) (**1**) [29]. This CCP comprises neutral mixed-valent μ_3-oxo trinuclear $[Mn^{II}Mn^{III}_2O(is)_6]$ clusters bridged by hmta into a linear 1D chain as shown in Figure 1. To model its magnetic behavior, an approximation considered two trimers coupled through Mn^{III} and Mn^{II} ions (the Mn$\cdots$Mn distances between clusters via hmta linkers are equal to 6.310 Å). For **1**, the authors reported significant antiferromagnetic intracluster interactions between Mn spin centers with $2J_1 = +32.5$ K and $2J_2 = -16.8$ K, whereas the intercluster interactions through hmta spacers were found to be weakly ferromagnetic.

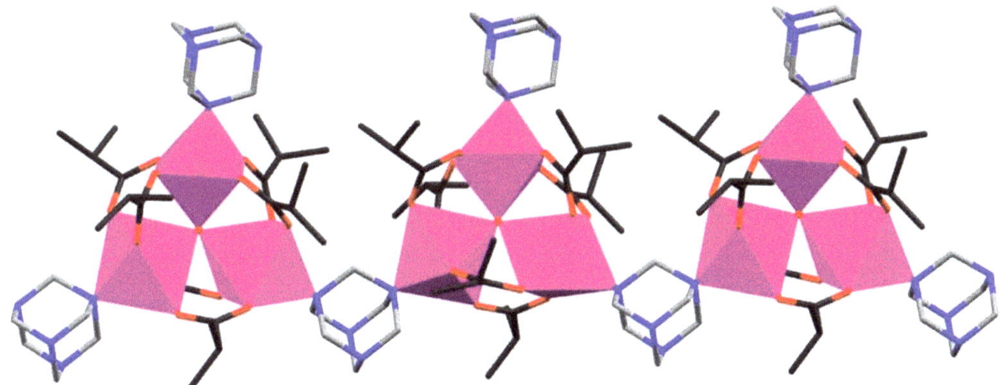

Figure 1. A linear 1D chain in $\{[Mn_3O(is)_6(hmta)_2]\cdot EtOH\}_n$ (**1**) CCP [29]. H atoms and solvent (EtOH) molecules are omitted for clarity. Color codes: C, gray; N, blue; O red sticks. C atoms in carboxylates are shown as black sticks, and Mn atoms are shown as magenta coordination polyhedra.

Combining the different pre-designed μ_3-oxo trinuclear pivalate complexes such as $[Fe_3O(piv)_6(H_2O)_3]piv\cdot 2(piv)$ and $[Mn_3O(piv)_6(hmta)_3]\cdot n\text{-PrOH}$ in one-pot reactions with the same spacer (hmta) lead to the formation of heteronuclear Fe/Mn-oxo 1D CCPs [30]. The solvothermal reaction of these precursors in MeCN at 120 °C for 4 h gave a heterometallic chain polymer $\{[Fe_2MnO(piv)_6(hmta)_2]\cdot 0.5(MeCN)\}_n$ (**2**), while refluxing in *n*-hexane resulted in the solvated heterometallic chain polymer $\{[Fe_2MnO(piv)_6(hmta)_2]\cdot Hpiv\cdot n\text{-}hexane\}_n$ (**3**). Magnetic studies of **2** and **3** indicated dominant antiferromagnetic interactions between the metal centers with significant intercluster interaction through hmta spacer with the following exchange parameters: $J_{Mn^{II}-Fe^{III}} = -17.4$ cm^{-1} and $J_{Fe^{III}-Fe^{III}} = -43.7$ cm^{-1} for **2**; and $J_{Mn^{II}-Fe^{III}} = -23.8$ cm^{-1} and $J_{Fe^{III}-Fe^{III}} = -53.4$ cm^{-1} for **3**. All intercluster exchange interactions were modeled using a molecular field model approximation to give $\lambda_{mf} = -0.219$ mol cm^{-3} (**2**) and $\lambda_{mf} = -0.096$ mol cm^{-3} (**3**) [30].

In comparison to Baca and Kögerler's approach, Kolotilov et al. [31–34] employed already preformed oxo-centered heterometallic trinuclear $[Fe_2MO(piv)_6(Hpiv)_3]$ (M^{II} = Co, Ni) pivalates to isolate a series of heterometallic 1D CCPs formulated as $\{[Fe_2CoO(piv)_6(bpe)]\cdot 0.5(bpe)\}_n$, $[Fe_2NiO(piv)_6(bpp)(dmf)]_n$, $\{[Fe_2NiO(piv)_6(pnp)(dmso)]\cdot 2.5(dmso)\}_n$, $[Fe_2NiO(piv)_6(pnp)(H_2O)]_n$, $\{[(Fe_2NiO(piv)_6)_4(Et\text{-}4\text{-}ppp)_6]\cdot 3(def)\}_n$, and $[Fe_2NiO(piv)_6(bpt)_{1.5}]_n$ (where bpe = 1,2-bis(4-pyridyl)ethylene, bpp = 1,3-bis(4-pyridyl)propane, pnp = 2,6-bis(4-pyridyl)-4-(1-naphtyl)pyridine, Et-4-ppp = 4-(*N*,*N*-diethylamino)phenyl-bis-2,6-(4-pyridyl)pyridine, bpt = 3,6-bis(3-pyridyl)-1,2,4,5-tetrazine).

The presence of three potential donor metal sites in the oxo-centered trinuclear carboxylate species makes this building block very attractive and useful for the construction of 2D CCPs. The first homometallic 2D $\{[Fe_3O(ac)_6(H_2O)_3][Fe_3O(ac)_{7.5}]_2\cdot 7(H_2O)\}_n$ cluster-based layer has been prepared by Long and coworkers [35] in 2007 from the reaction of a trinuclear iron acetate, $[Fe_3O(ac)_6(H_2O)_3]Cl\cdot 6H_2O$, $FeCl_3\cdot 4H_2O$ and sodium acetate in MeCN/water solution at room temperature in 4 months. The main feature of it is the formation of the "star" anionic layer of acetate-bridged $[Fe_3O(ac)_6]^+$ clusters that contain in its channels cationic guest-trimer $[Fe_3O(ac)_6(H_2O)_3]^+$ clusters. In 2009, Pavlishchuk et al. [36] reacted simple salt precursors, namely iron(III) nitrate nonahydrate and manganese(II) nitrate hexahydrate, with formic acid under heating to isolate the first heterometallic 2D CCP

$\{[Fe_3O(fa)_6][Mn(fa)_3(H_2O)_3]\cdot3.5(Hfa)\}_n$. This compound consists of trinuclear $[Fe_3O(fa)_6]^+$ units linked by mononuclear $[Mn(fa)_3(H_2O)_3]^-$ bridges into 2D honeycomb layers.

By refluxing the presynthesized pivalate complex compounds $[Mn_3O(piv)_6(hmta)_3]\cdot n$-PrOH and $[Fe_3O(piv)_6(H_2O)_3]piv\cdot2(Hpiv)$ in a hot toluene solution for 6 h, the new 2D heterometallic coordination polymer $\{[Fe_2MnO(piv)_6(hmta)_{1.5}]\cdot toluene\}_n$ (**4**) can be prepared [30]. In contrast to the above-mentioned 1D CCPs (**1–3**), in **4** hmta ligands, connect neighboring heterometallic $\{[Fe_2Mn(\mu_3\text{-}O)(piv)_6]$ pivalate clusters into a 2D corrugated layer as shown in Figure 2. The formed framework accommodates guest toluene molecules. In **4**, the exchange interactions between Mn^{II} and Fe^{III} were found to be antiferromagnetic ($J_{Mn^{II}-Fe^{III}} = -13.3$ cm^{-1}; $J_{Fe^{III}-Fe^{III}} = -35.4$ cm^{-1}), with significant intercluster interactions through hmta ($\lambda_{mf} = -0.051$ mol cm^{-3}). A series of 2D heterometallic pivalate CCPs built from $[Fe_2MO(piv)_6]$ (M^{II} = Co, Ni) clusters bridged by different polydentate polypyridyl-type linkers has also been reported by other groups [31,36–38].

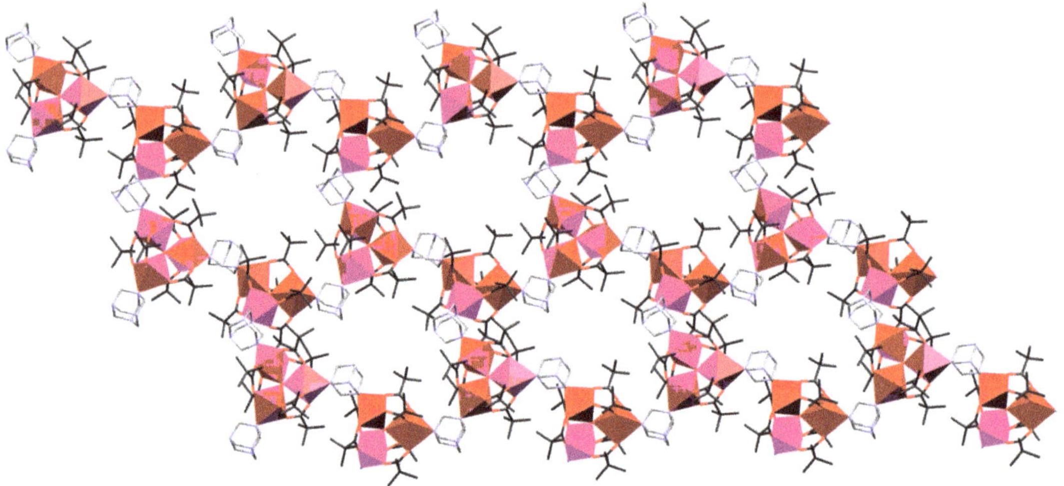

Figure 2. A heterometallic 2D layer in $\{[Fe_2MnO(piv)_6(hmta)_{1.5}]\cdot toluene\}_n$ (**4**) CCP [30]. H atoms and solvent toluene molecules are omitted for clarity. Color codes: C, gray; N, blue; O red sticks and Fe/Mn atoms are shown as brown/magenta coordination polyhedra. C atoms in carboxylates are shown as black sticks.

In 2014, Baca and Kögerler et al. [39] reported the first 3D cluster-based coordination polymers, $\{[Fe_3O(piv)_6(4,4'\text{-}bpy)_{1.5}](OH)\cdot0.75(dcm)\cdot8(H_2O)\}_n$ (**5**) and $[Fe_2CoO(piv)_6(bpe)_{0.5}(pyz)]_n$ (**6**). Relative to the 1D and 2D CCPs, 3D structures based on Mn/Fe trinuclear oxo-clusters are rare: up to now, only three compounds were reported [39,40]. CCPs **5** and **6** consist of μ_3-oxo-centered cationic homometallic $[Fe^{III}_3(\mu_3\text{-}O)(piv)_6]^+$ or neutral heterometallic $[Fe^{III}_2Co^{II}(\mu_3\text{-}O)(piv)_6]$ coordination clusters bridged by different *N,N'*-donor ligands: 4,4′-bipyridine (4,4′-bpy) in case of **5**, and 1,2′-bis(4-pyridyl)ethylene (bpe) and pyrazine (pyz) in case of **6**. They were prepared in a "one-pot" solvothermal reaction in dichloromethane from $[Fe_6O_2(OH)_2(piv)_{12}]$ and organic spacers, and, additionally, cobalt(II) pivalate was added in **6**. It is worth noting that the mutual arrangement of three-connected $[M_3(\mu_3\text{-}O)(piv)_6]$ nodes linked by a linear spacer determines the topology of the final CCPs. Here, neighboring μ_3-oxo trinuclear clusters are mutually perpendicular, and the pair of clusters may be regarded as a pseudo-tetrahedral four-connected binodal building block, and as a result, 3D porous networks are formed. A 6-fold interpenetrated network with rare (8.3)-c (**etc**) topology can be observed in **5**, and a three-fold interpenetrated network with (10.3)-b (**ths**) topology in **6** (Figure 3). Magnetic studies of **5** and **6** point to both ferro- and antiferromagnetic intra- and intercluster exchange interactions between the isotropic Fe^{III} and/or the strongly anisotropic (octahedrally coordinated) Co^{II}

spin centers. In particular, the $\chi_m T$ value of 7.54 cm³ K mol⁻¹ at 290 K, significantly smaller than the expected spin-only value of 13.1 cm³ K mol⁻¹ for three isolated $S = 5/2$ centers ($g_{iso} = 2.0$) for **5**, indicates dominant antiferromagnetic exchange interactions mediated by the central μ_3-O within {Fe₃(μ_3-O)} unit ($J_1 = -0.1$ cm⁻¹ and $J_2 = -27.0$ cm⁻¹) and the 4,4′-bpy bridges ($\lambda_{mf} = -0.609$ mol cm⁻³, the Fe···Fe distances via 4,4′-bpy are equal to 11.347 and 11.380 Å). For **6**, both contributions of the orbital momentum of Co^{II} center in {Fe₂Co} unit and intracluster (ferromagnetic) and intercluster (ferromagnetic) coupling within {Fe₂Co} triangular unit and between them have been considered via the magnetochemical computational framework CONDON [41,42]. The exchange interaction parameters are $J_{Co^{II}-Fe^{III}} = +55.0$ cm⁻¹, $J_{Fe^{III}-Fe^{III}} = -122.0$ cm⁻¹ and $\lambda_{mf} = +1.163$ mol cm⁻³ (through the pyz ligand the M···M distances are 7.096 and 7.142 Å, and through the bpe spacer these distances equal 13.603 Å). Subsequently, a 3D CCP, {[NH₄]₂[Fe₉O₃(ac)₂₃(H₂O)]}ₙ, with triangular [Fe₃(μ_3-O)(ac)₆]⁺ cations and acetate as linkers between the units has been reported by Bu et al. [40]. Its structure exhibits a 4-fold interpenetrating 3D network with a rare **eta-c4** net topology.

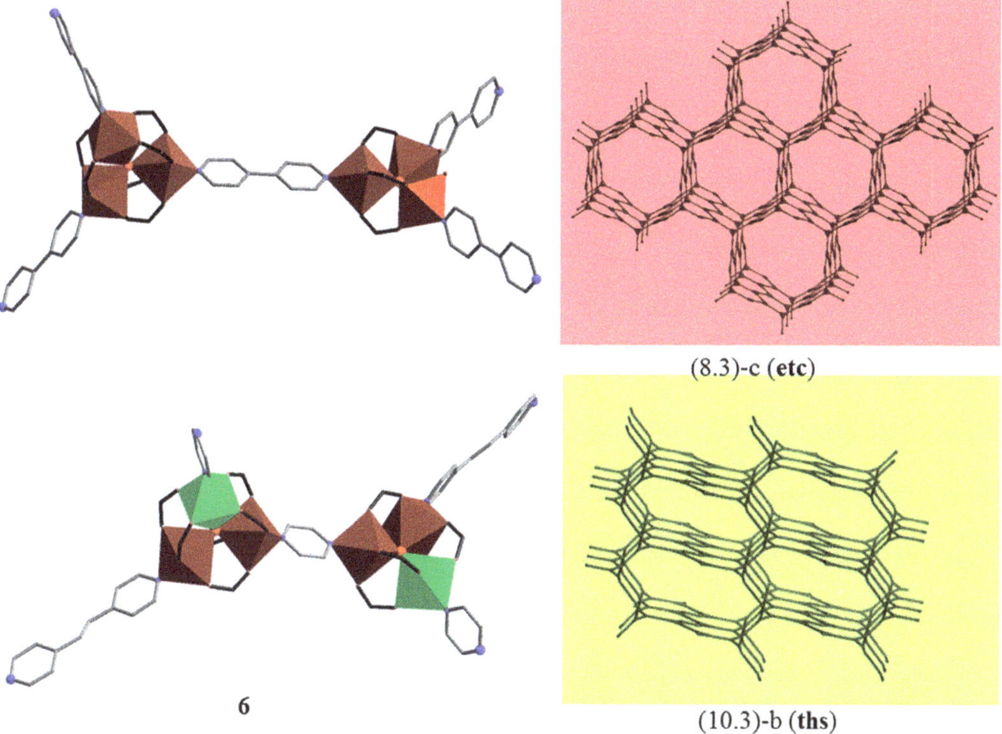

Figure 3. 3D layers in {[Fe₃O(piv)₆(4,4′-bpy)₁.₅](OH)·0.75(dcm)·8(H₂O)}ₙ (**5**) and [Fe₂CoO(piv)₆(bpe)₀.₅(pyz)]ₙ (**6**) CCPs [39]. H atoms and CMe₃ groups are omitted for clarity. Color codes: Fe, brown; Co, green coordination polyhedra; C, gray sticks; N, blue; O red spheres. C atoms in carboxylates are shown as black sticks.

3. Oxo-Tetranuclear Mn/Fe-Based CCPs

Oxo-tetranuclear Mn and Fe carboxylate complexes often called "butterfly" clusters, possess a well-known {M₄(μ_3-O)₂} core. The core comprises an inner "body" metal atoms doubly bridged by two μ_3-oxo atoms, connected outwards to the remaining "wing" metal atoms. The core can also be considered as two edge-sharing {M₃(μ_3-O)} triangular units. Although the core can be considered as potential [M₄O₂(O₂CR)ₓ(L)ᵧ] nodes, only 1D

CCPs from tetranuclear {Mn$_4$(μ_3-O)$_2$} and {Fe$_4$(μ_3-O)$_2$} carboxylate oxo-clusters have been reported thus far [29,43–45].

Christou and coworkers [44] first reported a 1D {Mn$_4$O$_2$}-based benzoate-bridged CCP formulated as [Mn$_4$O$_2$(ba)$_6$(dbm)$_2$(4,4′-bpy)]$_n$ (where Hba = benzoic acid; dbm = dibenzoylmethane) in 1994. Since then, only a few more {Mn$_4$O$_2$}-CCPs were reported. A 1D CCP, [Mn$_4$O$_2$(is)$_6$(bpm)(EtOH)$_4$]$_n$ (**7**), composed of mixed-valence tetranuclear [Mn^{II}$_2$Mn^{III}$_2$(μ_3-O)$_2$(is)$_6$(EtOH)$_4$] isobutyrate building blocks bridged by 2,2′-bipyrimidine (bpm) linkers (Figure 4), has been identified in 2008 [29] (Table 1). The addition of an ethanol solution of bpm to Mn(II) isobutyrate in tetrahydrofuran solution directly results in the formation of **7**. The temperature dependence of χT in **7** indicates dominant antiferromagnetic interactions inside the {Mn$_4$} units with a value of 13.46 cm^3 K mol^{-1} at 300 K that is close to the value expected for an uncoupled {Mn^{II}$_2$Mn^{III}$_2$} system (χT = 14.75 cm^3 K mol^{-1} for g = 2.00).

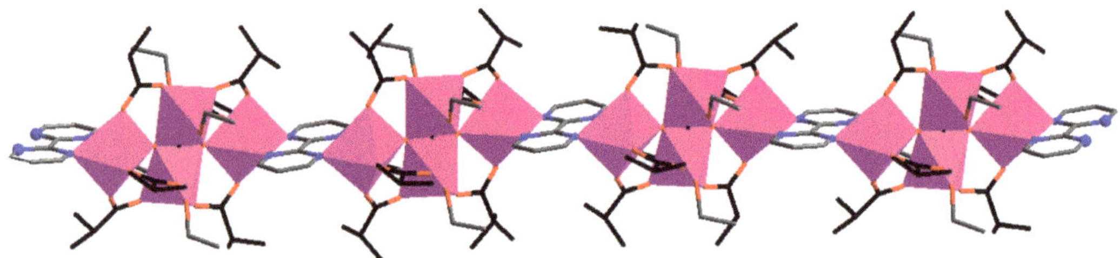

Figure 4. A linear 1D chain in [Mn$_4$O$_2$(is)$_6$(bpm)(EtOH)$_4$]$_n$ (**7**) CCP [29]. H atoms are omitted for clarity. Color codes: C, gray; O red sticks; N, blue spheres. C atoms in carboxylates are shown as black sticks, and Mn atoms are shown as magenta coordination polyhedra.

The first tetranuclear {Fe$_4$O$_2$}-based coordination polymer [Fe$_4$O$_2$(piv)$_8$(hmta)]$_n$ (**8**) reported in 2011 by Baca and coworkers [43] was isolated from the reaction of the μ_3-oxo trinuclear pivalate complex [Fe$_3$O(piv)$_6$(H$_2$O)$_3$]piv·2(Hpiv) with hmta in MeCN. In **8**, tetranuclear [Fe$_4$(μ_3-O)$_2$(piv)$_8$] clusters are bridged by hmta ligands forming zigzag chains (Figure 5). Magnetochemical analysis of the {Fe$_4$} butterfly-like CCP **8** indicates that the interactions between body-body Fe^{III} ions were antiferromagnetic with a least-squares fit yielding J_{bb} = −22 cm^{-1}.

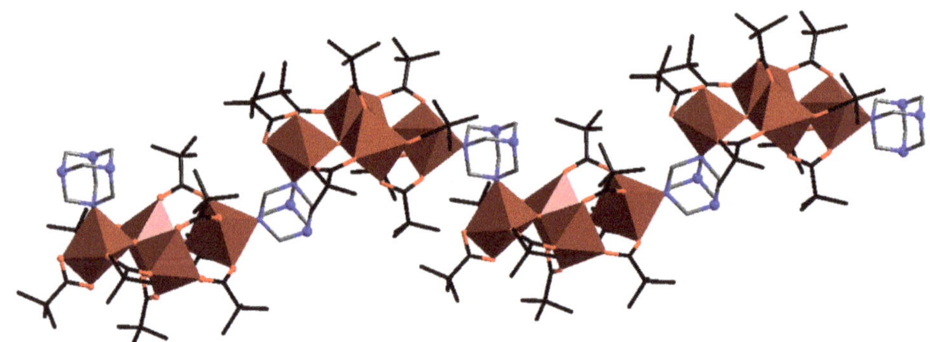

Figure 5. A zigzag 1D chain in [Fe$_4$O$_2$(piv)$_8$(hmta)]$_n$ (**8**) CCP [43]. H atoms are omitted for clarity. Color codes: C, gray; O red sticks; N, blue spheres. C atoms in carboxylates are shown as black sticks, and Fe atoms are shown as brown coordination polyhedra.

4. Oxo-Hexanuclear Mn/Fe CCPs

One of the fascinating Mn oxo-carboxylate clusters used as building blocks in the construction of CCPs are hexanuclear clusters $[Mn_6O_2(O_2CR)_{10}(L)_4]$ (where L = neutral monodentate ligand) with a central $\{Mn_6O_2\}$ core. The central mixed-valence $[Mn^{II}_4Mn^{III}_2(\mu_4\text{-}O)_2]^{10+}$ core consists of six Mn centers, two Mn^{III} and four Mn^{II} ions, arranged as two edge-sharing flattened Mn_4 tetrahedra, with a $\mu_4\text{-}O^{2-}$ ion in the center of each tetrahedron. Peripheral ligation is provided by bridging carboxylate groups. Within the core, two central Mn atoms are in the formal oxidation state +3, and the four terminal Mn atoms are in the lower oxidation state +2. Careful choice of appropriate exo-bidentate spacer ligands with the selection of starting precursors and media solution allowed interlinking the $\{Mn_6(\mu_4\text{-}O)_2\}$ oxo-clusters into 1D coordination polymeric chains, 2D layers or 3D frameworks.

The first 1D coordination polymers $[Mn_6O_2(piv)_{10}(Hpiv)_2(4,4'\text{-}bpy)]_n$ based on $\{Mn_6O_2\}$ pivalate building blocks bridged by 4,4'-bipyridine (4,4'-bpy) were obtained by Yamashita et al. in 2002 [46]. Assembling of $\{Mn_6O_2\}$ building clusters into coordination networks by shorter linkers was executed later. In 2008, Chen's group isolated the propionate-bridged $\{[Mn_6O_2(O_2CEt)_{10}(H_2O)_4]\cdot2(EtCO_2H)\}_n$ CCP [47] and Tasiopoulos's group in 2010 reported the acetato-bridged $\{[Mn_6O_2(ac)_{10}(H_2O)_4]\cdot2.5(H_2O)\}_n$ CCP [48].

A new series of 1D CCPs based on the $\{Mn_6(\mu_4\text{-}O)_2\}$ pivalate and isobutyrate clusters, namely $\{[Mn_6O_2(piv)_{10}(Hpiv)(EtOH)(na)]\cdot EtOH\cdot H_2O\}_n$ (**9**), $[Mn_6O_2(piv)_{10}(Hpiv)_2(en)]_n$ (**10**), $\{[Mn_6O_2(is)_{10}(pyz)_3]\cdot2(H_2O)\}_n$ (**11**), $[Mn_6O_2(is)_{10}(pyz)(MeOH)_2]_n$ (**12**), and $\{[Mn_6O_2(is)_{10}(pyz)_{1.5}(H_2O)]\cdot0.5(H_2O)\}_n$ (**13**), $\{[Mn_6O_2(is)_{10}(His)(EtOH)(bpea)]\cdot His\}_n$ (**14**), (where na = nicotinamide, en = ethyl nicotinate, pyz = pyrazine, bpea = 1,2-bis(4-pyridyl)ethane) have been reported [49,50] (Table 1). All CCPs, except **10**, have been prepared by the reaction of simple Mn^{II} pivalate or isobutyrate salts with the appropriate bridging ligand in different solvents: MeCN solution in case of **9** and **13**, MeCN/EtOH (1:1 *v:v*) solution for **11**, MeCN/MeOH (1:1 *v:v*) solution for **12**, and thf/EtOH solution (1:1 *v:v*) for **14**. CCP **10** was isolated from the reaction of $[Mn_6O_2(O_2CCMe_3)_{10}(thf)_4]$ with ethyl nicotinate in decane. The spacer length and spacer flexibility have an important effect on the complex structures in these systems. The shorter pyz ligand connects neighboring $\{Mn_6(\mu_4\text{-}O)_2\}$ clusters to form 1D zigzag chains in **11** (intercluster Mn$\cdots$Mn: 7.288 Å), linear chains in **12** (intercluster Mn$\cdots$Mn: 7.443 Å), and ladder-like chains in **13** (intercluster Mn$\cdots$Mn: 7.415 and 7.480 Å). The na group links the $\{Mn_6(\mu_4\text{-}O)_2\}$ clusters into linear chains in **10** (intercluster Mn$\cdots$Mn: 8.690 Å), and finally, bpea bridges the $\{Mn_6(\mu_4\text{-}O)_2\}$ clusters (intercluster Mn$\cdots$Mn: 9.239 Å) in a unique meander-type chain as shown in Figure 6 in **14**. Magnetochemical analysis of **9**, **11**, and **14** finds significant intercluster antiferromagnetic interactions through the shorter pyz and na linkers ($\lambda_{mf} = -1.131$ mol cm^{-3} (**9**), -1.508 mol cm^{-3} (**11**)), and the intercluster interaction via long bpea is negligible in **14**. Within $\{Mn_6\}$ fragments, the intracluster interactions are antiferromagnetic with the best-fit parameters $J_1 = -1.61$ cm^{-1}, $J_2 = J_3 = -0.89$ cm^{-1} and $J_4 = -1.23$ cm^{-1}. Here J_1 is the exchange coupling constant describing the nearest-neighbor $Mn^{III}\cdots Mn^{III}$ interactions through two μ_4-O centers, $J_2 = J_3$—for $Mn^{III}\cdots Mn^{II}$ interactions mediated by μ_4-O and μ_3-O centers and a carboxylate (J_2) or a μ_4-O center and a carboxylate (J_3), and J_4—for $Mn^{II}\cdots Mn^{II}$ interactions via a μ_4-O center and a carboxylate group. We note that Ovcharenko et al. [51] linked the polynuclear $\{Mn_6\}$ pivalate fragments by nitronyl nitroxide radical (2,4,4,5,5-pentamethyl-4,5-dihydro-1*H*-imidazolyl-3-oxide-1-oxyl (NIT-Me) to isolate 1D molecular magnet $\{[Mn_6O_2(piv)_{10}(thf)_2(NIT-Me)][Mn_6O_2(piv)_{10}(thf)_2(dcm)(NIT-Me)]\}_n$ with $T_c = 3.5$ K.

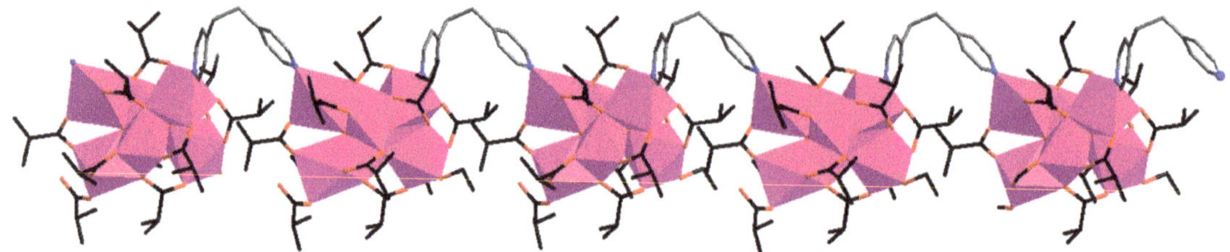

Figure 6. 1D meander-like chain in $\{[Mn_6O_2(is)_{10}(His)(EtOH)(bpea)]\cdot His\}_n$ (**14**) CCP [49]. H atoms and solvate His are omitted for clarity. Color codes: C, gray; O red sticks; N, blue spheres. C atoms in carboxylates are shown as black sticks, and Mn atoms are shown as magenta coordination polyhedra.

In contrast to hexanuclear $\{Mn_6(\mu_4\text{-}O)_2\}$ building blocks, the use of hexanuclear $\{Fe_6(\mu_3\text{-}O)_2\}$ building blocks in the design of CCPs is fairly scarce. Until now, only two such compounds have been published [52]. 1D zigzag chain coordination polymers $[Fe_6O_2(O_2CH_2)(piv)_{12}(diox)]_n$ (**15**) (Figure 7) and $[Fe_6O_2(O_2CH_2)(piv)_{12}(4,4'\text{-}bpy)]_n$ (**16**) (where diox = 1,4-dioxane; $4,4'$-bpy = 4,4′-bipyridine) have been prepared from smaller pre-designed μ_3-oxo trinuclear $[Fe_3O(piv)_6(H_2O)_3]piv\cdot 2(Hpiv)$ or hexanuclear $[Fe_6O_2(OH)_2(piv)_{12}]$ species by slow diffusion of MeCN into its solution in 1,4-dioxane (**15**) or heating starting $[Fe_6O_2(OH)_2(piv)_{12}]$ and $4,4'$-bpy in CH_2Cl_2/MeCN (**16**). Although the $\{Fe_6\}$ building blocks in **15** and **16** show the general similarity (Fe$\cdots$Fe distances differ less than 0.015 Å from their averages), the observed magnetic low-field susceptibility displays strong differences, mainly from inter-$\{Fe_6\}$ cluster coupling mediated either by the diox (closest Fe$\cdots$Fe contact: 7.171 Å) or the $4,4'$-bpy (11.39 Å) linkers. The magnetism of the $[Fe_6O_2(O_2CH_2)(piv)_{12}]$ clusters was modeled assuming that the $\{Fe_6\}$ cluster consists of two identical isosceles $Fe_3(\mu_3\text{-}O)$ triangles with three exchange energies (J_{1-3}) between the spin-only $(S = 5/2;\ {}^6A_{1g})$ Fe^{III} centers. CCPs **15** and **16** exhibit dominant antiferromagnetic interactions among the Fe^{III} ions within the $\{Fe_6\}$ clusters (J_1 and J_2 between Fe^{III} ions in $\{Fe_3O\}$ triangles and J_3—between the triangles) and possible antiferromagnetic intercluster interaction (λ_{mf}) with the best-fit $J_1 = -30.82\ \text{cm}^{-1}$, $J_2 = -14.43\ \text{cm}^{-1}$, $J_3 = -11.75\ \text{cm}^{-1}$ and $\lambda_{mf} = -11.6\ \text{mol cm}^{-3}$ for **15**, and $J_1 = -32.77\ \text{cm}^{-1}$, $J_2 = -14.11\ \text{cm}^{-1}$, $J_3 = -10.57\ \text{cm}^{-1}$ and $\lambda_{mf} = -0.12\ \text{mol cm}^{-3}$ for **16**.

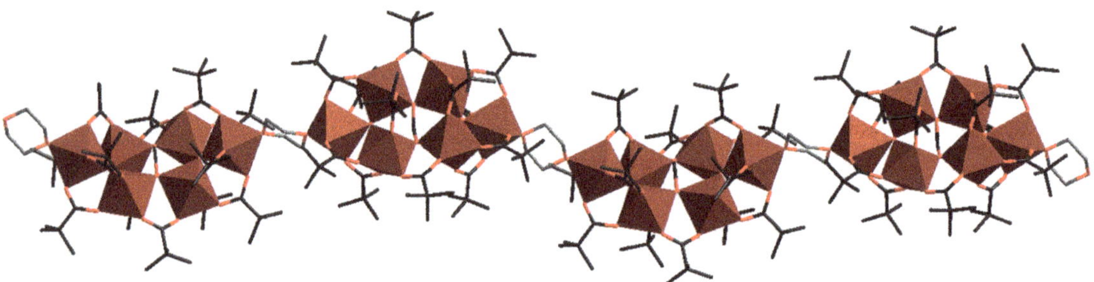

Figure 7. A 1D chain in $[Fe_6O_2(O_2CH_2)(piv)_{12}(diox)]_n$ (**15**) [52]. H atoms are omitted for clarity. Color codes: C, gray; O red sticks. C atoms in carboxylates are shown as black sticks, and Fe atoms are shown as brown coordination polyhedra.

Using isonicotinamide (ina), Ovcharenko and co-workers in 2013 [53] prepared the first 2D pivalate CCPs based on a $\{Mn_6O_2\}$ carboxylate motif, $\{[Mn_6O_2(piv)_{10}(ina)_2]\cdot 3(Me_2CO)\}_n$ and $\{[Mn_6O_2(piv)_{10}(ina)_2]\cdot 2(EtOAc)\}_n$. Employing the same ina spacer, an anhydrous $[Mn_6O_2(piv)_{10}(ina)_2]_n$ (**17**) CCP was obtained from the simple manganese(II) pivalate and ina in a mixture of MeCN/EtOH in 2014 [54] (Figure 8a). Similar to Ovcharenko's CCPs, $\{Mn_6\}$ pivalate clusters are bridged via ina spacer ligands in a 2D polymeric (4,4) layer with a Mn$^{II}\cdots$Mn^{II} distance of 9.279 and 9.357 Å. Two other 2D CCPs $\{[Mn_6O_2(piv)_{10}(adt\text{-}$

4)$_2$]·2(thf)}$_n$ (**18**) and {[Mn$_6$O$_2$(is)$_{10}$(adt-4)$_2$]·thf·3(EtOH)}$_n$ (**19**) were prepared from the reaction of a simple [Mn(piv)$_2$] salt (**18**) or a hexanuclear [Mn$_6$(is)$_{12}$(His)$_6$] isobutyrate cluster (**19**) and the bent, semi-rigid *N,N′*-donor aldrithiol (adt-4) spacer ligands in thf. Interestingly, four semi-rigid aldrithiol ligands joint the adjacent {Mn$_6$} clusters in a different way: if in **18**, a 2D polymeric layer with grid topology is formed, in **19**, an intriguing bilayer framework occurs (Figure 8b). In **19**, a pair of adt-4 ligands connect the neighboring {Mn$_6$} isobutyrate clusters to form linear chains with a Mn^{II}···Mn^{II} distance of 10.537 Å between neighboring {Mn$_6$} units. Another pair of adt-4 ligands double cross-link neighboring {Mn$_6$} clusters from differently oriented chains (Mn^{II}···Mn^{II}: 10.873 Å) resulting in a rare 2D bilayer (Figure 8b). Thus, the hexanuclear [Mn$_6$O$_2$(is)$_{10}$] isobutyrate clusters here serve as 3-connected T-shaped nodes and represent the first example of a CCP with such bilayer topology. Magnetic susceptibility measurements for **17–19** indicate a net of antiferromagnetic intracluster and intercluster exchange interactions, resulting in singlet ground states. An intracluster coupling scheme defines four types of exchange energies in {Mn$_6$} clusters with the best-fit $J_{Mn^{III}-Mn^{III}} = -15.4$ cm^{-1}, $J_{Mn^{II}-Mn^{II}} = -1.9$ cm^{-1}, $J_{Mn^{III}-Mn^{II}} = -2.8$ cm^{-1}, and $J_{Mn^{II}-Mn^{II},ext} = +5.5$ cm^{-1}, resulting in an overall $S = 0$ ground state for each {Mn$_6$} cluster. The molecular field parameter results in $\lambda = -1.1$ mol cm^{-3}, indicating antiferromagnetic intercluster coupling.

Table 1. Cluster-based coordination polymers (CCPs) built up from Fe/Mn-oxo pivalates and isobutyrates.

Code	Formulae	Oxo Building Block	Linker [1]	Dimensionality	Refs
1	{[Mn$_3$O(is)$_6$(hmta)$_2$]·EtOH}$_n$	{Mn$^{III}_2$Mn^{II}O}	hmta	1D	[29]
2	{[Fe$_2$MnO(piv)$_6$(hmta)$_2$]·0.5(MeCN)}$_n$	{Fe$^{III}_2$Mn^{II}O}	hmta	1D	[30]
3	{[Fe$_2$MnO(piv)$_6$(hmta)$_2$]·Hpiv·*n*-hexane}$_n$	{Fe$^{III}_2$Mn^{II}O}	hmta	1D	[30]
4	{[Fe$_2$MnO(piv)$_6$(hmta)$_{1.5}$]·toluene}$_n$	{Fe$^{III}_2$Mn^{II}O}	hmta	2D	[30]
5	{[Fe$_3$O(piv)$_6$(4,4′-bpy)$_{1.5}$](OH)·0.75(dcm)·8(H$_2$O)}$_n$	{Fe$^{III}_3$O}	4,4′-bpy	3D	[39]
6	[Fe$_2$CoO(piv)$_6$(bpe)$_{0.5}$(pyz)]$_n$	{Fe$^{III}_2$Co^{II}O}	bpe, pyz	3D	[39]
7	[Mn$_4$O$_2$(is)$_6$(bpm)(EtOH)$_4$]$_n$	{Mn$^{III}_2$Mn$^{II}_2$O$_2$}	bpm	1D	[29]
8	[Fe$_4$O$_2$(piv)$_8$(hmta)]$_n$	{Fe$^{III}_4$O$_2$}	hmta	1D	[43]
9	{[Mn$_6$O$_2$(piv)$_{10}$(Hpiv)(EtOH)(na)]·EtOH·H$_2$O}$_n$	{Mn$^{III}_2$Mn$^{II}_4$O$_2$}	na	1D	[49]
10	[Mn$_6$O$_2$(piv)$_{10}$(Hpiv)$_2$(en)]$_n$	{Mn$^{III}_2$Mn$^{II}_4$O$_2$}	en	1D	[50]
11	{[Mn$_6$O$_2$(is)$_{10}$(pyz)$_3$]·2(H$_2$O)}$_n$	{Mn$^{III}_2$Mn$^{II}_4$O$_2$}	pyz	1D	[49]
12	[Mn$_6$O$_2$(is)$_{10}$(pyz)(MeOH)$_2$]$_n$	{Mn$^{III}_2$Mn$^{II}_4$O$_2$}	pyz	1D	[50]
13	{[Mn$_6$O$_2$(is)$_{10}$(pyz)$_{1.5}$(H$_2$O)]·0.5(H$_2$O)}$_n$	{Mn$^{III}_2$Mn$^{II}_4$O$_2$}	pyz	1D	[50]
14	{[Mn$_6$O$_2$(is)$_{10}$(His)(EtOH)(bpea)]·His}$_n$	{Mn$^{III}_2$Mn$^{II}_4$O$_2$}	bpea	1D	[49]
15	[Fe$_6$O$_2$(O$_2$CH$_2$)(piv)$_{12}$(diox)]$_n$	{Fe$^{III}_6$O$_2$}	diox	1D	[52]
16	[Fe$_6$O$_2$(O$_2$CH$_2$)(piv)$_{12}$(4,4′-bpy)]$_n$	{Fe$^{III}_6$O$_2$}	4,4′-bpy	1D	[52]
17	[Mn$_6$O$_2$(piv)$_{10}$(ina)$_2$]$_n$	{Mn$^{III}_2$Mn$^{II}_4$O$_2$}	ina	2D	[54]
18	{[Mn$_6$O$_2$(piv)$_{10}$(adt-4)$_2$]·2(thf)}$_n$	{Mn$^{III}_2$Mn$^{II}_4$O$_2$}	adt-4	2D	[54]
19	{[Mn$_6$O$_2$(is)$_{10}$(adt-4)$_2$]·thf·3(EtOH)}$_n$	{Mn$^{III}_2$Mn$^{II}_4$O$_2$}	adt-4	2D	[54]

[1] hmta = hexamethylenetetramine; 4,4′-bpy = 4,4′-bipyridine; bpe = 1,2′-bis(4-pyridyl)ethylene; pyz = pyrazine; bpm = 2,2′-bipyrimidine; na = nicotinamide; en = ethyl nicotinate; bpea = 1,2-bis(4-pyridiyl)ethane; diox = 1,4-dioxane; ina = isonicotinamide; adt-4 = aldrithiol.

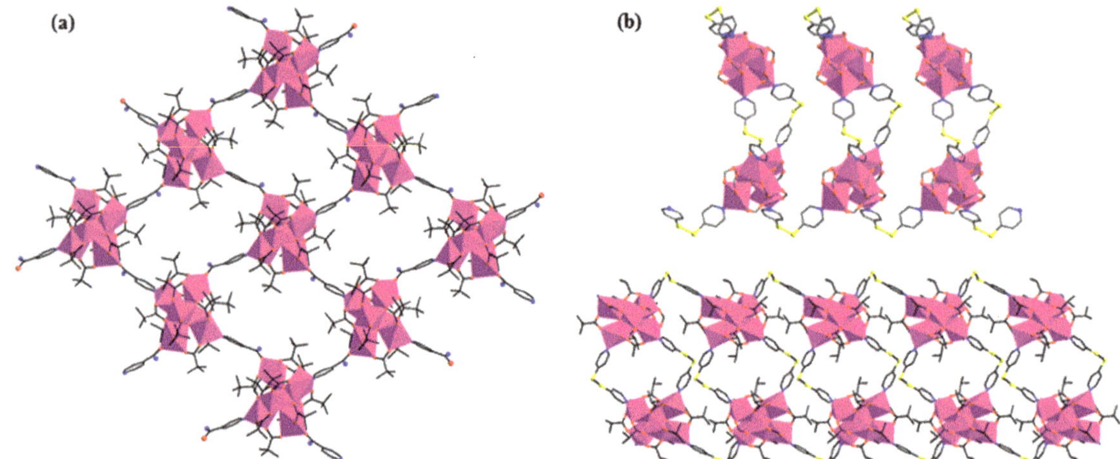

Figure 8. (**a**) A 2D (4,4) layer in $[Mn_6O_2(piv)_{10}(ina)_2]_n$ (**17**) CCP. (**b**) A rare T-shaped bilayer in $\{[Mn_6O_2(is)_{10}(adt\text{-}4)_2]\cdot thf\cdot3(EtOH)\}_n$ (**19**) CCP [54]. H atoms and solvent molecules are omitted for clarity. Color codes: C, gray sticks; O red; N, blue; S, yellow spheres. Mn atoms are shown as magenta coordination polyhedra.

The first example of 3D oxo-hexanuclear $\{Mn_6\}$ pivalate CCP appeared in 2004 when Ovcharenko et al. reported a honeycomb $[Mn_6O_2(piv)_{10}(NIT\text{-}Me)_2]_n$ motif [51]. The second 3D oxo-hexanuclear $[Mn_6O_2(O_2CCH_2C_6H_5)_{10}(pyz)_2]_n$ CCP was synthesized in 2012 by Ghosh and coworkers who used pyrazine (pyz) linker to connect $\{Mn_6\}$ phenyl acetate clusters into a diamond-like framework with a (6,4) topology [55].

5. Conclusions

Polynuclear $Mn^{II,III}/Fe^{III}$-oxo pivalate and isobutyrate clusters recommend themselves as extremely versatile building blocks. Their ancillary coordinating ligands are sufficiently flexible to allow for the formation of a wide variety of structurally diverse and topologically interesting CCPs. Using metal-containing carboxylate building units ranging from trinuclear $\{M_3O\}$ to hexanuclear $\{M_6O_2\}$ cores, 1D, 2D and 3D CCPs have successfully been isolated. We note that this robust and versatile materials platform, based on oxo-carboxylate coordination cluster structures, is particularly attractive in the field of "intelligent" multifunctional materials, including magnetic sensors and spintronic devices since the targeted molecular systems allow for a combination of their intrinsic stability and physical quantum characteristics with unique structural features. However, despite their prominent role at the forefront of magnetic materials research, significant efforts remain focused on developing and understanding their complex magnetic characteristics as well as establishing magneto-structural correlations for these systems.

Funding: This research in part was supported by the State Program of the National Agency for Research and Development of R. Moldova, grant number ANCD 20.80009.5007.15.

Acknowledgments: Svetlana Baca is grateful to the Alexander von Humboldt Foundation for Georg Forster Research Award.

References

1. Zhang, W.-X.; Liao, P.-Q.; Lin, R.-B.; Wei, Y.-S.; Zeng, M.-H.; Chen, X.-M. Metal cluster-based functional porous coordination polymers. *Coord. Chem. Rev.* **2015**, *293–294*, 263–278. [CrossRef]
2. Kang, X.-M.; Tang, M.-H.; Yang, G.-L.; Zhao, B. Cluster/cage-based coordination polymers with tetrazole derivatives. *Coord. Chem. Rev.* **2020**, *422*, 213424. [CrossRef]

3.   Andruh, M. Oligonuclear complexes as tectons in crystal engineering: Structural diversity and magnetic properties. *Chem. Commun.* **2007**, 2565–2577. [CrossRef]

4.   Roubeau, O.; Clérac, R. Rational assembly of high-spin polynuclear magnetic complexes into coordination networks: The case of a [Mn$_4$] single-molecule magnet building block. *Eur. J. Inorg. Chem.* **2008**, 4325–4342. [CrossRef]

5.   Jeon, I.R.; Clerac, R. Controlled association of single-molecule magnets (SMMs) into coordination networks: Towards a new generation of magnetic materials. *Dalton Trans.* **2012**, *41*, 9569–9586. [CrossRef]

6.   Shieh, M.; Liu, Y.-H.; Lia, Y.-H.; Lina, R.Y. Metal carbonyl cluster-based coordination polymers: Diverse syntheses, versatile network structures, and special properties. *Cryst. Eng. Comm.* **2019**, *21*, 7341–7364. [CrossRef]

7.   Batten, S.R.; Neville, S.M.; Turner, D.R. *Coordination Polymers: Design, Analysis and Application*; Royal Society of Chemistry: Cambridge, UK, 2008; pp. 1–471.

8.   Robson, R. A net-based approach to coordination polymers. *J. Chem. Soc. Dalton Trans.* **2000**, 3735–3744. [CrossRef]

9.   Tranchemontagne, D.J.; Mendoza-Cortes, J.L.; O'Keeffe, M.; Yaghi, O.M. Secondary building units, nets and bonding in the chemistry of metal-organic frameworks. *Chem. Soc. Rev.* **2009**, 1257–1283. [CrossRef] [PubMed]

10.   Eddaoudi, M.; Moler, D.B.; Li, H.; Chen, B.; Reineke, T.M.; O'Keeffe, M.; Yaghi, O.M. Modular chemistry: Secondary building units as a basis for the design of highly porous and robust metal-organic carboxylate frameworks. *Acc. Chem. Res.* **2001**, *34*, 319–330. [CrossRef]

11.   Perry, J.J.; Permana, J.A.; Zaworotko, M.J. Design and synthesis of metal–organic frameworks using metal–organic polyhedra as supermolecular building blocks. *Chem. Soc. Rev.* **2009**, *38*, 1400–1417. [CrossRef] [PubMed]

12.   Baca, S.G. The design of coordination networks from polynuclear Mn(II, III) and Fe(III) oxo-carboxylate clusters. In *Advances in Chemistry Research*; Taylor, J.C., Ed.; Nova Science Publishers: New York, NY, USA, 2018; Volume 43, pp. 81–130.

13.   Fang, X.; Kögerler, P. A polyoxometalate-based manganese carboxylate cluster. *Chem. Commun.* **2008**, 3396–3398. [CrossRef] [PubMed]

14.   Vaz, M.G.F.; Andruh, M. Molecule-based magnetic materials constructed from paramagnetic organic ligands and two different metal ions. *Coord. Chem. Rev.* **2021**, *427*, 213611. [CrossRef]

15.   Ratera, I.; Veciana, J. Playing with organic radicals as building blocks for functional molecular materials. *Chem. Soc. Rev.* **2012**, *41*, 303–349. [CrossRef]

16.   Kahn, O. Chemistry and physics of supramolecular magnetic materials. *Acc. Chem. Res.* **2000**, *33*, 647–657. [CrossRef] [PubMed]

17.   Mccleverty, J.A.; Ward, M.D. The role of bridging ligands in controlling electronic and magnetic properties in polynuclear complexes. *Acc. Chem. Res.* **1998**, *31*, 842–851. [CrossRef]

18.   Farrusseng, D. (Ed.) *Metal–Organic Frameworks. Applications from Catalysis to Gas Storage*; Wiley-VCH Verlag & Co. KgaA: Weinheim, Germany, 2011; p. 392.

19.   Winpenny, R.E.P. (Ed.) *Single-Molecule Magnets and Related Phenomena*; Structure and Bonding; Springer: Berlin/Heidelberg, Germany, 2006; Volume 122.

20.   Gatteschi, D.; Sessoli, R.; Villain, J. *Molecular Nanomagnets*; Oxford University Press: New York, NY, USA, 2006.

21.   Gatteschi, D.; Sessoli, R.; Cornia, A. Single-molecule magnets based on iron(III) oxo clusters. *Chem. Commun.* **2000**, 725–732. [CrossRef]

22.   Moushi, E.E.; Stamatatos, T.C.; Wernsdorfer, W.; Nastopoulos, V.; Christou, G.; Tasiopoulos, A.J. A Family of 3D coordination polymers composed of Mn$_{19}$ magnetic units. *Angew. Chem.* **2006**, *118*, 7886–7889. [CrossRef]

23.   Moushi, E.E.; Stamatatos, T.C.; Wernsdorfer, W.; Nastopoulos, V.; Christou, G.; Tasiopoulos, A.J. A Mn$_{17}$ octahedron with a giant ground-state spin: Occurrence in discrete form and as multidimensional coordination polymers. *Inorg. Chem.* **2009**, *48*, 5049–5051. [CrossRef]

24.   Moushi, E.E.; Stamatatos, T.C.; Wernsdorfer, W.; Nastopoulos, V.; Christou, G.; Tasiopoulos, A.J. 1-D coordination polymers consisting of a high-spin Mn$_{17}$ octahedral unit. *Polyhedron* **2009**, *28*, 1814–1817. [CrossRef]

25.   Hessel, L.W.; Romers, C. The crystal structure of "anhydrous manganic acetate". *Recl. Trav. Chim. Pays-Bas* **1969**, *88*, 545–552. [CrossRef]

26.   Albores, P.; Rentschler, E. Structural and magnetic characterization of a µ-1,5-dicyanamide-bridged iron basic carboxylate [Fe$_3$O(O$_2$C(CH$_3$)$_3$)$_6$] 1D chain. *Inorg. Chem.* **2008**, *47*, 7960–7962. [CrossRef]

27.   Long, D.L.; Kögerler, P.; Farrugia, L.J.; Cronin, L. Linking chiral clusters with molybdate building blocks: From homochiral helical supramolecular arrays to coordination helices. *Chem. Asian J.* **2006**, *1*, 352–357. [CrossRef]

28.   Wu, Q.L.; Han, S.D.; Wang, Q.L.; Zhao, J.P.; Ma, F.; Jiang, X.; Liu, F.C.; Bu, X.H. Divalent metal ions modulated strong frustrated M(II)–Fe(III)$_3$O (M = Fe, Mn, Mg) chains with metamagnetism only in a mixed valence iron complex. *Chem. Commun.* **2015**, *51*, 15336–15339. [CrossRef] [PubMed]

29.   Baca, S.G.; Malaestean, I.L.; Keene, T.D.; Adams, H.; Ward, M.D.; Hauser, J.; Neels, A.; Decurtins, S. One-dimensional manganese coordination polymers composed of polynuclear cluster blocks and polypyridyl linkers: Structures and properties. *Inorg. Chem.* **2008**, *47*, 11108–11119. [CrossRef]

30.   Dulcevscaia, G.M.; Filippova, I.G.; Speldrich, M.; Leusen, J.; Kravtsov, V.C.; Baca, S.G.; Kögerler, P.; Liu, S.X.; Decurtins, S. Cluster-based networks: 1D and 2D coordination polymers based on {MnFe$_2$(µ$_3$-O)}-type clusters. *Inorg. Chem.* **2012**, *51*, 5110–5117. [CrossRef] [PubMed]

31. Polunin, R.A.; Kiskin, M.A.; Cador, O.; Kolotilov, S.V. Coordination polymers based on trinuclear heterometallic pivalates and polypyridines: Synthesis, structure, sorption and magnetic properties. *Inorg. Chim. Acta* **2012**, *380*, 201–210. [CrossRef]

32. Lytvynenko, A.S.; Kiskin, M.A.; Dorofeeva, V.N.; Mishura, A.M.; Titov, V.E.; Kolotilov, S.V.; Eremenko, I.L.; Novotortsev, V.M. Redox-active porous coordination polymer based on trinuclear pivalate: Temperature-dependent crystal rearrangement and redox-behavior. *J. Solid State Chem.* **2015**, *223*, 122–130. [CrossRef]

33. Sotnik, S.A.; Polunin, R.A.; Kiskin, M.A.; Kirillov, A.M.; Dorofeeva, V.N.; Gavrilenko, K.S.; Eremenko, I.L.; Novotortsev, V.M.; Kolotilov, S.V. Heterometallic coordination polymers assembled from trigonal trinuclear Fe$_2$Ni-pivalate blocks and polypyridine spacers: Topological diversity, sorption, and catalytic properties. *Inorg. Chem.* **2015**, *54*, 5169–5181. [CrossRef] [PubMed]

34. Lytvynenko, A.S.; Polunin, R.A.; Kiskin, M.A.; Mishura, A.M.; Titov, V.E.; Kolotilov, S.V.; Novotortsev, V.M.; Eremenko, I.L. Electrochemical and electrocatalytic characteristics of coordination polymers based on trinuclear pivalates and heterocyclic bridging ligands. *Theor. Exp. Chem.* **2015**, *51*, 54–61. [CrossRef]

35. Zheng, Y.Z.; Tong, M.L.; Xue, W.; Zhang, W.X.; Chen, X.M.; Grandjean, F.; Long, G.J. A "Star" antiferromagnet: A polymeric iron(III) acetate that exhibits both spin frustration and long-range magnetic ordering. *Angew. Chem. Int. Ed.* **2007**, *46*, 6076–6080. [CrossRef]

36. Lytvynenko, A.S.; Kolotilov, S.V.; Cador, O.; Gavrilenko, K.S.; Golhen, S.; Ouahab, L.; Pavlishchuk, V.V. Porous 2D coordination polymeric formate built up by Mn(II) linking of Fe$_3$O units: Influence of guest molecules on magnetic properties. *Dalton Trans.* **2009**, 3503–3509. [CrossRef]

37. Polunin, R.A.; Kolotilov, S.V.; Kiskin, M.A.; Cador, O.; Mikhalyova, E.A.; Lytvynenko, A.S.; Golhen, S.; Ouahab, L.; Ovcharenko, V.I.; Eremenko, I.L.; et al. Topology control of porous coordination polymers by building block symmetry. *Eur. J. Inorg. Chem.* **2010**, 5055–5057. [CrossRef]

38. Dorofeeva, V.N.; Kolotilov, S.V.; Kiskin, M.A.; Polunin, R.A.; Dobrokhotova, Z.V.; Cador, O.; Golhen, S.; Ouahab, L.; Eremenko, I.; Novotortsev, V. 2D Porous honeycomb polymers versus discrete nanocubes from trigonal trinuclear complexes and ligands with variable topology. *Chem. Eur. J.* **2012**, *18*, 5006–5012. [CrossRef] [PubMed]

39. Botezat, O.; van Leusen, J.; Kravtsov, V.C.; Filippova, I.G.; Hauser, J.; Speldrich, M.; Hermann, R.P.; Krämer, K.W.; Liu, S.-X.; Decurtins, S.; et al. Interpenetrated (8,3)-c and (10,3)-b metal–organic frameworks based on {Fe^{III}$_3$} and {Fe^{III}$_2$Co^{II}} pivalate spin clusters. *Cryst. Growth. Des.* **2014**, *14*, 4721–4728. [CrossRef]

40. Zhao, J.P.; Han, S.D.; Jiang, X.; Xu, J.; Chang, Z.; Bu, X.H. A three dimensional magnetically frustrated metal–organic framework via the vertices augmentation of underlying net. *Chem. Commun.* **2015**, *51*, 4627–4630. [CrossRef] [PubMed]

41. Speldrich, M.; van Leusen, J.; Kögerler, P. CONDON 3.0: An updated software package for magnetochemical analysis - all the way to polynuclear actinide complexes. *J. Comput. Chem.* **2018**, *39*, 2133–2145. [CrossRef] [PubMed]

42. van Leusen, J.; Speldrich, M.; Schilder, H.; Kögerler, P. Comprehensive insight into molecular magnetism via CONDON: Full vs. Effective Models. *Coord. Chem. Rev.* **2015**, *289–290*, 137–148. [CrossRef]

43. Baca, S.G.; Filippova, I.G.; Keene, T.D.; Botezat, O.; Malaestean, I.L.; Stoeckli-Evans, H.; Kravtsov, V.C.; Chumacov, I.; Liu, S.-X.; Decurtins, S. Iron(III)-pivalate-based complexes with tetranuclear {Fe$_4$(μ_3-O)$_2$}$^{8+}$ cores and N-donor ligands: Formation of cluster and polymeric architectures. *Eur. J. Inorg. Chem.* **2011**, *3*, 356–367. [CrossRef]

44. Wang, S.; Tsai, H.L.; Folting, K.; Martin, J.M.; Hendrickson, D.N.; Christou, G. Covalent linkage of [Mn$_4$O$_2$(O$_2$CPh)$_6$(dbm)$_2$] into a dimer and a one-dimensional polymer (dbmH = dibenzoylmethane). *Chem. Commun.* **1994**, 671–673. [CrossRef]

45. Huang, D.; Zhang, X.; Ma, C.; Chen, H.; Chen, C.; Liu, Q.; Zhang, C.; Liao, D.; Li, L. Aggregation of tetranuclear Mn building blocks with alkali ion. Syntheses, crystal structures and chemical behaviors of monomer, dimer and polymer containing [Mn$_4$(μ_3-O)$_2$]$^{8+}$ units. *Dalton Trans.* **2007**, 680–688. [CrossRef]

46. Nakata, K.; Miyasaka, H.; Sugimoto, K.; Ishii, T.; Sugiura, K.; Yamashita, M. Construction of a one-dimensional chain composed of Mn$_6$ clusters and 4,4′-bipyridine linkers: The first step for creation of "Nano-Dots-Wires". *Chem. Lett.* **2002**, *31*, 658–659. [CrossRef]

47. Ma, C.-B.; Hu, M.Q.; Chen, H.; Chen, C.N.; Liu, Q.T. Aggregation of hexanuclear, mixed-valence manganese oxide clusters linked by propionato ligands to form a one-dimensional polymer [Mn$_6$O$_2$(O$_2$CEt)$_{10}$(H$_2$O)$_4$]$_n$. *Eur. J. Inorg. Chem.* **2008**, 5274–5280. [CrossRef]

48. Moushi, E.E.; Tasiopoulos, A.J.; Manos, M.J. Synthesis and structural characterization of a metal cluster and a coordination polymer based on the [Mn$_6$(μ_4-O)$_2$]$^{10+}$ unit. *Bioinorg. Chem. Appl.* **2010**, 367128. [CrossRef]

49. Malaestean, I.L.; Kravtsov, V.C.; Speldrich, M.; Dulcevscaia, G.; Simonov, Y.A.; Lipkowski, J.; Ellern, A.; Baca, S.G.; Kögerler, P. One-dimensional coordination polymers from hexanuclear manganese carboxylate clusters featuring a {Mn^{II}$_4$Mn^{III}$_2$(μ_4-O)$_2$} core and spacer linkers. *Inorg. Chem.* **2010**, *49*, 7764–7772. [CrossRef]

50. Darii, M.; Filippova, I.G.; Hauser, J.; Decurtins, S.; Liu, S.-X.; Kravtsov, V.C.; Baca, S.G. Incorporation of hexanuclear Mn(II,III) carboxylate clusters with a {Mn$_6$O$_2$} core in polymeric structures. *Crystals* **2018**, *8*, 100. [CrossRef]

51. Ovcharenko, V.; Fursova, E.; Romanenko, G.; Ikorskii, V. Synthesis and structure of heterospin compounds based on the [Mn$_6$(O)$_2$Piv$_{10}$]-cluster unit and nitroxide. *Inorg. Chem.* **2004**, *43*, 3332–3334. [CrossRef] [PubMed]

52. Baca, S.G.; Secker, T.; Mikosch, A.; Speldrich, M.; van Leusen, J.; Ellern, A.; Kögerler, P. Avoiding magnetochemical over-parametrization, exemplified by one-dimensional chains of hexanuclear iron(III) pivalate clusters. *Inorg. Chem.* **2013**, *52*, 4154–4156. [CrossRef] [PubMed]

53. Fursova, E.Y.; Ovcharenko, V.I.; Bogomyakov, A.S.; Romanenko, G.V. Structure of the complex of hexanuclear manganese pivalate with isonicotinamide. *J. Struct. Chem.* **2013**, *54*, 164–167. [CrossRef]
54. Malaestean, I.L.; Ellern, A.; van Leusen, J.; Kravtsov, V.C.; Kögerler, P.; Baca, S.G. Cluster-based networks: Assembly of a (4,4) layer and a rare T-shaped bilayer from [Mn$^{III}_2$Mn$^{II}_4$O$_2$(RCOO)$_{10}$] coordination clusters. *Cryst. Eng. Comm.* **2014**, *16*, 6523–6525. [CrossRef]
55. Kar, P.; Haldar, R.; Gómez-García, C.J.; Ghosh, A. Antiferromagnetic porous metal–organic framework containing mixed-valence [Mn$^{II}_4$Mn$^{III}_2$(μ_4-O)$_2$]$^{10+}$ units with catecholase activity and selective gas adsorption. *Inorg. Chem.* **2012**, *51*, 4265–4273. [CrossRef]

Article

The Impact of Structural Defects on Iodine Adsorption in UiO-66

John Maddock [1], Xinchen Kang [1], Lifei Liu [2], Buxing Han [2], Sihai Yang [1,*] and Martin Schröder [1,*]

[1] Department of Chemistry, University of Manchester, Manchester M13 9PL, UK; john.maddock@manchester.ac.uk (J.M.); kangxinchen@iccas.ac.cn (X.K.)

[2] Beijing National Laboratory for Molecular Sciences, CAS Key Laboratory of Colloid, Interface and Chemical Thermodynamics, Institute of Chemistry, Chinese Academy of Science, Beijing 100190, China; liulifei@iccas.ac.cn (L.L.); hanbx@iccas.ac.cn (B.H.)

* Correspondence: sihai.yang@manchester.ac.uk (S.Y.); M.Schroder@manchester.ac.uk (M.S.); Tel.: +44-161-275-1066 (S.Y.); +44-161-306-9119 (M.S.)

Abstract: Radioactive I_2 (iodine) produced as a by-product of nuclear fission poses a risk to public health if released into the environment, and it is thus vital to develop materials that can capture I_2 vapour. Materials designed for the capture and storage of I_2 must have a high uptake capacity and be stable for long-term storage due the long half-life of ^{129}I. UiO-66 is a highly stable and readily tuneable metal-organic framework (MOF) into which defect sites can be introduced. Here, a defective form of UiO-66 (UiO-66-FA) was synthesised and the presence of missing cluster moieties confirmed using confocal fluorescence microscopy and gas sorption measurements. The uptake of I_2 vapour in UiO-66-FA was measured using thermal gravimetric analysis coupled mass spectrometry (TGA-MS) to be 2.25 g g^{-1}, almost twice that (1.17 g g^{-1}) of the pristine UiO-66. This study will inspire the design of new efficient I_2 stores based upon MOFs incorporating structural defects.

Keywords: metal organic frameworks; defect; iodine adsorption; UiO-66

Citation: Maddock, J.; Kang, X.; Liu, L.; Han, B.; Yang, S.; Schröder, M. The Impact of Structural Defects on Iodine Adsorption in UiO-66. *Chemistry* **2021**, *3*, 525–531. https://doi.org/10.3390/chemistry3020037

Academic Editor: Catherine Housecroft

Received: 1 March 2021
Accepted: 8 April 2021
Published: 12 April 2021

Publisher's Note: MDPI stays neutral with regard to jurisdictional claims in published maps and institutional affiliations.

1. Introduction

Nuclear power is responsible for approximately 21% of the UK's energy production as of 2020 [1]. The products produced as a result of the fission of uranium pose a danger to the environment and public health, and so materials that can capture and store such fission products are of significant interest. ^{131}I and ^{129}I are volatile fission products with half-lives of 8 days and 1.57×10^7 years, respectively [2]. ^{131}I poses a serious risk to human health as it has been linked to the occurrence of thyroid cancer [3]. ^{129}I is less hazardous due to its low energy beta and gamma emissions, but poses a long-term environmental risk due to bioaccumulation [4]. It is vital also to prevent I_2 escaping into the environment as it can spread over a wide area due to its high solubility in water [5]. The capture of I_2 has been investigated using a wide variety of porous materials such as aerogels [6], zeolites [7], porous organic polymers [8] and covalent–organic framework [9] materials. The main drawbacks to using these materials are that they can be non-specific for I_2, have low uptakes, or have amorphous structures that prevent determination of preferred binding sites, thus restricting an understanding of the mechanism of action of the material.

Metal–organic framework (MOF) materials are often highly porous and crystalline and are well-known for their tuneable structures, potential high chemical stability and high surface areas [10]. The tunability of MOFs has allowed them to be used in a wide range of applications including catalysis [11], molecular separations [12] and the capture of gases such as CO_2 [13] and SO_2 [14]. The storage of I_2 by MOFs has been reported previously with uptakes reaching as high as 7.35 g g^{-1} in the case of the ionic liquid-doped material PCN-333 [15]. Proof-of-concept studies have also shown that MOFs can be used for the long-term storage of I_2 using glass sintering [16] and pressure-induced amorphization [17].

Functionalisation of the MOF structure is a common route to increasing I_2 adsorption and relies on the introduction of electron donating [18,19] or reactive groups [20] onto the ligands. Doping MOFs with Ag [21] and Cu [22] ions can also improve the I_2 uptake of the host MOF. Detailed studies into the adsorption mechanism of I_2 have also been carried out and highlight the importance of the structure of the MOF [23,24] and of metal cluster nodes for the efficient capture of I_2 [25]. The introduction of structural defects has been shown to increase the catalytic activity of MOFs [26,27] and increase the uptake of CO_2 by UiO-66 [28]. However, the impact of such defects on the adsorption of I_2 has not been investigated previously and we were interested to determine whether this was an appropriate methodology for improved I_2 adsorption (Figure S1, Supplementary Materials).

UiO-66 was chosen in this work due its high stability and the established synthesis and routes to the preparation of defective derivatives. Coordination modulation by undertaking the synthesis of UiO-66 in the presence of formic acid can introduce defects into UiO-66. The competitor ligand formate binds to metal clusters in place of the bridging terephthalate linker to produce missing linker defects. If there is sufficient missing linker, defects will occur within the overall stable geometry with certain metal clusters absent (Figure S2) [29]. We have analysed the presence of missing cluster defects within a sample of defect UiO-66, designated UiO-66-FA, using Brunauer–Emmett–Teller (BET) surface area analysis, thermal gravimetric analysis and confocal fluorescence microscopy. The uptake of I_2 has been measured and confirmed using Raman spectroscopy and thermal gravimetric analysis coupled with mass spectrometry. The presence of structural defects results in a nearly 100% enhancement in the adsorption capacity of I_2 within UiO-66-FA compared to pristine UiO-66.

2. Results and Discussion

UiO-66 was synthesised by dissolving $ZrCl_4$ and terephthalic acid in DMF and heating the solution to 120 °C for 24 h. UiO-66-FA was synthesised using the same general procedure but using a mixture of DMF and formic acid as solvent. Powder X-ray diffraction (PXRD) analysis of as-synthesised UiO-66 and UiO-66-FA (Figure 1) confirmed the phase purity of both samples [30,31]. The PXRD pattern of UiO-66-FA shows two additional Bragg peaks between 5 and 7° attributed to the presence of *reo* regions within the *fcu* structure caused by missing cluster defects [29,32]. These Bragg peaks are also broad, highlighting the disorder of these defect sites throughout the parent structure of UiO-66.

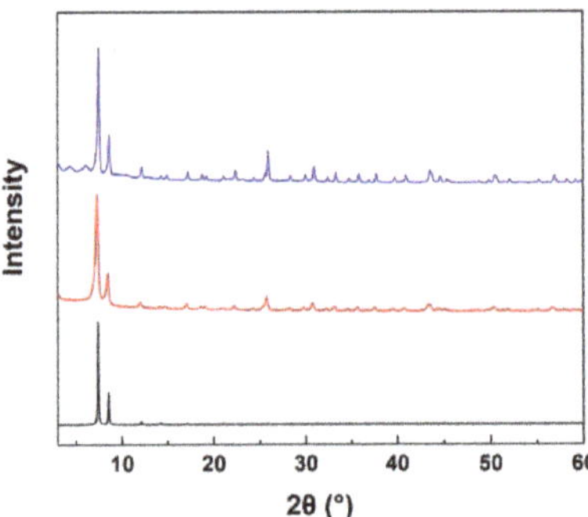

Figure 1. PXRD (powder X-ray diffraction) patterns for UiO-66 synthesised with (blue) and without formic acid (red). Simulated PXRD for UiO-66 (black) [27].

Confocal fluorescence microscopy (CFM) can be used to visualise mesoporous defects within electro-synthesised MOFs [27]. In these systems, Lewis acid sites found in defect sites and boundaries catalyse the formation of a fluorescent oligomer from furfuryl alcohol, the monomer of which is not fluorescent (Figure S3). On exposure of UiO-66 and UiO-66-FA to furfuryl alcohol it was noted that the UiO-66-FA sample had a darker colour than the sample exposed to UiO-66 (Figure S4). This suggested that a more fluorescent oligomer was being produced by UiO-66-FA as the oligomer has a dark brown colour compared to the colourless furfuryl alcohol. This conclusion was supported by CFM which showed a uniform spread of high intensity fluorescence throughout the sample of UiO-66-FA exposed to furfuryl alcohol. The sample of UiO-66 exposed to furfuryl alcohol shows weaker fluorescence which is not evenly distributed across the sample (Figure 2). The relatively small amount of fluorescence and its location for the UiO-66 sample can be explained by the presence of Lewis acid sites at the edges of the crystals, with the high intensity fluorescence across the sample of UiO-66-FA confirming the presence of an increased number of defects within the structure of UiO-66-FA compared to UiO-66.

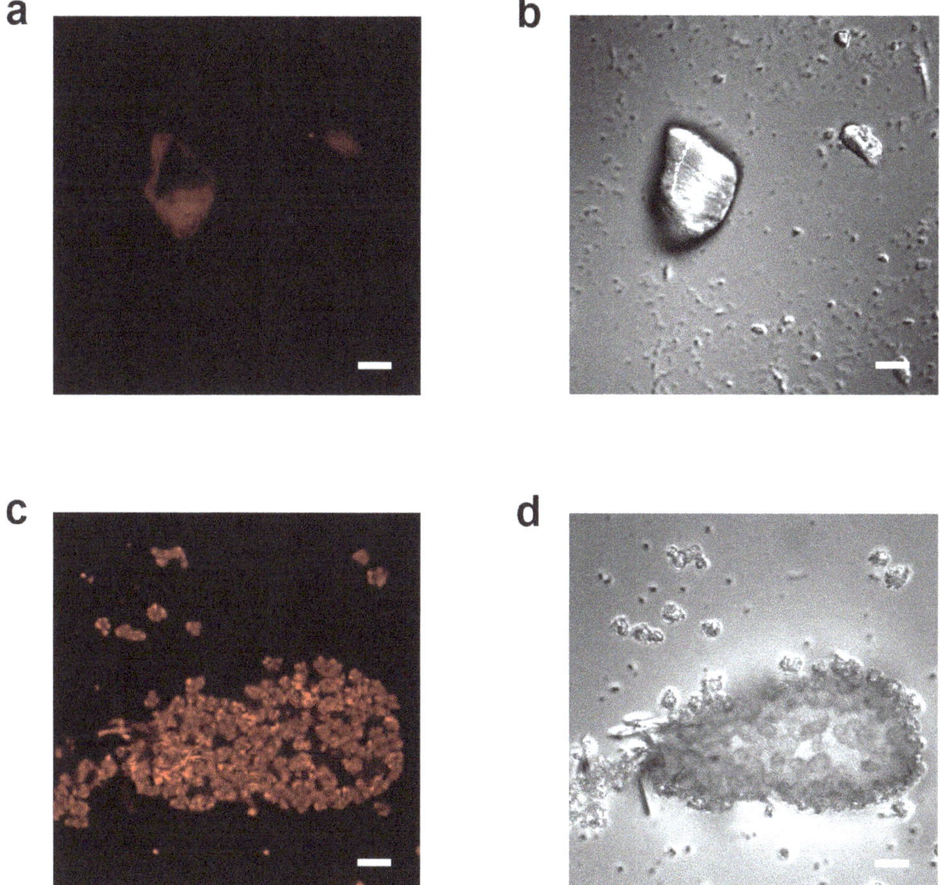

Figure 2. CFM (confocal fluorescence microscopy) and micrograph images: (**a**) fluorescence micrograph of UiO-66; (**b**) micrograph of UiO-66; (**c**) fluorescence micrograph of UiO-66-FA; (**d**) micrograph of UiO-66-FA. Scale bars are 10 μm, 10 μm, 5 μm and 5 μm, respectively.

UiO-66-FA shows a higher BET surface area than UiO-66, 1705 and 1170 m^2 g^{-1}, respectively, as determined from the N_2 adsorption isotherm, consistent with the removal of linkers and clusters to form defect sites in UiO-66-FA. The pore size distribution data (Figure S5) confirm that larger pores are present in UiO-66-FA with significant peaks above 8 Å radius attributed to the large pores created due to missing cluster moieties. In contrast, UiO-66 only shows pores of less than 9 Å radius [30]. The calculated micropore volume also supports the formation of larger pores with 0.30 cm^3 g^{-1} in UiO-66 and 0.73 cm^3 g^{-1} in UiO-66-FA. PXRD, CFM and BET results confirm the presence of defects in UiO-66-FA.

I_2 adsorption was carried out in a sealed flask by heating for 3 days the activated, desolvated solid MOF sample with solid I_2, each inside an open glass vessel. The MOF sample was then removed from the vessel for further analysis. TGA analysis of the I_2-loaded samples showed a drop in mass between 100 and 200 °C attributed to loss of I_2 as monitored by mass spectrometry (Figure 3). The I_2 uptake over three repeat cycles gave an average uptake for UiO-66 and UiO-66-FA of 1.17 and 2.25 g g^{-1}, respectively, consistent with the increased porosity due to defects in the structure of UiO-66-FA. The weight drop observed at around 500 °C is linked to decomposition of the ligand and the percentage drop for UiO-66-FA is less than that of UiO-66, reflecting fewer ligands present in the UiO-66-FA. The Raman spectrum of solid I_2 shows a peak at 180 cm^{-1} assigned to the υI-I stretching vibration. This peak is shifted in both I_2-loaded UiO-66 samples (Figure 4), and the presence of this single peak rules out the presence of triiodide that would produce a peak between 110 and 140 cm^{-1} [33]. Other observable features in the complete Raman spectra (Figure S6) include peaks at 1600 cm^{-1}, 1150 cm^{-1} and 850 cm^{-1} assigned to the C-C bonds in the aromatic ring of the ligand [34]. The overlapping peaks seen at 1400 cm^{-1} are due to the COO stretching vibration of the linker overlapping with another aromatic ring Raman peak. There is no change to these peaks upon I_2 adsorption in both UiO-66 and UiO-66-FA, confirming that the linker remains intact after I_2 adsorption. I_2 has strong interaction with unsaturated zirconium clusters and increased access to the framework of UiO-66-FA results in a high I_2 uptake.

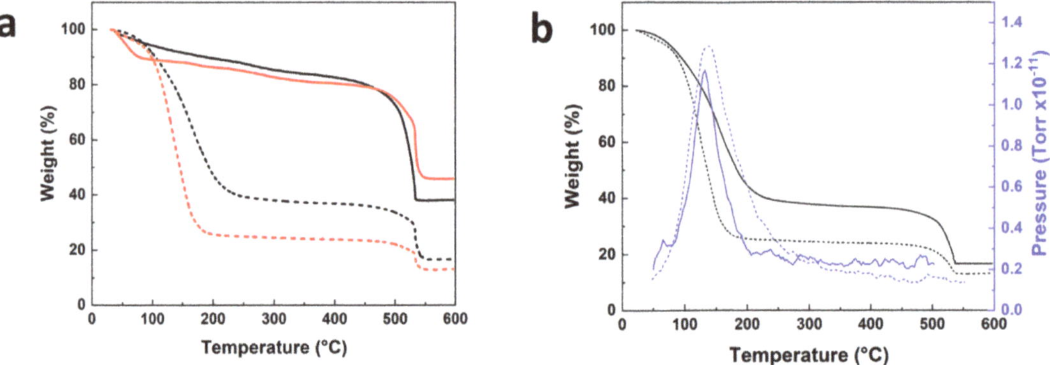

Figure 3. (**a**) TGA for UiO-66 (black) and UiO-66-FA (red); dashed lines indicate I_2-loaded samples. (**b**) TGA-MS results (black line) for I_2-loaded UiO-66 (solid line) and I_2-loaded UiO-66-FA (dashed).

The cycling of I_2 adsorption was carried out to show that even when clusters are removed from UiO-66 the structure remains stable on the adsorption and desorption of I_2 (Figure 5). The decrease in I_2 uptake observed after each cycle in derivatives of UiO-66 has been observed previously [25,35]. However, this drop in uptake appears to be less for UiO-66-FA, which could be due to the openness of the structure reducing the impact of I_2 removal. Samples of MOF were monitored by PXRD on removal of I_2 from loaded material (Figures S7 and S8). The PXRD patterns show little change in the structure or crystallinity throughout the cycling experiments, confirming that defects within UiO-66-FA do not decrease its stability on I_2 loading.

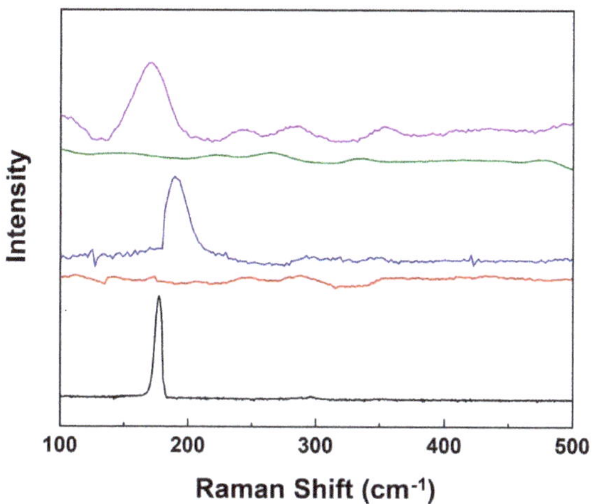

Figure 4. Raman spectra of solid I_2 (black), activated UiO-66 (blue), I_2-loaded UiO-66 (red), activated UiO-66-FA (green) and I_2-loaded UiO-66-FA (purple). The complete spectra are shown in Figure S5.

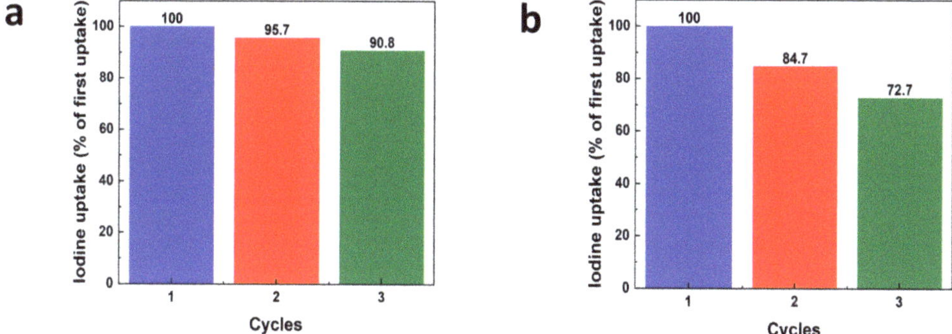

Figure 5. I_2 uptakes compared to the first cycle for (**a**) UiO-66 and (**b**) UiO-66-FA. First cycle (blue), second cycle (red) and third cycle (green). PXRD of samples after I_2 desorption are shown in Figures S7 and S8.

In summary, the impact of structural defects on I_2 uptake capacity has been studied in UiO-66 and UiO-66-FA. The presence of defects in UiO-66-FA was verified by BET surface area, TGA and CFM analysis. Defects caused by missing clusters within the structure of UiO-66 results in an increase I_2 adsorption from 1.17 to 2.25 g g^{-1} for UiO-66 and UiO-66-FA, respectively, and the overall increased porosity of the latter also contributes to higher I_2 uptake. Cycling of I_2 loading in UiO-66-FA confirms that I_2 uptake can be increased without compromising the stability of the MOF, an approach that can be applied potentially to other capture systems.

Supplementary Materials: The following are available online at https://www.mdpi.com/article/10.3390/chemistry3020037/s1, Figure S1: Colour change observed in UiO-66-FA with loading of I_2. Figure S2: Structure of UiO-66 and of UiO-66-FA. Figure S3: Reaction scheme for the synthesis of the fluorescent oligomer from furfuryl alcohol. Figure S4: Sample of UiO-66 and UiO-66-FA in furfuryl alcohol after oligomerization reaction. Figure S5: N_2 adsorption isotherm and pore size distribution of UiO-66 and UiO-66-FA. Figure S6: Raman spectra of I_2, UiO-66, I_2-loaded UiO-66, UiO-66-FA, and I_2-loaded UiO-66-FA. Figure S7: PXRD of UiO-66 after removal of captured I_2. Figure S8: PXRD of UiO-66-FA after removal of captured I_2.

Author Contributions: J.M. and X.K.: all experimental work; J.M. and L.L.: validation experiments; B.H., S.Y. and M.S.: design of project, supervision and overall strategy of project. All authors have contributed to the preparation of the manuscript. All authors have read and agreed to the published version of the manuscript.

Funding: We thank EPSRC (EP/I011870) and the Royal Society (IC170327), National Natural Science Foundation of China (21733011, 21890761), the University of Manchester and Institute of Chemistry, Chinese Academy of Sciences for funding. This project has received funding from the European Research Council (ERC) under the European Union's Horizon 2020 research and innovation programme (grant agreement No 742401, *NANOCHEM*). X.K. is supported by Royal Society Newton International Fellowship.

Conflicts of Interest: The authors declare no conflict of interest.

References

1. World Nuclear Association. Nuclear Power in the United Kingdom. Available online: https://www.world-nuclear.org/information-library/country-profiles/countries-t-z/united-kingdom.aspx (accessed on 27 June 2020).
2. Audi, G.; Bersillon, O.; Blachot, J.; Wapstra, A.H. The NUBASE evaluation of nuclear and decay properties. *Nucl. Phys. A* **2003**, *729*, 3–128. [CrossRef]
3. Shakhtarin, V.V.; Tsyb, A.F.; Stepanenko, V.F.; Orlov, M.Y.; Kopecky, K.J.; Davis, S. Iodine deficiency, radiation dose, and the risk of thyroid cancer among children and adolescents in the Bryansk region of Russia following the Chernobyl power station accident. *Int. J. Epidemiol.* **2003**, *32*, 584–591. [CrossRef]
4. Satoh, Y.; Ueda, S.; Kakiuchi, H.; Ohtsuka, Y.; Hisamatsu, S.J. Concentrations of iodine-129 in coastal surface sediments around spent nuclear fuel reprocessing plant at Rokkasho, Japan, during and after its test operation. *Radioanal. Nucl. Chem.* **2019**, *322*, 2019–2024. [CrossRef]
5. Ioannides, K.; Stamoulis, K.; Papachristodoulou, C.J. Environmental radioactivity measurements in north–western Greece following the Fukushima nuclear accident. *Radioanal. Nucl. Chem.* **2013**, *298*, 1207–1213. [CrossRef]
6. Riley, B.J.; Chun, J.; Ryan, J.V.; Matyáš, J.; Li, X.S.; Matson, D.W.; Sundaram, S.K.; Strachan, D.M.; Vienna, J.D. Chalcogen-based aerogels as a multifunctional platform for remediation of radioactive iodine. *RSC Adv.* **2011**, *1*, 1704–1715. [CrossRef]
7. Chibani, S.; Chebbi, M.; Lebègue, S.; Cantrel, L.; Badawi, M. Impact of the Si/Al ratio on the selective capture of iodine compounds in silver-mordenite: A periodic DFT study. *Phys. Chem. Chem. Phys.* **2016**, *18*, 25574–25581. [CrossRef]
8. Su, K.; Wang, W.; Li, B.; Yuan, D. Azo-bridged calix[4]resorcinarene-based porous organic frameworks with highly efficient enrichment of volatile iodine. *ACS Sustain. Chem. Eng.* **2018**, *6*, 17402–17409. [CrossRef]
9. Wang, C.; Wang, Y.; Ge, R.; Song, X.; Xing, X.; Jiang, Q.; Lu, H.; Hao, C.; Guo, X.; Gao, Y.; et al. A 3D covalent organic framework with exceptionally high iodine capture capability. *Chem. Eur. J.* **2018**, *24*, 585–589. [CrossRef]
10. Furukawa, H.; Cordova, K.E.; O'Keeffe, M.; Yaghi, O.M. The chemistry and applications of metal-organic frameworks. *Science* **2013**, *341*, 1230444. [CrossRef] [PubMed]
11. Caratelli, C.; Hajek, J.; Cirujano, F.G.; Waroquier, M.; Xamena, F.X.L.I.; van Speybroeck, V. Nature of active sites on UiO-66 and beneficial influence of water in the catalysis of Fischer esterification. *J. Catal.* **2017**, *352*, 401–414. [CrossRef]
12. Herm, Z.R.; Bloch, E.D.; Long, J.R. Hydrocarbon Separations in metal–organic frameworks. *Chem. Mater.* **2014**, *26*, 323–338. [CrossRef]
13. Ding, M.; Flaig, R.W.; Jiang, H.L.; Yaghi, O.M. Carbon capture and conversion using metal–organic frameworks and MOF-based materials. *Chem. Soc. Rev.* **2019**, *48*, 2783–2828. [CrossRef]
14. Smith, G.L.; Eyley, J.E.; Han, X.; Zhang, X.; Li, J.; Jacques, N.M.; Godfrey, H.G.W.; Argent, S.P.; McPherson, L.J.M.; Teat, S.J.; et al. Reversible coordinative binding and separation of sulfur dioxide in a robust metal–organic framework with open copper sites. *Nat. Mater.* **2019**, *18*, 1358–1365. [CrossRef]
15. Tang, Y.; Huang, H.; Li, J.; Xue, W.; Zhong, C. IL-induced formation of dynamic complex iodide anions in IL@MOF composites for efficient iodine capture. *J. Mater. Chem. A* **2019**, *7*, 18324–18329. [CrossRef]
16. Sava, D.F.; Garino, T.J.; Nenoff, T.M. Iodine confinement into metal–organic frameworks (mofs): Low-temperature sintering glasses to form novel glass composite material (GCM) alternative waste forms. *Ind. Eng. Chem. Res.* **2012**, *51*, 614–620. [CrossRef]
17. Chapman, K.W.; Sava, D.F.; Halder, G.J.; Chupas, P.J.; Nenoff, T.M. Trapping guests within a nanoporous metal-organic framework through pressure-induced amorphization. *J. Am. Chem. Soc.* **2011**, *133*, 18583–18585. [CrossRef] [PubMed]
18. Munn, A.S.; Millange, F.; Frigoli, M.; Guillou, N.; Falaise, C.; Stevenson, V.; Volkringer, C.; Loiseau, T.; Cibin, G.; Walton, R.I. Iodine sequestration by thiol-modified MIL-53(Al). *CrystEngComm* **2016**, *18*, 8108–8114. [CrossRef]
19. Falaise, C.; Volkringer, C.; Facqueur, J.; Bousquet, T.; Gasnot, L.; Loiseau, T. Capture of iodine in highly stable metal–organic frameworks: A systematic study. *Chem. Commun.* **2013**, *49*, 10320–10322. [CrossRef]
20. Marshall, R.J.; Griffin, S.L.; Wilson, C.; Forgan, R.S. Postsynthetic bromination of UiO-66 analogues: Altering linker flexibility and mechanical compliance. *Chem. Eur. J.* **2016**, *22*, 4870–4877. [CrossRef] [PubMed]
21. Chen, J.; Gao, Q.; Zhang, X.; Liu, Y.; Wang, P.; Jiao, Y.; Yang, Y. Nanometer mixed-valence silver oxide enhancing adsorption of ZIF-8 for removal of iodide in solution. *Sci. Total Environ.* **2019**, *646*, 634–644. [CrossRef]

22. Qi, B.; Liu, Y.; Zheng, T.; Gao, Q.; Yan, X.; Jiao, Y.; Yang, Y.J. Highly efficient capture of iodine by Cu/MIL-101. *Solid State Chem.* **2018**, *258*, 49–55. [CrossRef]

23. Zhang, X.; da Silva, I.; Godfrey, H.G.W.; Callear, S.K.; Sapchenko, S.A.; Cheng, Y.; Vitórica-Yrezábal, I.; Frogley, M.D.; Cinque, G.; Tang, C.C.; et al. Confinement of iodine molecules into triple-helical chains within robust metal–organic frameworks. *J. Am. Chem. Soc.* **2017**, *139*, 16289–16296. [CrossRef]

24. Gallis, D.F.S.; Ermanoski, I.; Greathouse, J.A.; Chapman, K.W.; Nenoff, T.M. Iodine gas adsorption in nanoporous materials: A combined experiment–modeling study. *Ind. Eng. Chem. Res.* **2017**, *56*, 2331–2338. [CrossRef]

25. Chen, P.; He, X.; Pang, M.; Dong, X.; Zhao, S.; Zhang, W. Iodine capture using Zr-based metal–organic frameworks (Zr-MOFs): Adsorption performance and mechanism. *ACS Appl. Mater. Interfaces* **2020**, *12*, 20429–20439. [CrossRef]

26. Chaemchuen, S.; Luo, Z.; Zhou, K.; Mousavi, B.; Phatanasri, S.; Jaroniec, M.; Verpoort, F. Defect formation in metal–organic frameworks initiated by the crystal growth-rate and effect on catalytic performance. *J. Catal.* **2017**, *354*, 84–91. [CrossRef]

27. Kang, X.; Lyu, K.; Li, L.; Li, J.; Kimberley, L.; Wang, B.; Liu, L.; Cheng, Y.; Frogley, M.D.; Rudić, S.; et al. Integration of mesopores and crystal defects in metal-organic frameworks via templated electrosynthesis. *Nat. Commun.* **2019**, *10*, 1–9. [CrossRef] [PubMed]

28. Wu, H.; Chua, Y.S.; Krungleviciute, V.; Tyagi, M.; Chen, P.; Yildirim, T.; Zhou, W. Unusual and highly tunable missing-linker defects in zirconium metal–organic framework UIO-66 and their important effects on gas adsorption. *J. Am. Chem. Soc.* **2013**, *135*, 10525–10532. [CrossRef]

29. Shearer, G.C.; Vitillo, J.G.; Bordiga, S.; Svelle, S.; Olsbye, U.; Lillerud, K.P. Functionalizing the defects: Postsynthetic ligand exchange in the metal organic framework UiO-66. *Chem. Mater.* **2016**, *28*, 7190–7193. [CrossRef]

30. Koutsianos, A.; Kazimierska, E.; Barron, A.R.; Taddei, M.; Andreoli, E. A new approach to enhancing the CO_2 capture performance of defective UiO-66 via post-synthetic defect exchange. *Dalton Trans.* **2019**, *48*, 3349–3359. [CrossRef]

31. Cavka, J.H.; Jakobsen, S.; Olsbye, U.; Guillou, N.; Lamberti, C.; Bordiga, S.; Lillerud, K.P. A new zirconium inorganic building brick forming metal organic frameworks with exceptional stability. *J. Am. Chem. Soc.* **2008**, *130*, 13850–13851. [CrossRef] [PubMed]

32. DeStefano, M.R.; Islamoglu, T.; Garibay, S.J.; Hupp, J.T.; Farha, O.K. Room-temperature synthesis of UiO-66 and thermal modulation of densities of defect sites. *Chem. Mater.* **2017**, *29*, 1357–1361. [CrossRef]

33. Svensson, P.H.; Kloo, L. Synthesis, structure, and bonding in polyiodide and metal iodide—iodine systems. *Chem. Rev.* **2003**, *103*, 1649–1684. [CrossRef] [PubMed]

34. Schwan, J.; Ulrich, S.; Batori, V.; Ehrhardt, H.; Silva, S.R.P. Raman spectroscopy on amorphous carbon films. *J. Appl. Phys.* **1996**, *80*, 440–447. [CrossRef]

35. Wang, Z.; Huang, Y.; Yang, J.; Li, Y.; Zhuang, Q.; Gu, J. The water-based synthesis of chemically stable Zr-based MOFs using pyridine-containing ligands and their exceptionally high adsorption capacity for iodine. *Dalton Trans.* **2017**, *46*, 7412–7420. [CrossRef] [PubMed]

chemistry

MDPI

Article

Fluorescent Detection of Carbon Disulfide by a Highly Emissive and Robust Isoreticular Series of Zr-Based Luminescent Metal Organic Frameworks (LMOFs)

Ever Velasco [1,†], Yuki Osumi [1,†], Simon J. Teat [2], Stephanie Jensen [3], Kui Tan [4], Timo Thonhauser [3] and Jing Li [1,*]

[1] Department of Chemistry and Chemical Biology, Rutgers University, 123 Bevier Road, Piscataway, NJ 08854, USA; eov2@chem.rutgers.edu (E.V.); yto1@scarletmail.rutgers.edu (Y.O.)

[2] Advanced Light Source Lawrence Berkeley National Laboratory, Berkeley, CA 94720, USA; sjteat@lbl.gov

[3] Department of Physics and Center for Functional Materials, Wake Forest University, Winston-Salem, NC 27109, USA; jensensj@wfu.edu (S.J.); thonhauser@wfu.edu (T.T.)

[4] Department of Materials Science & Engineering, University of Texas at Dallas, Richardson, TX 75080, USA; kuitan@utdallas.edu

* Correspondence: jingli@rutgers.edu

† These authors contributed equally to this work.

Abstract: Carbon disulfide (CS_2) is a highly volatile neurotoxic species. It is known to cause atherosclerosis and coronary artery disease and contributes significantly to sulfur-based pollutants. Therefore, effective detection and capture of carbon disulfide represents an important aspect of research efforts for the protection of human and environmental health. In this study, we report the synthesis and characterization of two strongly luminescent and robust isoreticular metal organic frameworks (MOFs) $Zr_6(\mu_3\text{-}O)_4(OH)_8(tcbpe)_2(H_2O)_4$ (here termed **1**) and $Zr_6(\mu_3\text{-}O)_4(OH)_8(tcbpe\text{-}f)_2(H_2O)_4$ (here termed **2**) and their use as fluorescent sensors for the detection of carbon disulfide. Both MOFs demonstrate a calorimetric bathochromic shift in the optical bandgap and strong luminescence quenching upon exposure to carbon disulfide. The interactions between carbon disulfide and the frameworks are analyzed by in-situ infrared spectroscopy and computational modelling by density functional theory. These results reveal that both the Zr metal node and organic ligand act as the preferential binding sites and interact strongly with carbon disulfide.

Keywords: metal organic frameworks (MOFs); coordination polymers (CP); luminescent sensing; carbon disulfide

Citation: Velasco, E.; Osumi, Y.; Teat, S.J.; Jensen, S.; Tan, K.; Thonhauser, T.; Li, J. Fluorescent Detection of Carbon Disulfide by a Highly Emissive and Robust Isoreticular Series of Zr-Based Luminescent Metal Organic Frameworks (LMOFs). *Chemistry* **2021**, *3*, 327–337. https://doi.org/10.3390/chemistry3010024

Received: 23 December 2020
Accepted: 17 February 2021
Published: 1 March 2021

1. Introduction

Metal organic frameworks (MOFs) are an extremely versatile class of crystalline, permanently porous inorganic-organic materials that have gained significant attention over the past two decades. These materials are constructed from the self-assembly process of metal ions or metal clusters and organic linkers to form coordinatively bonded extended and porous network structures. The ability to control the nature of the metal ions and linkers make MOFs a promising class of materials for applications in gas storage and separation, catalysis, luminescent sensing, and other areas [1–9].

The incorporation of either luminescent metal nodes or organic linkers into frameworks make it possible to design and construct luminescent MOFs (LMOFs). As a subcategory of MOFs, LMOFs have been extensively studied for the luminescent detection of chemical species, as LED phosphors, and so on [10–16]. As chemical sensors, they serve as promising candidates for the detection of a wide range of molecules whereby analyte interactions alter the optical properties of the LMOF. These changes in their optical properties may manifest as either quenching or enhancement in luminescence intensity, and/or shifts in emission energy. Such effects are governed by electron or energy transfer processes that occur between the analyte and the framework [5,17]. LMOF-based sensors offer rapid and sensitive detection limits resulting in a promising and simple alternative for current sensing technology that requires the use of expensive instrumentation.

Carbon disulfide (CS_2) is a highly volatile neurotoxic species produced in the manufacturing of viscose rayon fibers, the production of agricultural pesticides, as a vulcanizing accelerant, and as a chemical by-product of syngas produced from biomass [18–20]. CS_2 is a nonpolar organic solvent that can dissolve a wide range of chemicals; however, it is also known to cause atherosclerosis and coronary artery disease [21]. In addition to its detrimental effects on human health, environmentally, CS_2 is known to contribute significantly to sulfur-based pollutants as it readily oxidizes into carbonyl sulphide and sulfur dioxide that contribute to the formation of acid rain [22]. Despite the toxicity of CS_2 there has been limited work on the luminescent detection of CS_2 and less for the capture of this chemical using MOFs [23]. While MOFs have been reported previously for the luminescent sensing of CS_2, they are limited to magnesium, indium, calcium, and lead-based coordination polymers which often suffer from framework instability [24–27]. In addition, the exact interactions between the frameworks and CS_2 were not investigated in depth in these studies. To the best of our knowledge, the current work represents the first example that makes use of highly robust zirconium-based MOFs for the detection of CS_2. Our study is not only on the luminescent sensing performance of the two Zr-LMOFs but also on the molecular interactions that occur in the MOF-analyte system with the help of in-situ Fourier-transform infrared (FTIR) spectroscopic experiments and density functional theory (DFT) calculations.

2. Materials and Methods

2.1. Materials

All reagents were used directly from commercially available sources without further purification. Both ligands were synthesized according to previously published methods with slight variations [28,29]. Their synthesis is outlined in Schemes 1 and 2, and detailed in the Supporting Information.

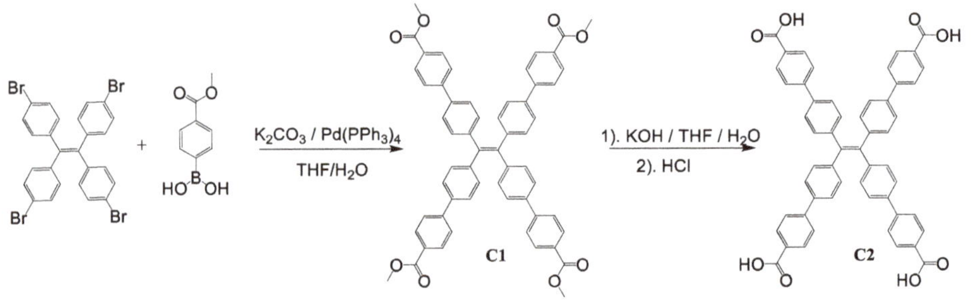

Scheme 1. Synthesis of H$_4$tcbpe.

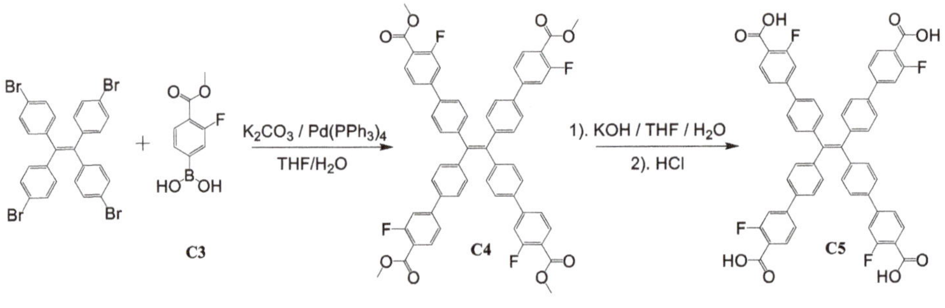

Scheme 2. Synthesis of H$_4$-tcbpe-F.

2.2. Synthesis of 1 and 2

$Zr_6(\mu_3\text{-}O)_4(OH)_8(tcbpe)_2(H_2O)_4$ (here termed **1**). A solution containing 35.7 mg of $ZrOCl_2 \cdot xH_2O$ was suspended in 3 mL of DMA, to which 1.7 mL of formic acid was added and sonicated for 10 min until fully dissolved. To this solution was added 16.7 mg of H_4tcbpe and it was sonicated for an additional 10 min. The suspension was then placed in a pre-heated 120 °C oven for 48 h. Reactions were filtered to afford (20 mg 53.1% yield based on the inorganic salt of) pale green needle-shaped crystals suitable for single crystal X-ray diffraction.

$Zr_6(\mu_3\text{-}O)_4(OH)_8(tcbpe\text{-}f)_2(H_2O)_4$ (here termed **2**). Similarly, 35.7 mg of $ZrOCl_2 \cdot xH_2O$, 3 mL of DMA, and 1.7 mL of formic acid were added into a 20 mL scintillation vial and sonicated until dissolved. Upon full dissolution, 18.5 mg of H_4tcbpe-F was added into the vial and it was sonicated for 10 min. The cloudy suspension was placed into a preheated oven at 120 °C for 48 h. Pale green needle-shaped single crystals were collected after filtration (20 mg, 52% yield based on the inorganic salt).

2.3. Powder X-ray Diffraction (PXRD)

Powder X-ray diffraction measurements were taken at room temperature from 3–35° (2θ) with a scan speed of $2°\ min^{-1}$ using a copper k_α radiation source ($\lambda = 1.5406$ Å) from a Rigaku Ultima-IV X-ray diffractometer (Billerica, MA, USA).

2.4. Thermogravimetric Analysis (TGA)

The thermal decomposition analysis was performed using a TA instruments Q5000IR analyzer (TA Instruments, New Castle, DE, USA) under a flow of nitrogen and sample purge rate at 10 mL/min and 12 mL/min, respectively. About 10 mg of sample was loaded on the platinum pan and gradually increased from room temperature to 600 °C with a heating rate of $10\ °C\ min^{-1}$.

2.5. Diffuse Reflectance Measurement(UV-VIS)

Absorption spectra at room temperature were collected with a Shimadzu UV-3600 (Kyoto, Japan) from 200–800 nm. Samples were prepared by evenly loading the powders in between two quartz glass plates. The collected reflectance data were then converted through the Kubelka–Munk function, $\alpha/S = (1 - R)^2/2R$ (α is absorption coefficient, S is scattering coefficient, and R is reflectance).

2.6. Photoluminescence Spectra (PL)

Photoluminescence data were recorded at room temperature using a Duetta fluorescence and absorbance spectrometer from Horiba scientific (Kyoto, Japan). Fluorescence measurements of single crystals of **1** or **2** were collected at different states. Since **1** and **2** undergo mechanochromic shifts in emission that resemble those of outgassed samples, the fluorescence spectra were collected in both the as-made and outgassed states. To collect fluorescent measurements for as-made samples, reactions were filtered off, washed with DMF and immediately measured. Outgassed samples were dried prior to treatment with the specified solvent exchanged measurements without grinding.

2.7. Internal Quantum Yield Measurements (QY)

Internal quantum yield measurements were measured on a C9920-02 absolute quantum yield system from Hamamatsu Photonics (Shizuoka, Japan). The light source is a 150 W xenon monochromatic with a 3.3 inch integrating sphere. Either sodium salicylate (360 nm excitation) or cerium-doped yttruim aluminum garnet (YAG:Ce, 450 nm excitation) were used as standards for QY measurements with QY values of 60% and 95% at their respective wavelengths.

2.8. Chemical Stability Experiments

The chemical stability of **1** and **2** was explored: in brief, approximately 20 mg of **1** or **2** was immersed into 10 mL of different solutions of varying pH and allowed to set for 48 h. Samples were collected through vacuum filtration, washed with 3 aliquots of water (20 mL) and 3 aliquots of N',N'-Dimethylformamide (10 mL). After washing, samples were collected and used to collect PXRD patterns.

2.9. Fluorescent Sensing Experiments

Fluorescent measurements for the detection of CS_2 were performed using 5 mg of **1** and **2** previously ground in an agate mortar and pestle with 2 mL of DMF. The resulting slurry was then pipetted into a cuvette. Grinding in DMF allows for the formation of fine powder without affecting the luminescence profile of the as-made samples (mechanochromic behavior if ground dry, Figure S19). The PL spectra of the outgassed samples may be replicated after extensive grinding. Under continuous stirring, CS_2 was added in and allowed to stir for 5 min in between sample measurements of increasing concentration of CS_2.

2.10. In-Situ Infrared Spectroscopy (FTIR)

In-situ IR measurements were performed on a Nicolet 6700 FTIR spectrometer (Waltham, MA, USA) using a liquid N_2-cooled mercury cadmium telluride (MCT-A) detector. A vacuum cell is placed in the sample compartment of the infrared spectrometer with the sample at the focal point of the beam. The samples (~5 mg) were gently pressed onto KBr pellets and placed into a cell that was connected to a vacuum line for evacuation. The samples of **1** and **2** were activated by evacuation overnight at 150 °C, respectively, and then cooled back to room temperature for CS_2 vapor exposure measurement.

2.11. Computational Methods

We explored the interactions of the CS_2 with the LMOFs by looking at the charge rearrangement upon binding of the CS_2 by subtracting CS_2 and framework charge densities from the CS_2 + framework charge density. This allowed us to see how the guest molecules rearranged the charge density of the system. The charge densities were obtained by first calculating the optimized structures in the Vienna Ab initio Simulation Package (VASP) [30,31] at the DFT level with the vdW-DF exchange-correlation functional [32–34] in order to capture the long range van der Waals interactions between the CS_2 and the framework. Optimizations were performed until SCF loops reached an energy convergence of 1×10^{-4} eV and forces were below 1 meV/Å for each atom. Only the Γ-point was considered with a plane-wave energy cutoff of 600 eV. Once the optimizations were completed, the charge density images were created using the aforementioned methods.

3. Results and Discussion

Here, we report the synthesis of $Zr_6(\mu_3\text{-}O)_4(OH)_8(\text{tcbpe})_2(H_2O)_4$ (**1**) and $Zr_6(\mu_3\text{-}O)_4(OH)_8(\text{tcbpe-F})_2(H_2O)_4$ (**2**) through modified conditions [35] using $ZrOCl_2\cdot xH_2O$, $H_4\text{tcbpe}$ (or $H_4\text{tcbpe-F}$), DMA (N,N'-dimethylacetamide) and formic acid at 120 °C for 48 h. Compound **1** crystallizes in an orthorhombic crystal system, space group *Cmcm* with $Zr_6O_4(OH)_8(H_2O)_4$ secondary-building units (SBUs) that are bridged together by 4-connected organic linkers. The overall structure is a two-fold interpenetrated network with *scu* topology (Figure 1a–d). Single crystals for **2** were not solved but the PXRD patterns of as-made **1**, as-made **2**, and the simulated diffraction pattern of **1** are in good agreement, suggesting it is isoreticular to **1** (Figure 2a). Both compounds inherit flexibility from their interpenetrated network where the nature of the solvent within the pores can alter the structure slightly. Compound **1** was previously reported under packing with solvents such as DMF, EtOH, Et$_2$O, MeOH, and toluene; here we include the single crystal structure for the altered packing with DMA molecules within the pores (CCDC: 2005129). The chemical stability of **1** and **2** in water and acidic/basic aqueous solutions at room temperature was confirmed by PXRD analysis (Figure 2b and Figure S1, Electronic Supporting Information (ESI)). The thermal stability of the two compounds was evaluated by thermogravimetric (TG) analysis. Both compounds remained highly crystalline upon heating at 150 and 200 °C for 6 h, as confirmed by PXRD patterns (Figures S2 and S3).

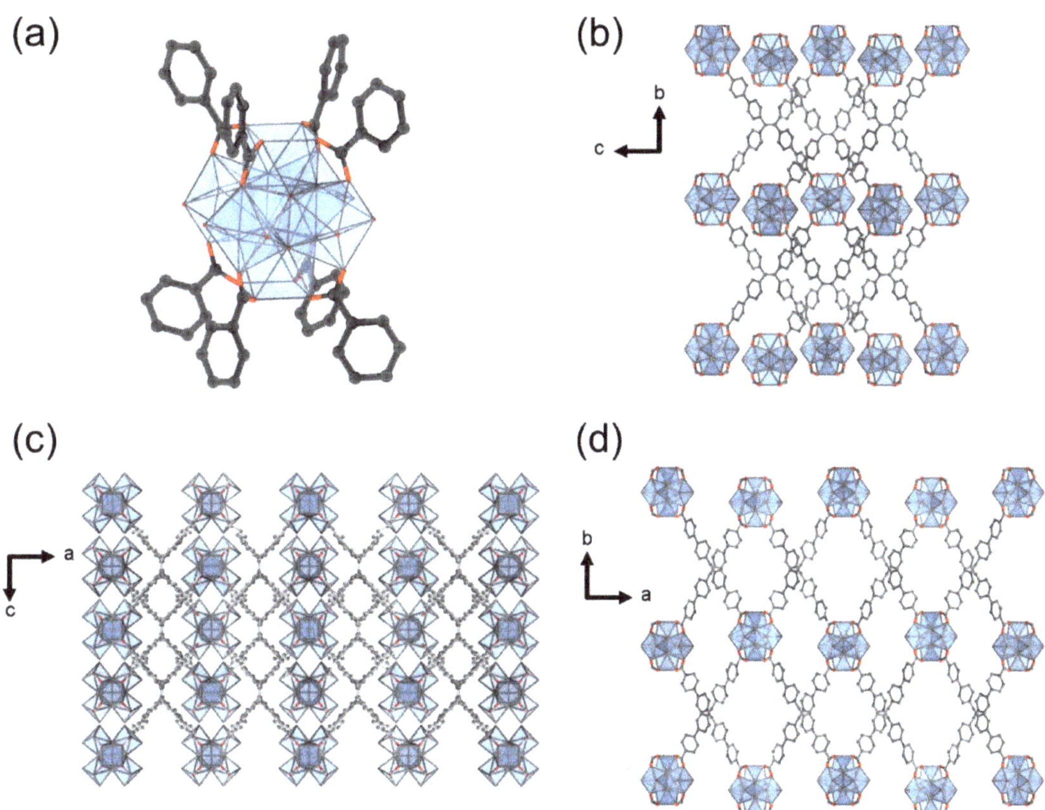

Figure 1. Crystal structure of **1**. (**a**) 8-connected zirconium SBU, (**b**) two-fold interpenetrated nets along the crystallographic *a*-axis, (**c**) *b*-axis, and (**d**) *c*-axis.

As demonstrated in previous work, the incorporation of a strongly emissive ligand with aggregation induced-emission (AIE) properties into a framework results in highly emissive LMOFs [28,29,36–38]. The optical properties of **1** and **2** were analyzed by UV-Vis and photoluminescence (PL) spectroscopy. Under the excitation of UV light (365 nm) **1** has an emission centered at 457 nm (blue color) with a relatively high internal quantum yield (IQY) of 68%. Upon outgassing, **1** undergoes a bathochromic shift and emits at 532 nm (green color) under 455 nm excitation (Figure S4, ESI). An IQY value of 69% is achieved. Similarly, the as-made sample of **2** exhibits an emission peak centered at 492 nm under 365 nm excitation and an IQY value of 48.7%. Upon fully outgassing, **2** displays an emission centered at 557 nm under 455 nm excitation with a significantly increased IQY value of 73.6% (Figure S5, ESI). The emission profiles of the individual ligands used in the assembly of **1** and **2** resemble the emission profiles of the outgassed versions of both materials (Figure S6, ESI). The observed shift in emission of **1** and **2** corresponds well with their optical band gap shifts from 2.72 eV to 2.44 eV (Figure S7, ESI) and from 2.53 eV to 2.42 eV (Figure S8, ESI) in the as-made and outgassed samples, respectively. To outgas samples and retain permanent porosity we referred to a previously published technique that exploits the use of organic solvents with low dipole moments and surface tension due to the structural sensitivity upon outgassing [39]. Briefly, samples were solvent exchanged with dimethylformamide, dichloromethane and finally hexane without letting samples dry in between steps. Small amounts of samples were removed and studied by TGA to ensure complete removal of solvents in between solvent exchange steps (Figure S10, ESI). Using this solvent exchange strategy, N_2 isotherms collected at 77 K indicate that **1** has a Brunauer-Emmett-Teller (BET) surface area of ~527 m^2/g with a pore size of ~6 Å (Figure S11, ESI). The pore size distribution analysis suggests that **1** is microporous, with a pore diameter of ~5.8 Å. Furthermore, the flexible nature of the framework is seen as the structure undergoes significant peak shifting upon outgassing, and this flexible nature will be explained later (Figure S12, ESI). Both **1** and **2** are stable in either aqueous, acidic, or high boiling point solvents which prompted us to study their sensing performance under these solvent systems. Upon the addition of CS_2 to **1** and **2** we noted the following changes: i) significant decrease in luminescence intensity; ii) large bathochromic shift in the optical band gap; iii) slight structural alternations upon packing of CS_2 that manifest as shift in peaks of their respective PXRD patterns. Subsequent photoluminescence studies reveal that **1** and **2** are good sensors for the detection of CS_2. The addition of CS_2 at various concentrations to suspensions of **1** (Figure 2c) in DMA led to gradual decrease in the overall luminescent intensity associated with a color change (Figure 2e,f). We then tested the quenching efficiency of **1** and **2** in low concentrations of CS_2 (in the range of 0 to 350 µM, Figures S13 and S14, ESI). From this information we applied a Stern–Volmer analysis (Figures S15 and S16, ESI), where we observed a linear decrease in luminescence intensity with regards to the concentration of analyte. From the Stern–Volmer plots, we calculated the limit of detection (LOD) for **1** and **2** (based on 3σ/m) to be 2.89 ppm and 2.58 ppm, respectively. As seen in Figure 2d and Figure S9, ESI, **1** and **2** exhibit a significant reduction in optical bandgap associated with color change upon exposure to CS_2. As-made samples of **1** and **2** possess band gaps of 2.72 eV and 2.53 eV that shift to 2.38 eV and 2.36 eV, respectively. To understand the PL quenching mechanism, we investigated the possibility of both energy transfer and electron transfer processes. The optical absorption spectrum of CS_2 and the emission profiles of **1** and **2** show no spectral overlap (Figure 3a and Figure S17, ESI), clearly suggesting that there is essentially no energy transfer [40]. The quenching is likely due to the electron transfer as described in detail in our previous studies [40–43]. Previous work has demonstrated that the highest occupied molecular orbital (HOMO) and lowest unoccupied molecular orbital (LUMO) energy levels of a LMOF can be altered through the functionalization of the organic linker [44] which can improve the efficiency of the electron transfer process. In this study, we selected fluorine for the ligand functionalization as the process can be carried out by Suzuki cross coupling, and as it is small enough not to cause a severe steric effect resulting in a different structure [45]. Indeed, the overall connectivity and topology of the resulting MOF was retained thus generating an isoreticular structure. However, the results from our sensing experiments suggest that fluorine functionalization did not provide a noticeable improvement to the overall quenching performance. To demonstrate the selectivity of the MOF, we also tested the change in luminescence of sample **1** when suspended in other common volatile organic compounds (VOCs) including dichloromethane, ethyl acetate, acetone, methanol, benzene, and toluene. In all cases, very little quenching was observed (Figure S20).

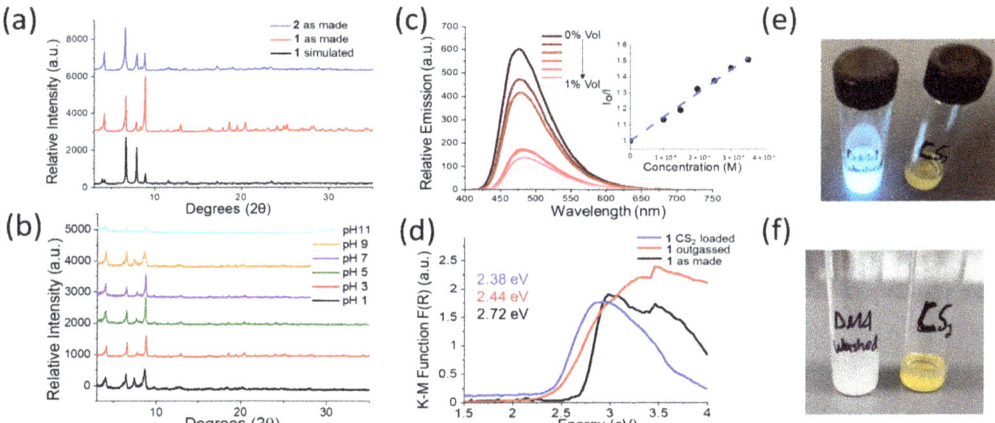

Figure 2. (**a**) PXRD patterns of simulated **1** compared to the experimental patterns of **1** and **2**. (**b**) PXRD patterns of **1** after exposure to different pH values. (**c**) Luminescence quenching profile of **1** when exposed to increasing concentrations of CS_2 in DMA (inset: the Stern–Volmer plot). (**d**) Optical band gaps of **1** in different forms. A shift in optical band gap of **1** after exposure to CS_2 was observed. (**e,f**) Images of **1** before and after exposure to CS_2: under UV excitation (**e**) and under daylight (**f**).

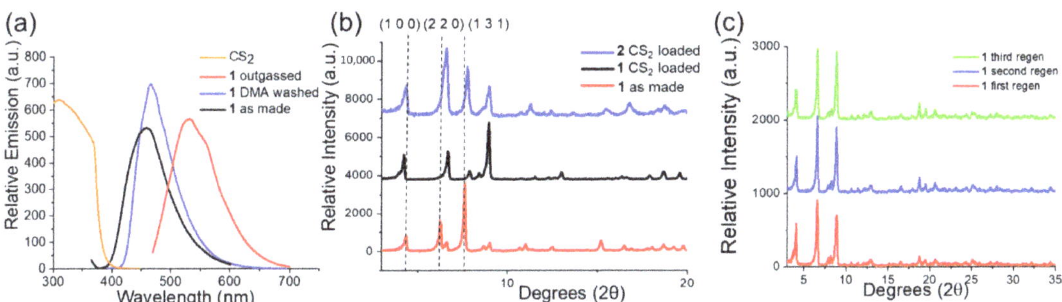

Figure 3. (**a**) The optical absorption spectrum of pure CS_2 and emission spectra of **1** in different forms (as-made, DMA washed and outgassed). (**b**) PXRD patterns of the as-made **1** and CS_2-loaded **1** and **2**. (**c**) The PXRD patterns of **1** after three cycles of regeneration upon exposure to CS_2, 1 M HCl, and DMF.

In addition, PXRD analysis on both **1** and **2** recovered after exposure to CS_2 indicates structural changes related to the packing of the interpenetrated nets as the shifts in the diffraction peaks correspond mainly to the (2 2 0) and (1 3 1) planes (Figure 3b). The PXRD patterns of the as-made samples of **1** and **2** are nearly identical with no obvious shifts in the peak positions, suggesting that, even with the addition of the fluorine atom to the organic linker, the packing of the interpenetrated nets remains the same and that the shifting between the two nets is a direct result from the loading of CS_2. Exposing CS_2-loaded samples of **1** to dilute HCl reverts the structures to their original as-made form with respect to both structural and luminescent profiles, and samples maintain their crystallinity for up to three cycles (Figure 3c). To understand the analyte interactions that occur within the framework at the molecular level and identify potential interaction sites within the system we used in-situ IR spectroscopy to probe the interaction of **1** and **2** upon exposure of CS_2 in the vapor phase. The IR measurements were conducted to monitor changes in the vibrational modes of **1** or **2** before and after exposure to 3 Torr of CS_2.

The spectra were collected in-situ during the exposure to CS_2 and desorption under evacuation of gas phase. Carbon dioxide, a close relative to CS_2, is known to interact strongly with Zr-MOFs through the Zr metal nodes. [46] The spectroscopic results here on CS_2 also reveal that the molecule has a prominent interaction with bridged μ_3-OH bonds from the Zr_6 SBU. In addition, it also interacts with the C-H bonds of the organic linkers in both **1** and **2** (Figure 4a and Figure S18, ESI), as evidenced by following observations. First, upon exposure to CS_2, the vibrational bands of both **1** and **2** at 3674 cm^{-1}, corresponding to stretching modes $\nu(\mu_3$-OH) of the Zr_6 node [47], show an obvious red shift in the difference spectra (Figure 4a). Furthermore, the in plane and out of plane C-H vibrational modes of the organic linkers at 1200–600 cm^{-1} are significantly perturbed. In addition to these changes in the vibrational bands of **1** and **2**, the spectra of adsorbed CS_2 collected under desorption also show notable differences with respect to those of free CS_2 (Figure 4a and Figure S17, ESI). Specifically, the hot band ($\nu_1 + \nu_3$) exhibits a much sharper peak centered at 2164 cm^{-1} [48]. As seen in the desorption profiles, the application of a static vacuum is capable of desorbing CS_2 from the framework, indicative of a relatively weak interaction. Thus, the in-situ IR study not only confirms the interaction of CS_2 with the framework but also reveals that the hydrogen atoms of the ligand may play a role in the binding of CS_2. To help elucidate the interactions between CS_2 and the MOF we also employed DFT calculations to analyze the electronic structure of CS_2-loaded **1**. Within the structure of **1**, there are two distinct types of "pockets" near the Zr-SBU where the interactions between CS_2 and the framework may occur. As seen in Figure 4b, pocket **I** sits between two of the coordinately saturated sites occupied by two organic ligands where Zr-O-C bonds predominate, and pocket **II** sits between two Zr atoms and contains two hydroxyl functional groups where Zr-O-H bonds predominate. The charge rearrangement diagram shows a significant increase in electron density near the zirconium node (Figure 4c) at an isolevel of 2×10^{-4} eV/Å, which indicates a weak bond. It reveals an electron density contribution from CS_2 onto the Zr atom, the oxygen atoms directly bonded to Zr, and the hydrogen atoms of the linker closest to that site. However, it is important to note that the most preferential binding interaction of CS_2 and the framework occurs in pocket **II** where the Zr-O-C bonds predominate. While the in-situ IR measurements reveal that there is a shift in the vibrational modes of the Zr-OH bond of pocket **I**, the computational calculations find pocket **II** to be a more favorable binding site. The sulfur atom binds in a head-on fashion and the zirconium node experiences a significant loss in electron density with a calculated binding energy between 300–500 meV. However, it would be feasible to assume that exposure to excess CS_2 will first occupy all the most preferential binding sites and then occupy the remaining Zr-OH bonds of SBU within the framework, which explains the observed IR shift.

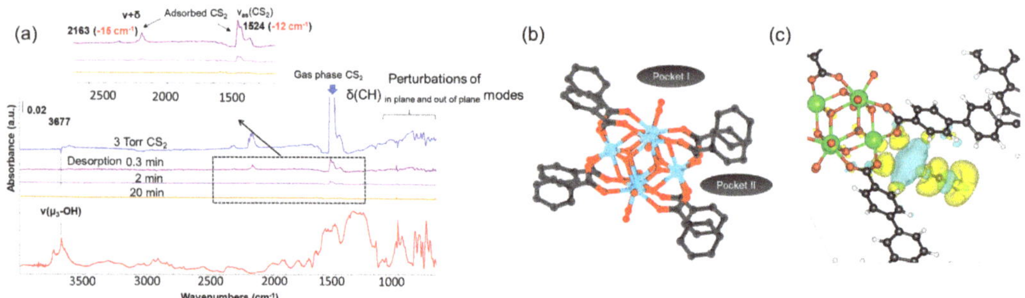

Figure 4. (**a**) difference IR spectra of **1** during adsorption (~3 min, top) and desorption (middle three) of CS_2, referenced to activated MOF in vacuum (bottom). The CS_2 gas signal of asymmetric stretching band $\nu3$ at 1560–1400 cm^{-1} is not fully shown due to its prohibitively high intensity. (**b**) The two distinct pockets in the MOF structure that are surrounded by Zr-O bonds. (**c**) Charge rearrangement diagram showing an increase in electron density (yellow) and loss of electron density (cyan) upon CS_2 interaction at an isolevel of 2×10^{-4} eV.

4. Conclusions

In conclusion, we have synthesized and characterized two robust and isoreticular Zr-based LMOFs with underlying *scu* topology and two-fold interpenetration. Both **1** and **2** are strongly emissive with high PL quantum yields of 68% (69%) and 49% (74%), respectively in as-made (outgassed) form. Investigation of their use as fluorescent sensors for the detection of CS_2 shows that both compounds exhibit a bathochromic shift in their optical absorption spectra and significantly reduced emission intensity upon exposure to CS_2. Luminescent sensing experiments reveal that both are capable of sensing CS_2 in solutions of DMA with detection limits of 2.89 ppm and 2.58 ppm, respectively. Upon adsorption of CS_2, subtle shifts in the PXRD patterns imply that the packing of the two interpenetrated nets is altered. To confirm the molecular interactions occurring between CS_2 and the framework, in-situ IR experiments were performed upon exposure to CS_2. These results confirm that CS_2 interacts with the OH- groups from the Zr-SBU and the C-H groups from the organic linker. DFT calculations further corroborate this conclusion and suggest that the preferential binding sites for CS_2 occur closest to the oxygen atoms from the carboxylic acid group that coordinates directly to the Zr-SBU. This work provides another example where luminescence-based sensing can be achieved via host-guest interactions that occur between the analyte and Zr-based MOFs. Our work demonstrates a new host-guest interaction that may be useful for the future development of highly sensitive MOF materials capable of sensing CS_2 in the vapor phase.

Supplementary Materials: The following are available online https://www.mdpi.com/2624-8549/3/1/24/s1. CCDC 2005129 contains the supplementary crystallographic data for this manuscript. This data may be obtained free of charge via www.ccdc.cam.ac.uk/data_request/cif, by email-ing data_request@ccdc.cam.ac.uk, or by contacting The Cambridge Crystallographic Data Centre, 12 Union Road, Cambridge CB2 1EZ, UK; fax: +44-1223-336033.

Author Contributions: The manuscript was written through contributions of all authors. E.V. and Y.O. the synthesis and characterization of the ligands and complexes, and wrote the first draft; S.J.T., solved the crystal structure of compound 1; K.T. contributed the in-situ FTIR measurements; S.J. and T.T. contributed the DFT calculations. J.L. revised the writing; The data was analyzed by the specific author who collected the corresponding data. E.V. and J.L. reviewed the work; J.L. supervised the team. All authors have read and agreed to the published version of the manuscript.

Funding: This research was funded by the U. S. Department of Energy, Office of Science, Office of Basic Energy Sciences, grant number DE-SC0019902. This research used resources of the Advanced Light Source, which is a DOE Office of Science User Facility under contract no. DE-AC02-05CH11231.

Institutional Review Board Statement: Not applicable.

Informed Consent Statement: Not applicable.

Data Availability Statement: Not applicable.

Conflicts of Interest: The authors declare no conflict of interest. The funders had no role in the design of the study; in the collection, analyses, or interpretation of data; in the writing of the manuscript, or in the decision to publish the results.

Sample Availability: Not available.

References

1. Li, J.; Wang, X.; Zhao, G.; Chen, C.; Chai, Z.; Alsaedi, A.; Hayat, T.; Wang, X. Metal-organic framework-based materials: Superior adsorbents for the capture of toxic and radioactive metal ions. *Chem. Soc. Rev.* **2018**, *47*, 2322–2356. [CrossRef] [PubMed]
2. Wang, H.; Li, J. Microporous Metal-Organic Frameworks for Adsorptive Separation of C5-C6 Alkane Isomers. *Acc. Chem. Res.* **2019**, *52*, 1968–1978. [CrossRef]
3. Adil, K.; Belmabkhout, Y.; Pillai, R.S.; Cadiau, A.; Bhatt, P.M.; Assen, A.H.; Maurin, G.; Eddaoudi, M. Gas/vapour separation using ultra-microporous metal-organic frameworks: Insights into the structure/separation relationship. *Chem. Soc. Rev.* **2017**, *46*, 3402–3430. [CrossRef] [PubMed]
4. Bao, Z.; Chang, G.; Xing, H.; Krishna, R.; Ren, Q.; Chen, B. Potential of microporous metal–organic frameworks for separation of hydrocarbon mixtures. *Energy Environ. Sci.* **2016**, *9*, 3612–3641. [CrossRef]

5. Lustig, W.P.; Mukherjee, S.; Rudd, N.D.; Desai, A.V.; Li, J.; Ghosh, S.K. Metal-organic frameworks: Functional luminescent and photonic materials for sensing applications. *Chem. Soc. Rev.* **2017**, *46*, 3242–3285. [CrossRef] [PubMed]

6. Cui, Y.; Yue, Y.; Qian, G.; Chen, B. Luminescent functional metal-organic frameworks. *Chem. Rev.* **2012**, *112*, 1126–1162. [CrossRef] [PubMed]

7. Zhang, Y.; Yuan, S.; Day, G.; Wang, X.; Yang, X.; Zhou, H.-C. Luminescent sensors based on metal-organic frameworks. *Coord. Chem. Rev.* **2018**, *354*, 28–45. [CrossRef]

8. Zhao, X.; Wang, Y.; Li, D.-S.; Bu, X.; Feng, P. Metal–Organic Frameworks for Separation. *Adv. Mater.* **2018**, *30*, 1705189. [CrossRef]

9. Zhai, Q.-G.; Bu, X.; Zhao, X.; Li, D.-S.; Feng, P. Pore Space Partition in Metal–Organic Frameworks. *Acc. Chem. Res.* **2017**, *50*, 407–417. [CrossRef] [PubMed]

10. Allendorf, M.D.; Bauer, C.A.; Bhakta, R.K.; Houk, R.J. Luminescent metal-organic frameworks. *Chem. Soc. Rev.* **2009**, *38*, 1330–1352. [CrossRef]

11. Lustig, W.P.; Li, J. Luminescent metal–organic frameworks and coordination polymers as alternative phosphors for energy efficient lighting devices. *Coord. Chem. Rev.* **2018**, *373*, 116–147. [CrossRef]

12. Zhou, X.-S.; Fan, R.-Q.; Du, X.; Hao, S.-E.; Fang, R.; Wang, P.; Xing, K.; Wang, A.N.; Yang, Y.-L.; Liu, Z.-G. Key effect of robust $\pi\cdots\pi$ stacking on AIE performance for supramolecular indium(III)–organic assemblies and application in PMMA-doped hybrid material. *Inorg. Chem. Commun.* **2018**, *90*, 39–44. [CrossRef]

13. Nagarkar, S.S.; Desai, A.V.; Ghosh, S.K. A fluorescent metal-organic framework for highly selective detection of nitro explosives in the aqueous phase. *Chem. Commun.* **2014**, *50*, 8915–8918. [CrossRef]

14. Mollick, S.; Mandal, T.N.; Jana, A.; Fajal, S.; Desai, A.V.; Ghosh, S.K. Ultrastable Luminescent Hybrid Bromide Perovskite@MOF Nanocomposites for the Degradation of Organic Pollutants in Water. *ACS Appl. Nano Mater.* **2019**, *2*, 1333–1340. [CrossRef]

15. Zhai, Z.-W.; Yang, S.-H.; Cao, M.; Li, L.-K.; Du, C.-X.; Zang, S.-Q. Rational Design of Three Two-Fold Interpenetrated Metal–Organic Frameworks: Luminescent Zn/Cd-Metal–Organic Frameworks for Detection of 2,4,6-Trinitrophenol and Nitrofurazone in the Aqueous Phase. *Cryst. Growth Des.* **2018**, *18*, 7173–7182. [CrossRef]

16. Dong, X.-Y.; Huang, H.-L.; Wang, J.-Y.; Li, H.-Y.; Zang, S.-Q. A Flexible Fluorescent SCC-MOF for Switchable Molecule Identification and Temperature Display. *Chem. Mater.* **2018**, *30*, 2160–2167. [CrossRef]

17. Hu, Z.; Deibert, B.J.; Li, J. Luminescent metal-organic frameworks for chemical sensing and explosive detection. *Chem. Soc. Rev.* **2014**, *43*, 5815–5840. [CrossRef]

18. Gargurevich, I.A. Hydrogen Sulfide Combustion: Relevant Issues under Claus Furnace Conditions. *Ind. Eng. Chem. Res.* **2005**, *44*, 7706–7729. [CrossRef]

19. Meng, X.; de Jong, W.; Pal, R.; Verkooijen, A.H.M. In bed and downstream hot gas desulphurization during solid fuel gasification: A review. *Fuel Process. Technol.* **2010**, *91*, 964–981. [CrossRef]

20. Glarborg, P.; Halaburt, B.; Marshall, P.; Guillory, A.; Troe, J.; Thellefsen, M.; Christensen, K. Oxidation of reduced sulfur species: Carbon disulfide. *J. Phys. Chem. A* **2014**, *118*, 6798–6809. [CrossRef]

21. Anstey, M.; Sun, A.; Paap, S.M.; Foltz, G.; Jaeger, C.; Hoette, T.; Ochs, M.E. *Inherently Safer Technology Gaps Analysis Study*; United States. Department of Energy: Washington, DC, USA, 2012.

22. Wayne, D.; Monnery, W.Y.S.; Behie, L.A. Modelling the modified claus process reaction furnace and the implications on plant design and recovery. *Can. J. Chem. Eng.* **1993**, *71*, 711–724.

23. McGuirk, C.M.; Siegelman, R.L.; Drisdell, W.S.; Runcevski, T.; Milner, P.J.; Oktawiec, J.; Wan, L.F.; Su, G.M.; Jiang, H.Z.H.; Reed, D.A.; et al. Cooperative adsorption of carbon disulfide in diamine-appended metal-organic frameworks. *Nat. Commun.* **2018**, *9*, 5133. [CrossRef]

24. Wu, Z.-F.; Tan, B.; Feng, M.-L.; Lan, A.-J.; Huang, X.-Y. A magnesium MOF as a sensitive fluorescence sensor for CS2 and nitroaromatic compounds. *J. Mater. Chem. A* **2014**, *2*, 6426–6431. [CrossRef]

25. Dong, Y. Tetranuclear cluster-based Pb(II)-MOF: Synthesis, crystal structure and luminescence sensing for CS2. *J. Mol. Struct.* **2018**, *1160*, 46–49. [CrossRef]

26. Wu, Z.-F.; Tan, B.; Feng, M.-L.; Du, C.-F.; Huang, X.-Y. A magnesium-carboxylate framework showing luminescent sensing for CS2 and nitroaromatic compounds. *J. Solid State Chem.* **2015**, *223*, 59–64. [CrossRef]

27. Wu, Z.-F.; Tan, B.; Velasco, E.; Wang, H.; Shen, N.-N.; Gao, Y.-J.; Zhang, X.; Zhu, K.; Zhang, G.-Y.; Liu, Y.-Y.; et al. Fluorescent In based MOFs showing "turn on" luminescence towards thiols and acting as a ratiometric fluorescence thermometer. *J. Mater. Chem. C* **2019**, *7*, 3049–3055. [CrossRef]

28. Wei, Z.; Gu, Z.Y.; Arvapally, R.K.; Chen, Y.P.; McDougald, R.N., Jr.; Ivy, J.F.; Yakovenko, A.A.; Feng, D.; Omary, M.A.; Zhou, H.C. Rigidifying fluorescent linkers by metal-organic framework formation for fluorescence blue shift and quantum yield enhancement. *J. Am. Chem. Soc.* **2014**, *136*, 8269–8276. [CrossRef] [PubMed]

29. Lustig, W.P.; Wang, F.; Teat, S.J.; Hu, Z.; Gong, Q.; Li, J. Chromophore-Based Luminescent Metal-Organic Frameworks as Lighting Phosphors. *Inorg. Chem.* **2016**, *55*, 7250–7256. [CrossRef]

30. Kresse, G.; Furthmüller, J. Efficient iterative schemes for \textit{ab initio} total-energy calculations using a plane-wave basis set. *Phys. Rev. B* **1996**, *54*, 11169–11186. [CrossRef] [PubMed]

31. Kresse, G.; Joubert, D. From ultrasoft pseudopotentials to the projector augmented-wave method. *Phys. Rev. B* **1999**, *59*, 1758–1775. [CrossRef]

32. Thonhauser, T.; Cooper, V.R.; Li, S.; Puzder, A.; Hyldgaard, P.; Langreth, D.C. Van der Waals density functional: Self-consistent potential and the nature of the van der Waals bond. *Phys. Rev. B* **2007**, *76*, 125112. [CrossRef]

33. Langreth, D.C.; Lundqvist, B.I.; Chakarova-Käck, S.D.; Cooper, V.R.; Dion, M.; Hyldgaard, P.; Kelkkanen, A.; Kleis, J.; Lingzhu, K.; Shen, L.; et al. A density functional for sparse matter. *J. Phys. Condens. Matter* **2009**, *21*, 084203. [CrossRef]

34. Kristian, B.; Valentino, R.C.; Kyuho, L.; Elsebeth, S.; Thonhauser, T.; Per, H.; Bengt, I.L. van der Waals forces in density functional theory: A review of the vdW-DF method. *Rep. Prog. Phys.* **2015**, *78*, 066501.

35. Chen, C.-X.; Wei, Z.-W.; Cao, C.-C.; Yin, S.-Y.; Qiu, Q.-F.; Zhu, N.-X.; Xiong, Y.-Y.; Jiang, J.-J.; Pan, M.; Su, C.-Y. All Roads Lead to Rome: Tuning the Luminescence of a Breathing Catenated Zr-MOF by Programmable Multiplexing Pathways. *Chem. Mater.* **2019**, *31*, 5550–5557. [CrossRef]

36. Hu, Z.; Huang, G.; Lustig, W.P.; Wang, F.; Wang, H.; Teat, S.J.; Banerjee, D.; Zhang, D.; Li, J. Achieving exceptionally high luminescence quantum efficiency by immobilizing an AIE molecular chromophore into a metal-organic framework. *Chem. Commun.* **2015**, *51*, 3045–3048. [CrossRef] [PubMed]

37. Wang, F.; Liu, W.; Teat, S.J.; Xu, F.; Wang, H.; Wang, X.; An, L.; Li, J. Chromophore-immobilized luminescent metal-organic frameworks as potential lighting phosphors and chemical sensors. *Chem. Commun. (Camb.)* **2016**, *52*, 10249–10252. [CrossRef]

38. Gong, Q.; Hu, Z.; Deibert, B.J.; Emge, T.J.; Teat, S.J.; Banerjee, D.; Mussman, B.; Rudd, N.D.; Li, J. Solution processable MOF yellow phosphor with exceptionally high quantum efficiency. *J. Am. Chem. Soc.* **2014**, *136*, 16724–16727. [CrossRef] [PubMed]

39. Ma, J.; Kalenak, A.P.; Wong-Foy, A.G.; Matzger, A.J. Rapid Guest Exchange and Ultra-Low Surface Tension Solvents Optimize Metal-Organic Framework Activation. *Angew. Chem. Int. Ed. Engl.* **2017**, *56*, 14618–14621. [CrossRef] [PubMed]

40. Hu, Z.; Tan, K.; Lustig, W.P.; Wang, H.; Zhao, Y.; Zheng, C.; Banerjee, D.; Emge, T.J.; Chabal, Y.J.; Li, J. Effective sensing of RDX via instant and selective detection of ketone vapors. *Chem. Sci.* **2014**, *5*, 4873–4877. [CrossRef]

41. Pramanik, S.; Hu, Z.; Zhang, X.; Zheng, C.; Kelly, S.; Li, J. A Systematic Study of Fluorescence-Based Detection of Nitroexplosives and Other Aromatics in the Vapor Phase by Microporous Metal–Organic Frameworks. *Chem. A Eur. J.* **2013**, *19*, 15964–15971. [CrossRef]

42. Hu, Z.; Lustig, W.P.; Zhang, J.; Zheng, C.; Wang, H.; Teat, S.J.; Gong, Q.; Rudd, N.D.; Li, J. Effective Detection of Mycotoxins by a Highly Luminescent Metal–Organic Framework. *J. Am. Chem. Soc.* **2015**, *137*, 16209–16215. [CrossRef] [PubMed]

43. Pramanik, S.; Zheng, C.; Zhang, X.; Emge, T.J.; Li, J. New Microporous Metal−Organic Framework Demonstrating Unique Selectivity for Detection of High Explosives and Aromatic Compounds. *J. Am. Chem. Soc.* **2011**, *133*, 4153–4155. [CrossRef] [PubMed]

44. Lustig, W.P.; Shen, Z.; Teat, S.J.; Javed, N.; Velasco, E.; O'Carroll, D.M.; Li, J. Rational design of a high-efficiency, multivariate metal–organic framework phosphor for white LED bulbs. *Chem. Sci.* **2020**, *11*, 1814–1824. [CrossRef]

45. Pang, J.; Yuan, S.; Qin, J.; Liu, C.; Lollar, C.; Wu, M.; Yuan, D.; Zhou, H.-C.; Hong, M. Control the Structure of Zr-Tetracarboxylate Frameworks through Steric Tuning. *J. Am. Chem. Soc.* **2017**, *139*, 16939–16945. [CrossRef]

46. Grissom, T.G.; Driscoll, D.M.; Troya, D.; Sapienza, N.S.; Usov, P.M.; Morris, A.J.; Morris, J.R. Molecular-Level Insight into CO_2 Adsorption on the Zirconium-Based Metal–Organic Framework, UiO-66: A Combined Spectroscopic and Computational Approach. *J. Phys. Chem. C* **2019**, *123*, 13731–13738. [CrossRef]

47. Tan, K.; Jensen, S.; Feng, L.; Wang, H.; Yuan, S.; Ferreri, M.; Klesko, J.P.; Rahman, R.; Cure, J.; Li, J.; et al. Reactivity of Atomic Layer Deposition Precursors with OH/H2O-Containing Metal Organic Framework Materials. *Chem. Mater.* **2019**, *31*, 2286–2295. [CrossRef]

48. Mivehvar, F.; Lauzin, C.; McKellar, A.R.W.; Moazzen-Ahmadi, N. Infrared spectra of rare gas–carbon disulfide complexes: He–CS2, Ne–CS2, and Ar–CS2. *J. Mol. Spectrosc.* **2012**, *281*, 24–27. [CrossRef]

 *chemistry*

 MDPI

Article

Coordination Polymers Constructed from Semi-Rigid *N,N′*-Bis(3-pyridyl)terephthalamide and Dicarboxylic Acids: Effect of Ligand Isomerism, Flexibility, and Identity

Chia-Jou Chen [1], Chia-Ling Chen [1], Yu-Hsiang Liu [1], Wei-Te Lee [1], Ji-Hong Hu [1], Pradhumna Mahat Chhetri [2],* and Jhy-Der Chen [1],*

[1] Department of Chemistry, Chung-Yuan Christian University, Chung-Li 320, Taiwan; jelly4112@gmail.com (C.-J.C.); jialing41666@gmail.com (C.-L.C.); g10963021@cycu.edu.tw (Y.-H.L.); lku23y230230@gmail.com (W.-T.L.); joejoe860831@gmail.com (J.-H.H.)
[2] Department of Chemistry, Amrit Campus, Tribhuvan University, Kathmandu 44600, Nepal
* Correspondence: mahatp@gmail.com (P.M.C.); jdchen@cycu.edu.tw (J.-D.C.); Tel.: +886-3-265-3351(J.-D.C.)

Abstract: Reactions of the semi-rigid *N,N′*-bis(3-pyridyl)terephthalamide (**L**) with divalent metal salts in the presence of dicarboxylic acids afforded $[Cd(L)_{0.5}(1,2\text{-BDC})(H_2O)]_n$ (1,2-H_2BDC = benzene-1,2-dicarboxylic acid), **1**, $\{[Cd(L)_{1.5}(1,3\text{-BDC})(H_2O)]\cdot 5H_2O\}_n$ (1,3-H_2BDC = benzene-1,3-dicarboxylic acid), **2a**, $\{[Cd(1,3\text{-BDC})(H_2O)_3]\cdot 2H_2O\}_n$, **2b**, $\{[Cd(L)_{0.5}(1,4\text{-BDC})(H_2O)_2]\cdot H_2O\}_n$ (1,4-H_2BDC = benzene-1,4-dicarboxylic acid), **3**, and $[Cu(L)_{0.5}(5\text{-tert-IPA})]_n$ (5-tert-IPA = 5-tert-butylbenzene-1,3-dicarboxylic acid), **4**, which have been structurally characterized by single crystal X-ray diffraction. Complexes **1** and **3** are two-dimensional (2D) layers with the **bey** and the **hcb** topologies, and **2a** and **2b** are one-dimensional (1D) ladder and zigzag chain, respectively, while **4** shows a 3-fold interpenetrated three-dimensional (3D) net with the **cds** topology. The structures of these coordination polymers containing the semi-rigid **L** ligands are subject to the donor atom positions and the identity of the dicarboxylate ligands, which are in marked contrast to those obtained from the flexible bis-pyridyl-bis-amide ligands that form self-catenated nets. The luminescence of **1** and **3** and thermal properties of complexes **1**, **3**, and **4** are also discussed.

Keywords: coordination polymers; semi-rigid ligand; ligand isomerism; ligand flexibility

Citation: Chen, C.; Chen, C.; Liu, Y.; Lee, W.; Hu, J.; Chhetri, P.M.; Chen, J.-D. Coordination Polymers Constructed from Semi-Rigid *N,N′*-Bis(3-pyridyl)-terephthalamide and Dicarboxylic Acids: Effect of Ligand Isomerism, Flexibility, and Identity. *Chemistry* **2021**, *3*, 1–12. https://doi.org/10.3390/chemistry3010001

Received: 27 November 2020
Accepted: 21 December 2020
Published: 22 December 2020

Publisher's Note: MDPI stays neutral with regard to jurisdictional claims in published maps and institutional affiliations.

1. Introduction

Coordination polymers (CPs) constructed from metal salts and spacer ligands are coordination compounds with repeating coordination entities extending in 1, 2, or 3 dimensions [1]. As a result of their interesting structures and properties, CPs have been extensively studied by scientists during recent years [2,3]. Apart from metal ions, a wise choice of the spacer ligands involving the neutral and the auxiliary polycarboxylate anions are important in determining the structural types of CPs in a mixed ligand system, which are also subject to counter ions, solvents, temperature, metal to ligand ratio, etc.

The chemistry of CPs containing flexible bis-pyridyl-bis-amide (bpba) ligands has been investigated extensively during recent year [4]. The ligand-isomerism of the flexible bpba as well as the auxiliary dicarboxylate ligands have been found important in determining the structural diversity [5–8]. Cd(II) perchlorate CPs constructed from the flexible and isomeric *N,N′*-di(2-pyridyl)adipoamide, *N,N′*-di(3-pyridyl)adipoamide and *N,N′*-di(4-pyridyl)adipoamide exhibit a 1D zigzag chain, a 2D pleated sheet, and a 3D entangled framework, respectively [5], while the Cd(II) CPs comprising *N,N′*-di(3-pyridyl)adipoamide ligands are directed by the isomeric 1,2-, 1,3-, and 1,4-benzenedicarboxylate ligands, resulting in a 1D ladder chain, a 2D layer with loops and a 3D self-catenated net, respectively [6]. Reactions of the flexible *N,N′*-di(3-pyridyl)suberoamide with Cu(II) salts in the presence of the isomeric 1,2-, 1,3-, and 1,4-phenylenediacetic acids afforded CPs showing a 3D net with the 3,5T1 topology, a 5-fold interpenetrated 3D net of **cds** topology, and a 1D self-catenated network [7], while reactions of N,N'-di(3-pyridyl)suberoamide and Cd(II) salts with the isomeric phenylenediacetic acids afforded a loop-like 1D chain, a self-catenated net with the $(6^5 \cdot 8)$ topology and a 2D layer with the **sql** topology, respectively [8]. Probably, the self-catenated CPs can be achieved by the manipulation of the ligand-isomerism of the auxiliary polycarboxylate ligands.

We are investigating the CPs comprising the semi-rigid bpba and the dicarboxylate ligands, in an attempt to elucidate the effect of the rigidity of the neutral spacer on the structures of CPs supported by the dicarboxylate ligands [9]. In this report, N,N'-bis(3-pyridyl)terephthalamide (**L**), Figure 1, and divalent metal salts were used to react with the isomeric benzene-dicarboxylic acids as well as 5-tert-butylbenzene-1,3-dicarboxylic acid to investigate the effect of the donor atom position and the identity of the polycarboxylate ligands on the structural diversity of CPs with semi-rigid spacer. The syntheses and structures of $[Cd(\mathbf{L})_{0.5}(1,2\text{-BDC})(H_2O)]_n$ (1,2-H_2BDC = benzene-1,2-dicarboxylic acid), **1**, $\{[Cd(\mathbf{L})_{1.5}(1,3\text{-BDC})(H_2O)] \cdot 5H_2O\}_n$ (1,3-H_2BDC = benzene-1,3-dicarboxylic acid), **2a**, $\{[Cd(1,3\text{-BDC})(H_2O)_3] \cdot 2H_2O\}_n$, **2b**, $\{[Cd(\mathbf{L})_{0.5}(1,4\text{-BDC})(H_2O)_2] \cdot H_2O\}_n$ (1,4-H_2BDC = benzene-1,4-dicarboxylic acid), **3**, and $[Cu(\mathbf{L})_{0.5}(5\text{-tert-IPA})]_n$ (5-tert-IPA = 5-tert-butylbenzene-1,3-dicarboxylic acid), **4**, form the subject of this report. Thermal and luminescent properties of the available complexes are also discussed.

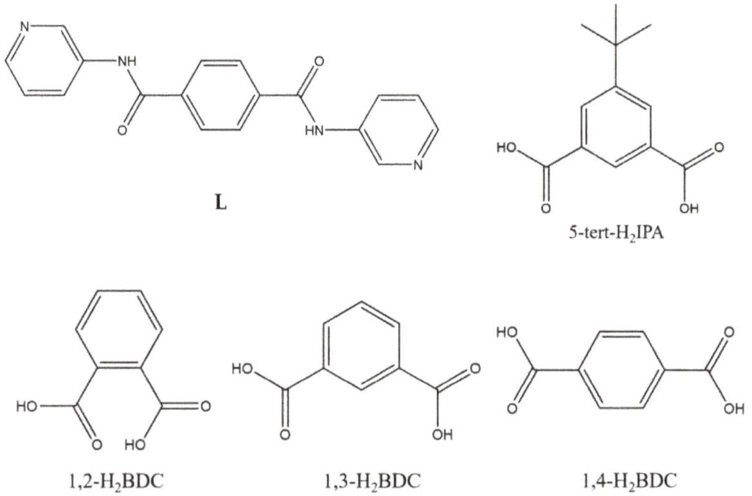

Figure 1. Structures of **L**, 5-tert-butylbenzene-1,3-dicarboxylic acid (5-tert-H$_2$IPA), benzene-1,2-dicarboxylic acid (1,2-H$_2$BDC), benzene-1,3-dicarboxylic acid (1,3-H$_2$BDC), and benzene-1,4-dicarboxylic acid (1,4-H$_2$BDC).

2. Materials and Methods

2.1. General Procedures

Elemental analyses were performed on a PE 2400 series II CHNS/O (PerkinElmer Instruments, Shelton, CT, USA). Solid state IR spectra were measured on a JASCO FT/IR-460 plus spectrometer (JASCO, Easton, MD, USA). Powder X-ray diffraction (PXRD) patterns were carried out on a Bruker D2 PHASER diffractometer (Bruker Corporation, Karlsruhe, Germany). Solid state emission spectroscopy was done using a Hitachi F-4500 spectrometer (Hitachi, Tokyo, Japan). TGA curves were obtained form an SII Nano Technology Inc. TGA/DTA 6200 analyzer (Seiko Instruments Inc., Torrance, CA, USA) in 30–800 °C under a nitrogen atmosphere.

2.2. Materials

The reagent $Cu(OAc)_2 \cdot H_2O$ was purchased from SHOWA Co. (Saitama, Japan), 5-tert-butylbenzene-1,3-dicarboxylic acid (5-tert-H_2IPA) from Aldrich Co. (Wyoming, IL, USA), benzene-1,2-dicarboxylic acid (1,2-H_2BDC), benzene-1,3-dicarboxylic acid (1,3-H_2BDC) and benzene-1,4-dicarboxylic acid (1,4-H_2BDC) from ACROS Co. (Pittsburgh, PA, USA), and $Cd(OAc)_2 \cdot 2H_2O$ from Fisher Scientific Co.(Pittsburgh, PA, USA). N,N'-bis(3-pyridyl)terephthalamide (**L**) was prepared according to published procedures [10].

2.3. Preparations

2.3.1. $[Cd(L)_{0.5}(1,2\text{-BDC})(H_2O)]_n$, **1**

A mixture of $Cd(OAc)_2 \cdot H_2O$ (0.027 g, 0.10 mmol), **L** (0.032 g, 0.10 mmol), benzene-1,2-dicarboxylic acid (1,2-H_2BDC) (0.017 g, 0.10 mmol), and 10 mL of 0.01 M NaOH solution in water was sealed in a 23 mL Teflon-lined stainless steel autoclave which was heated under autogenous pressure to 120 °C for two days and then the reaction system was cooled to room temperature at a rate of 2 °C per hour. Colorless crystals suitable for single-crystal X-ray diffraction were obtained which were washed with water and dried under vacuum. Yield: 0.022 g (49%). Anal. Calcd for $C_{17}H_{13}CdN_2O_6$ (MW = 453.69): C, 45.00; H, 2.89; N, 6.17%. Found: C, 44.99; H, 3.05; N, 6.16%. FT-IR (cm^{-1}): 3500(w), 3068(m), 1676(s), 1612(s), 1550(s), 1488(s), 1430(s), 1378(s), 1336(m), 1301(s), 1198(m), 1108(m), 800(m), 760(m), 731(m), 696(m).

2.3.2. $\{[Cd(L)_{1.5}(1,3\text{-BDC})(H_2O)] \cdot 5H_2O\}_n$, **2a**, and $\{[Cd(1,3\text{-BDC})(H_2O)_3] \cdot 2H_2O\}_n$, **2b**

Prepared by following the similar procedures for **1** except that 1,3- H_2BDC was reacted at 90 °C. Different colorless crystals with poor crystallinity were obtained, which were difficult to be separated manually. Careful test of the crystals showed that two complexes $\{[Cd(L)_{1.5}(1,3\text{-BDC})(H_2O)] \cdot 5H_2O\}_n$, **2a**, and $\{[Cd(1,3\text{-BDC})(H_2O)_3] \cdot 2H_2O\}_n$, **2b**, can be found, which were structurally characterized by using single crystal X-ray crystallography. Although the X-ray crystallography data of **2a** are humble, its structure can be clearly identified.

2.3.3. $\{[Cd(L)_{0.5}(1,4\text{-BDC})(H_2O)_2] \cdot H_2O\}_n$, **3**

Prepared by following the similar procedures for **1** except that benzene-1,4-dicarboxylic acid (1,4-H_2BDC) (0.017 g, 0.10 mmol) was used. Yield: 0.026 g (38%). Anal. Calcd for $C_{17}H_{15}CdN_2O_7 \cdot H_2O$ (MW = 489.74): C, 41.69; H, 3.50; N, 5.72%. Anal. Calcd for $C_{17}H_{15}CdN_2O_7$ (without cocrystallized water molecule) (MW = 453.69): C, 45.00; H, 2.89; N, 6.17%. Found: C, 44.25; H, 3.35; N, 6.85%. FT-IR (cm^{-1}): 3677(w), 3303(m), 1673(m), 1608(m), 1552(s), 1488(s), 1386(s), 1332(s), 1295(s), 1203(m), 1120(m), 1012(w), 885(w), 838(s), 806(m), 746(m), 698(m), 646(m), 528(w).

2.3.4. $[Cu(L)_{0.5}(5\text{-tert-IPA})]_n$, **4**

Prepared as described for **1** except $Cu(OAc)_2 \cdot H_2O$ (0.020 g, 0.10 mmol), **L** (0.032 g, 0.10 mmol), 5-tert-H_2IPA (0.021 g, 0.10 mmol) and 10 mL of 0.01 M NaOH solution in water were used. Yield: 0.014 g (32%). Anal. Calcd for $C_{21}H_{19}CuN_2O_5$ (MW = 442.92): C, 56.71; H, 4.31; N, 6.30%. Found: C, 56.08; H, 3.99; N, 6.46%. FT-IR (cm^{-1}): 3271(m), 3203(m), 3118(m), 3066(m), 2970(m), 1680(s), 1625(s), 1583(s), 1543(s), 1479(m), 1444(s), 1421(s), 1378(s), 1331(s), 1305(m), 1279(s), 1238(m), 1189(m), 1112(m), 1053(w), 1016(w), 865(w), 806(m), 773(m), 748(m), 721(s), 692(s), 634(w).

The experimental powder X-ray patterns of **1**, **3**, and **4** match quite well with their simulated ones, indicating the bulk purities, Figures S1–S3.

2.4. X-ray Crystallography

A Bruker AXS SMART APEX II CCD diffractometer (graphite-monochromated MoKα = 0.71073 Å) was used to collect the diffraction data of complexes **1–4** [11]. Data reduction was performed by standard methods with use of well-established computational procedures, involving Lorentz and polarization corrections and empirical absorption correction based on "multi-scan". Heavier atom positions were located by the direct method and the other atoms were found in alternating difference Fourier maps and least-square refinements. The hydrogen atoms except those of the water molecules were added by using the HADD command in SHELXTL 6.1012 [12]. Table 1 lists the crystal data for **1**, **2b**, **3**, and **4**, while those of **2a** are provided in the Supplementary Materials.

Table 1. Crystal data for complexes **1–4**.

	1	**2b**	**3**	**4**
Formula	$C_{17}H_{13}CdN_2O_6$	$C_8H_{14}CdO_9$	$C_{17}H_{17}CdN_2O_8$	$C_{21}H_{19}CuN_2O_5$
Formula weight	453.69	366.59	489.72	442.92
Crystal system	Monoclinic	Triclinic	Triclinic	Monoclinic
Space group	$P2_1/c$	$P\bar{1}$	$P\bar{1}$	$C2/c$
a, Å	14.6784(10)	10.2076(3)	6.34570(10)	21.600(3)
b, Å	5.8992(3)	11.1103(3)	9.4314(2)	10.6511(15)
c, Å	19.8917(11)	12.6791(3)	16.7255(3)	17.041(2)
$\alpha,^\circ$	90	102.1594(10)	100.8415(12)	90
$\beta,^\circ$	107.498(4)	102.2797(10)	95.6385(12)	100.715
$\gamma,^\circ$	90	103.5718(10)	104.0750(11)	90
V, Å^3	1642.74(17)	1313.41(6)	942.66(3)	3852.2(9)
Z	4	4	2	8
D_{calc}, Mg/m^3	1.834	1.854	1.725	1.527
F(000)	900	728	490	1824
μ(Mo K$_\alpha$), mm^{-1}	1.368	1.697	1.206	1.170
Range(2θ) for data collection, deg	2.91 to 52.00	3.42 to 56.63	4.70 to 56.63	3.84 to 56.52
Independent reflections	3173 [R(int) = 0.0794]	6502 [R(int) = 0.0306]	4703 [R(int) = 0.0344]	4756 [R(int) = 0.0847]
Data/restraints/parameters	3173/0/241	6502/0/348	4703/0/269	4756/492/285
quality-of-fit indicator [c]	1.004	1.003	1.034	1.059
Final R indices [I > 2σ(I)] [a,b]	R1 = 0.0436, wR2 = 0.0561	R1 = 0.0305, wR2 = 0.0965	R1 = 0.0303, wR2 = 0.0596	R1 = 0.0589, wR2 = 0.1310
R indices (all data)	R1 = 0.0886, wR2=0.0638	R1 = 0.0331, wR2 = 0.0994	R1 = 0.0418, wR2=0.0633	R1 = 0.0950, wR2=0.1486

[a] $R_1 = \Sigma||F_o| - |F_c||/\Sigma|F_o|$. [b] $wR_2 = [\Sigma_W(F_o^2 - F_c^2)^2/\Sigma w(F_o^2)^2]^{1/2}$. $w = 1/[\sigma^2(F_o^2) + (ap)^2 + (bp)]$, $p = [\max(F_o^2 \text{ or } 0) + 2(F_c^2)]/3$. a = 0.0195, b = 0.0000 for **1**; a = 0.0624, b = 1.6811 for **2b**; a = 0.0276, b = 0.1920 for **3**; a = 0.0634, b = 0.0000 for **4**. [c] quality-of-fit = $[\Sigma_W(|F_o^2| - |F_c^2|)^2/(N_{observed} - N_{parameters})]^{1/2}$.

3. Results

3.1. Structure of **1**

A single-crystal X-ray diffraction analysis shows that **1** crystallizes in the monoclinic space group $P2_1/c$. The asymmetric unit consists of one Cd(II) cation, one half **L** ligand, one 1,2-BDC^{2-} ligand, and one coordinated water molecule. Figure 2a depicts a drawing showing the coordination environment of Cd(II), which is seven-coordinated by six oxygen atoms from three independent 1,2-BDC^{2-} ligands, one water molecule [Cd-O = 2.291(3) − 2.524(3) Å] and one pyridyl nitrogen atom from the **L** ligands [Cd-N = 2.273(4) Å], resulting in a distorted pentagonal bipyramidal geometry. The metal atoms are linked by the 1,2-BDC^{2-} ligands to form 1D linear chains with various dinuclear units and the distances between the two Cd(II) metal centers are 5.0180(2) and 3.9650(2) Å for the large and small rings, respectively, Figure 2b. The 1D chains are further linked together by the **L** ligands to afford a 2D layer. If the Cd(II) cations are defined as 4-connected nodes and the 1,2-BDC^{2-} cations as 3-connected nodes, while the **L** ligands as linkers, the structure of complex **1** can be regarded as a 3,4-connected 2D net with the $(4^2.6^3.8)(4^2.6)$-**bey**; 3,4L83 topology, Figure 2c, determined by using ToposPro program [13].

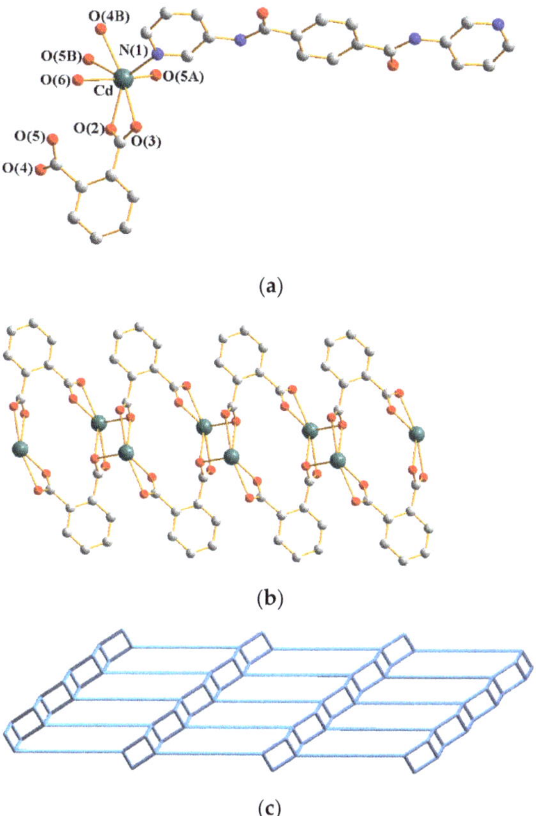

(a)

(b)

(c)

Figure 2. Figure 2. (**a**) Coordination environment of Cd(II) ion in **1**. Symmetry transformations used to generate equivalent atoms: (A) x, y + 1, z (B) −x + 1, −y + 1,− z + 1 (C) x, y − 1, z (D) −x + 2, −y + 4, −z + 1. (**b**) A drawing showing the 1D chain linked by the 1,2-BDC^{2-} ligands. (**c**) A drawing showing the 2D layer with the **bey** topology.

3.2. Structures of *2a* and *2b*

Crystals of **2a** conform to the triclinic space group *P*ī. The asymmetric unit consists of one Cd(II) cation, one and a half **L** ligands, one 1,3-BDC^{2-} ligand, one coordinated water molecule, and five cocrystallized water molecules. Figure 3a depicts a drawing showing the coordination environment about Cd(II), which is six-coordinated by three oxygen atoms from two 1,3-BDC^{2-} ligands [Cd-O = 2.017(6) − 2.220(6) Å], one coordinated water molecule [Cd-O = 2.074(6) Å], and two pyridyl nitrogen atoms from the two **L** ligands [Cd-N = 2.193(6) and 2.225(7) Å], resulting in a distorted octahedral geometry. The metal atoms are linked by the 1,3-BDC^{2-} ligands to form 1D ladder chains with bidentate bridging **L** as rungs and monodentate **L** dangling on the two supporting sides, Figure 3b. Interestingly, the 1D ladder chains are further reinforced by the **L** ligands through the π–π interactions (3.74, 3.86, and 3.73 Å) to afford a 2D layer, Figure 3c.

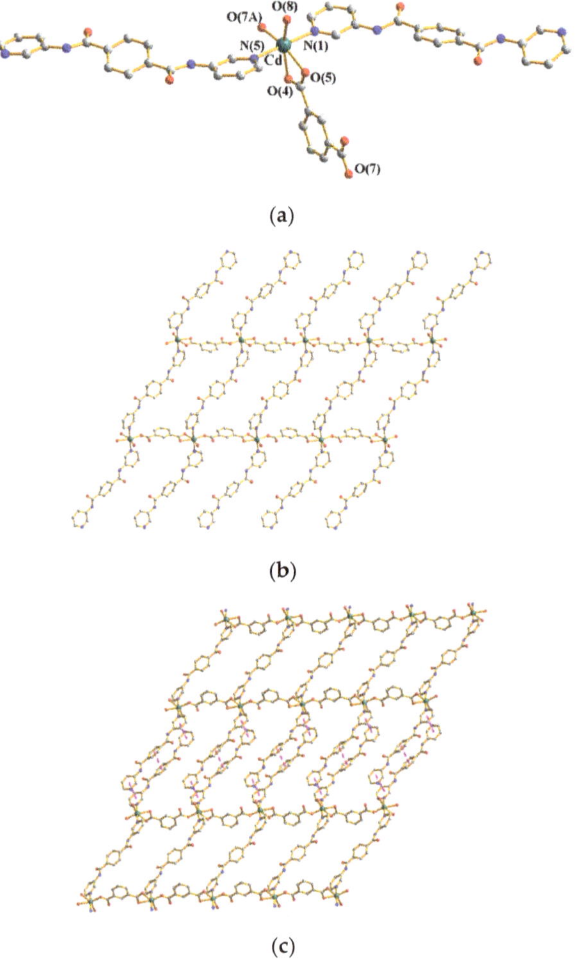

(a)

(b)

(c)

Figure 3. (**a**) Coordination environment of Cd(II) ion in **2a**. Symmetry transformations used to generate equivalent atoms: (A) x + 1, y, z. (**b**) A drawing showing the 1D ladder chain with dangling **L** ligands on the sides. (**c**) A drawing showing the 2D layer supported by the π–π interactions (purple dashed lines).

Crystals of **2b** conform to the triclinic space group $P\bar{1}$ and each asymmetric unit consists of two Cd(II) cations, two 1,3-BDC^{2-} ligands, six coordinated water, and four lattice water molecules. Figure 4a depicts a drawing showing the coordination environment of Cd(II), which is seven-coordinated by seven oxygen atoms from two 1,3-BDC^{2-} ligands, three water molecules [Cd-O = 2.277(3) − 2.423(2) Å], resulting in a distorted pentagonal bipyramidal geometry. The Cd(II) cations are further linked together by 1,3-BDC^{2-} to afford 1D chain, Figure 4b. It is noted that solvothermal reaction of Cd(CH$_3$COO)$_2$·2H$_2$O with 1,3-H$_2$BDC in DMF/H$_2$O afforded a 3D framework in which the 1,3-BDC^{2-} ligand bridge four Cd(II) ions [14], while replacement of 1,3-H$_2$BDC with 1,4-H$_2$BDC gave a 1D zigzag chain [14,15]. A comparison with the structure of **2b** indicates that the structures of the Cd(II)-BDC^{2-}-based complexes are subject to the donor atom positions of the BDC^{2-} ligands and the solvent system.

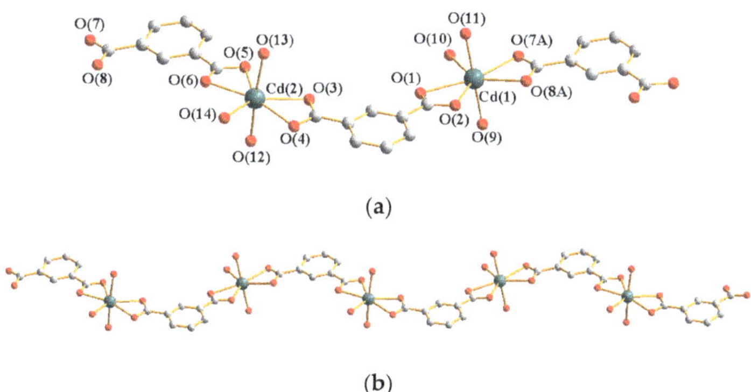

(a)

(b)

Figure 4. (**a**) Coordination environment of Cd(II) ion in **2b**. Symmetry transformations used to generate equivalent atoms: (A) x, y + 1, z − 1 (B) x, y − 1, z + 1 (C) −x + 1, −y + 1, −z. (**b**) A drawing showing the 1D chain.

3.3. Structure of 3

The structure of **3** was solved in the triclinic space group $P\bar{1}$ and each asymmetric unit consists of one Cd(II) cation, a half **L** ligand, two halves of a 1,4-BDC^{2-} ligand, two coordinated water, and one lattice water molecules. Figure 5a depicts a drawing showing the coordination environment of Cd(II), which is six-coordinated by five oxygen atoms from two 1,4-BDC^{2-} ligands, two water molecules [Cd-O = 2.2306(18) − 2.3644(17) Å] and one pyridyl nitrogen atom from the **L** ligand [Cd-N = 2.3445(19) Å], resulting in a distorted octahedral geometry. The Cd(II) cations are further linked together by 1,4-BDC^{2-} to afford 1D chains, Figure 5b, which are further linked by the **L** ligands to form a 2D layer. Considering the Cd(II) cations as 3-connected nodes and the organic ligands as linkers, the structure of complex **3** can be simplified as a 3-connected net with the 6^3-**hcb** topology, Figure 5c, determined by using ToposPro program [12].

3.4. Structure of 4

X-ray diffraction analysis reveals that crystals of **4** conform to the monoclinic space group $C2/c$ with each asymmetric unit comprising one Cu(II) cation, one 5-tert-IPA^{2-}, and one half **L**. The Cu(II) metal center, Figure 6a, is five-coordinated by four oxygen atoms from four independent 5-tert-IPA^{2-} ligands [Cu-O = 1.962(3) − 2.027(2) Å] and one pyridyl nitrogen atom from the **L** ligand [Cu-N = 2.190(3) Å], resulting in a distorted square pyramidal geometry (τ_5 = 0.006). Dinuclear paddlewheel Cu(II) units are linked by four 5-tert-IPA^{2-} ligands to form 1D looped chains, Figure 6b, with a distance of 2.7002(3) Å between the two Cu(II) ions, which are further linked by the **L** ligands to form a 3D framework. ToposPro program reveals that the structure of **4** can be regarded as a 4-connected net with the (6^5·8)-**cds** topology, Figure 6c, showing 3-fold interpenetration, Figure 6d, if the dinuclear Cu(II) units are considered as 4-connected nodes.

The structural topology of **4** is the same as that of [Zn(**L**)$_{0.5}$(5-tert-IPA)]$_n$ [9], indicating that the metal identity has no effect on the structural diversity of CPs constructed from the semi-rigid **L** and 5-tert-H$_2$IPA, which is in marked contrast to those from the flexible bpba ligands [6]. For example, reactions of Zn(II) and Cd(II) salts with N,N'-di(3-pyridyl)adipoamide and 1,2-H$_2$BDC afforded a 1D double-looped chain and a 2D layer, while with 1,3-H$_2$BDC gave a 3-fold interpenetrated **hcb** layers and a 1D ladder chain, respectively [6]. The metal effect on the structural diversity of CPs is thus subject to the flexibility of the bpba spacer ligands.

3.5. Ligand Conformation and Bonding Mode

The ligand conformation of **L** can be determined by considering the relative orientation of the two C=O groups, which forms cis and trans if the C=O groups appear in the same and the opposite direction, respectively [9]. On the other hand, the anti-anti, syn-anti, and syn-syn designations, based on the relative positions of the pyridyl nitrogen and amide oxygen atom, can also be shown [4,16]. The number of possible ligand conformations is thus quite less than that of the flexible bpba ligands with methylene carbon atoms as the spacer [4]. According to the descriptor for **L**, the ligand conformations of **L** in **1**, **2a**, and **3** are trans *anti-anti*, while that in **4** is trans *syn-syn*.

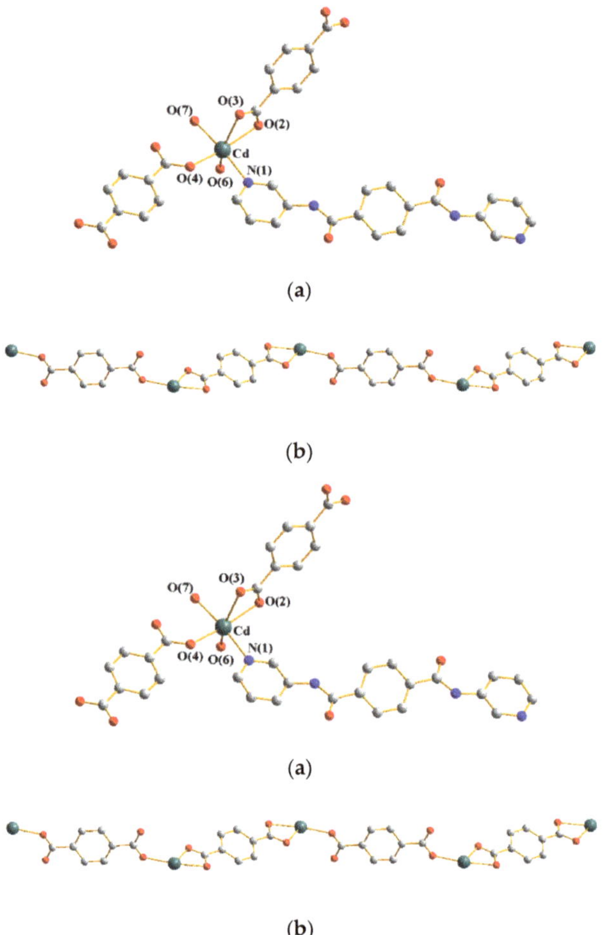

Figure 5. (**a**) Coordination environment of Cd(II) ion in **3**. Symmetry transformations used to generate equivalent atoms: (A) −x, −y + 2, −z + 2 (B) −x + 2, −y + 1, −z + 2 (C) −x + 3, −y + 2, −z + 1. (**b**) A drawing showing the 1D chain supported by the 1,4-BDC^{2-} ligands. (**c**) A drawing showing the 2D layer with the **hcb** topology.

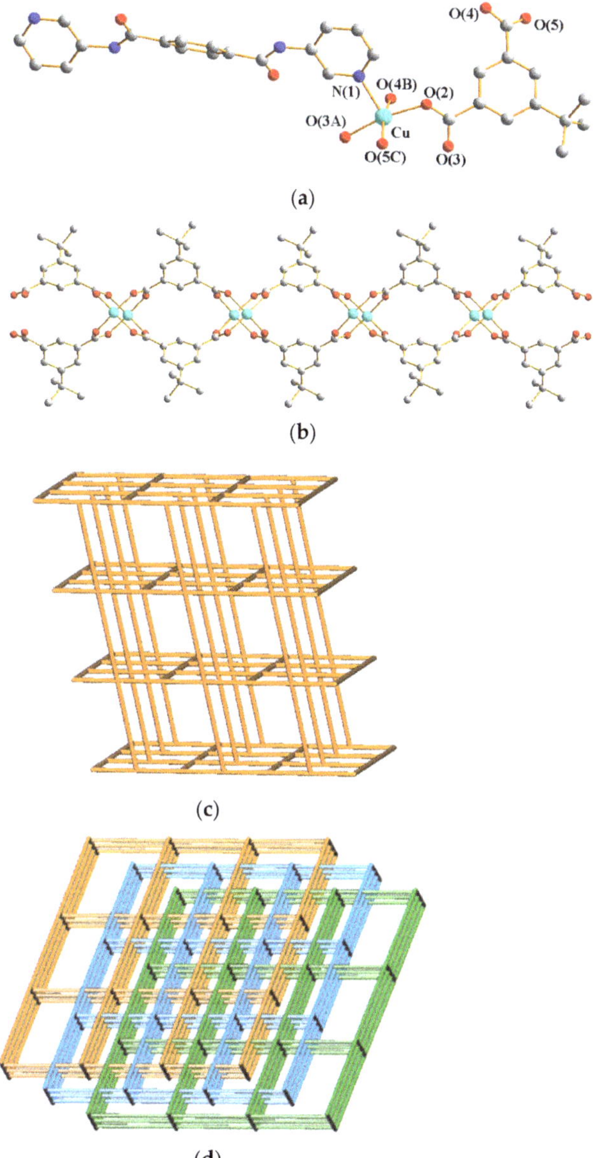

Figure 6. (**a**) Coordination environment of Cu(II) ion in **4**. Symmetry transformations used to generate equivalent atoms: (A) $-x + 1$, $-y$, $-z + 1$; (B) $-x + 1$, y, $-z + 1/2$; (C) x, $-y$, $z + 1/2$. (**b**) A drawing showing the paddlewheel type dinuclear units linked by the 5-tert-IPA^{2-} ligands. (**c**) A drawing showing the 3D net with the **cds** topology. (**d**) A drawing showing the 3-fold interpenetration.

Figure 7 shows the coordination modes for the dicarboxylate ligands. The 1,2-HBTC^{2-} ligands of **1** adopt μ_3-$\kappa^2 O,O'$:$\kappa O''\kappa^2 O''O'''$, coordination mode I, and the 1,3-BDC^{2-} ligands of **2a** and **2b** adopt μ_2-$\kappa^2 O,O'$:$\kappa O''$, coordination mode II, and μ_2-$\kappa^2 O,O'$:$\kappa^2 O''O'''$, coordination mode III, respectively, while the 1,4-BDC^{2-} ligands of **3** display μ_2-$\kappa^2 O,O'$:$\kappa^2 O''O'''$, coordination IV, and μ_2-κO:$\kappa O'$, coordination V, and the 5-tert-IPA^{2-} ligands of **4** adopt μ_4-$\kappa O,\kappa O'$:$\kappa O''\kappa O'''$, coordination mode VI.

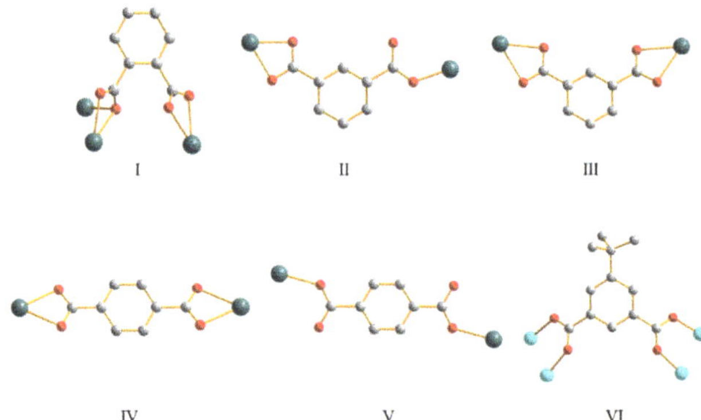

Figure 7. Coordination modes for the dicarboxylate ligands.

3.6. Ligand Flexibility and Isomeric Effect

Flexible bpba ligands containing different –$(CH_2)_n$– backbones are susceptible to the changes of the structural diversity because of the ability to adjust to the stereochemical requirements [4]. It has been shown that by manipulating the isomeric polycarboxylic acids, self-catenated CPs based on flexible bpba can be prepared [5–8]. In marked contrast, while the structural diversity of the Cd(II) CPs based on the semi-rigid **L** can be directed by the isomeric BDC^{2-} ligands, only low-dimensional networks of 2D layers and 1D ladder can be found, probably indicating the lack of the flexibility of **L** to adopt the proper conformation for the formation of a self-catenated net. The isomeric effect on the structural diversity of CPs in a mixed system thus relies on the flexibility of the neutral bpba ligands, not to mention the size and geometry of the metal ions.

3.7. Thermal Properties

Thermal gravimetric analyses (TGA) were carried out to examine the thermal decomposition of the compounds, Table 2. The samples were recorded from about 25 to 800 °C at 10 °C min^{-1} under a N_2 atmosphere. As shown in Figure S4, the TGA curve of **1** shows the gradual weight loss of one coordinated water molecule (calculated 3.97%, observed 4.76%) in 100–280 °C. The weight loss of 81.83% in 280–800 °C corresponds to the decomposition of half **L** and one 1,2-BDC^{2-} ligands (calculated 71.42%). TGA curve of **3**, Figure S5, shows the gradual weight loss of one cocrystallized water molecule (calculated 3.68%, observed 4.12%) in 40–95 °C. The weight loss of two coordinated water molecules (calculated 7.35%, observed 8.21%) in 100–350 °C. The weight loss of 66.41% in 360–706 °C corresponds to the decomposition of one and half **L** and one 1,4-BDC^{2-} ligands (calculated 66.19%). As shown in Figure S6, the TGA curve of **4** shows the weight loss of 81.98% in 250–800 °C corresponds to the decomposition of half **L** and one 5-tert-IPA^{2-} ligands (calculated 85.79%). It is noted that the starting decomposition temperature of the organic ligands of **4** is significantly lower than those of **1** and **3**, probably indicating that the 2D layers of **1** and **3** are more stable than the 3-fold interpenetrated net of **4**.

Table 2. Thermal properties of **1**, **3**, and **4**.

Complex	Weight Loss of Solvent, T, °C (Observed/Calc),%	Weight Loss of Ligand, T, °C (Observed/Calc),%
1	100–280 (4.76/3.97)	280–800 (81.83/71.42)
3	40–95 (4.12/3.68) 100–350(8.21/7.35)	360–706 (66.41/66.19)
4		250–800 (81.98/85.79)

3.8. Luminescent Properties

Table 3 summarizes the luminescent properties of **L**, 1,2-H$_2$BDC, 1,4-H$_2$BDC, **1**, and **3**. Figures S7–S11 depict the corresponding excitation/emission spectra, which were measured in the solid state at room temperature. The emissions of **L**, 1,2-H$_2$BDC and 1,4-H$_2$BDC appear in the range of 340–431 nm, which may be ascribed to the intraligand π^* → n or π^* → π transitions, while **1** and **3** in mixed systems show emissions at 416 and 434 nm upon excitations at 318 and 334 nm, respectively. Since oxidation or reduction of Cd(II) ion is not possible, it is thus not probable that the emissions of **1** and **3** are due to ligand-to-metal charge transfer (LMCT) or metal-to-ligand charge transfer (MLCT) [17]. Therefore, these emissions may be attributed to ligand-to-ligand charge transfer (LLCT). The blue- and red-shift of the emission wavelengths of **1** and **3** with respect to their corresponding organic ligands may be ascribed to the different dicarboxylate ligands that result in different structural types.

Table 3. The excitation and emission wavelengths of **L**,1,2-H$_2$BDC, 1,4-H$_2$BDC and complexes **1** and **3** in solid state.

Ligand.	λ_{ex} **(nm)**	λ_{em} **(nm)**	Complex	λ_{ex} **(nm)**	λ_{em} **(nm)**
L	276	431	**1**	318	416
1,2-H$_2$BDC	283	340	**3**	334	434
1,4-H$_2$BDC	277/331	384			

4. Conclusions

Divalent CPs constructed from the semi-rigid **L**, polycarboxylic acids, and metal salts have been synthesized successfully under hydrothermal reactions. Complexes **1** and **3** are 2D layers with the **bey** topology and the **hcb** topology, and **2a** and **2b** form a 1D ladder and a zigzag chain, respectively, while **4** is a 3-fold interpenetrated 3D nets with the **cds** topology. The structures of **1–4** are subject to the changes of the polycarboxylate ligands, indicating the effect of the ligand identity as well as the ligand isomerism on the structural diversity. While the flexible bpba are more likely to form self-catenated nets by the manipulation of the isomeric dicarboxylate ligands, the structural diversity is limited for semi-rigid **L**, presumably due to the lack of the flexibility to adopt the proper conformations for the formation of the entangled CPs.

Supplementary Materials: The following are available online at https://www.mdpi.com/2624-854 9/3/1/1/s1. Powder X-ray patterns (Figures S1–S3). TGA curves (Figures S4–S6). Emission Spectra (Figures S7–S11). X-ray data for **2a** (Tables S1–S3). Crystallographic data for **1, 2b, 3,** and **4** have been deposited with the Cambridge Crystallographic Data Centre, CCDC No. 2046035-2046038.

Author Contributions: Investigation, C.-J.C. and C.-L.C.; data curation, Y.-H.L. and J.-H.H.; validation, W.-T.L.; review and supervision, P.M.C. and J.-D.C. All authors have read and agreed to the published version of the manuscript.

Funding: This research was funded by Ministry of Science and Technology of the Republic of China, grant number MOST 108-2113-M-033-004.

Institutional Review Board Statement: Not applicable.

Informed Consent Statement: Not applicable.

Data Availability Statement: Data available in a publicly accessible repository.

Acknowledgments: We are grateful to the Ministry of Science and Technology of the Republic of China for support.

Conflicts of Interest: The authors declare no conflict of interest.

References

1. Batten, S.R.; Champness, N.R.; Chen, X.-M.; Garcia-Martinez, J.; Kitagawa, S.; Öhrström, L.; O'Keeffe, M.; Suh, M.P.; Reedijk, J. Terminology of metal–organic frameworks and coordination polymers. *Pure Appl. Chem.* **2013**, *85*, 1715–1724. [CrossRef]
2. Batten, S.R.; Neville, S.M.; Turner, D.R. *Coordination Polymers: Design, Analysis and Application*; Royal Society of Chemistry: Cambridge, UK, 2009.
3. Wales, D.J.; Grand, J.; Ting, V.P.; Burke, R.D.; Edler, K.J.; Bowen, C.R.; Mintova, S.; Burrows, A.D. Gas sensing using porous materials for automotive applications. *Chem. Soc. Rev.* **2015**, *44*, 4290–4321. [CrossRef] [PubMed]
4. Thapa, K.B.; Chen, J.-D. Crystal engineering of coordination polymers containing flexible bis-pyridyl-bis-amide ligands. *CrystEngComm* **2015**, *17*, 4611–4626. [CrossRef]
5. Hsu, Y.-F.; Hu, H.-L.; Wu, C.-J.; Yeh, C.-W.; Proserpio, D.M.; Chen, J.-D. Ligand isomerism-controlled structural diversity of cadmium(II) perchlorate coordination polymers containing dipyridyladipoamide ligands. *CrystEngComm* **2009**, *11*, 168–176. [CrossRef]
6. Cheng, P.-C.; Kuo, P.-T.; Liao, Y.-H.; Xie, M.-Y.; Hsu, W.; Chen, J.-D. Ligand-isomerism controlled structural diversity of Zn(II) and Cd(II) coordination polymers from mixed dipyridyladipoamide and benzenedicarboxylate ligands. *Cryst. Growth Des.* **2013**, *13*, 623–632. [CrossRef]
7. Lo, Y.-C.; Hsu, W.; He, H.-Y.; Hyde, S.-T.; Proserpio, D.M.; Chen, J.-D. Structural directing roles of isomeric phenylenediacetate ligands in the formation of coordination networks based on flexible *N,N*-di(3-pyridyl)suberoamide. *CrystEngComm* **2015**, *17*, 90–97. [CrossRef]
8. He, H.-Y.; Hsu, C.-H.; Chang, H.-Y.; Yang, X.-K.; Chhetri, P.M.; Proserpio, D.M.; Chen, J.-D. Self-catenated coordination polymers involving bis-pyridyl-bis-amide. *Cryst. Growth. Des.* **2017**, *17*, 1991–1998. [CrossRef]
9. Chen, C.-J.; Lee, W.-T.; Hu, J.-H.; Chhetri, P.M.; Chen, J.-D. Structural Diversity and Modification in Ni(II) Coordination Polymers: Peculiar Phenomenon of Reversible Structural Transformation between 1D Ladder and 2D Layer. *CrystEngComm* **2020**, *22*, 7565–7574. [CrossRef]
10. Rajput, L.; Singha, S.; Biradha, K. Comparative Structural Studies on Homologues of Amides and Reverse Amides: Unprecedented 4-fold Interpenetrated Quartz Network, New β-Sheet, and Two-Dimensional Layers. *Cryst. Growth Des.* **2007**, *7*, 2788–2795. [CrossRef]
11. Bruker AXS. *APEX2, V2008.6; SAD ABS V2008/1; SAINT+ V7.60A.; SHELXTL V6.14*; Bruker AXS Inc.: Madison, WI, USA, 2008.
12. Sheldrick, G.M. Crystal structure refinement with SHELXL. *Acta Cryst.* **2015**, *71*, 3–8.
13. Blatov, V.A.; Shevchenko, A.P.; Proserpio, D.M. Applied topological analysis of crystal structures with the program package ToposPro. *Cryst. Growth Des.* **2014**, *14*, 3576–3586. [CrossRef]
14. Zhang, F.; Bei, F.-L.; Cao, J.-M. Solvothermal Synthesis and Crystal Structure of Two Cd^{II} Coordination Polymers of Benzenedicarboxylic. *J. Chem. Cryst.* **2008**, *38*, 561–565. [CrossRef]
15. Ahmed, I.; Yunus, U.; Nadeem, M.; Bhatti, M.H.; Mehmood, M. Post synthetically modified compounds of Cd-MOF by L-amino acids for luminescent applications. *J. Solid State Chem.* **2020**, *287*, 121320. [CrossRef]
16. Chang, M.-N.; Yang, X.-K.; Chhetri, P.-M.; Chen, J.-D. Metal and Ligand Effects on the Construction of Divalent Coordination Polymers Based on bis-Pyridyl-bis-amide and Polycarboxylate Ligands. *Polymers* **2017**, *9*, 691. [CrossRef] [PubMed]
17. Sun, D.; Yan, Z.-H.; Blatov, V.A.; Wang, L.; Sun, D.-F. Syntheses, topological structures, and photoluminescences of six new Zn(II) coordination polymers based on mixed tripodal imidazole ligand and varied polycarboxylates. *Cryst. Growth Des.* **2013**, *13*, 1277–1289. [CrossRef]

 *chemistry*

Article

Neutral and Cationic Chelidonate Coordination Polymers with *N,N′*-Bridging Ligands

Rosa Carballo [1,*], Ana Belén Lago [2,*], Arantxa Pino-Cuevas [1], Olaya Gómez-Paz [1], Nuria Fernández-Hermida [1] and Ezequiel M. Vázquez-López [1]

[1] Departamento de Química Inorgánica, Instituto de Investigación Sanitaria Galicia Sur (IISGS), Universidade de Vigo, 36310 Vigo, Spain; aripc-27@hotmail.com (A.P.-C.); olgomez@uvigo.es (O.G.-P.); grupo_qi5@uvigo.es (N.F.-H.); ezequiel@uvigo.gal (E.M.V.-L.)

[2] Departamento de Química, Facultad de Ciencias, Sección Química Inorgánica, Universidad de la Laguna, 38206 La Laguna, Spain

[*] Correspondence: rcrial@uvigo.es (R.C.); alagobla@ull.edu.es (A.B.L.)

Abstract: The biomolecule chelidonic acid (H_2chel, 4-oxo-4H-pyran-2,6-dicarboxylic acid) has been used to build new coordination polymers with the bridging *N,N′*-ligands 4,4′-bipyridine (4,4-bipy) and 1,2-bis(4-pyridyl)ethane (bpe). Four compounds have been obtained as single crystals: 1D cationic coordination polymers $[M(4,4\text{-bipy})(OH_2)_4]^{2+}$ with chelidonate anions and water molecules in the second coordination sphere in $^1_\infty[Zn(4,4\text{-bipy})(H_2O)_4]$chel·$3H_2O$ (**2**) and in the two pseudopolymorphic $^1_\infty[Cu(4,4\text{-bipy})(H_2O)_4]$chel·$nH_2O$ (*n* = 3, **4a**; *n* = 6, **4b**), and the 2D neutral coordination polymers $^2_\infty[Zn(chel)(4,4\text{-bipy})(H_2O)]·2H_2O$ (**1**) and $^2_\infty[Zn(chel)(bpe)(H_2O)]·H_2O$ (**3**) where the chelidonate anion acts as a bridging ligand. The effects of the hydrogen bonds on the crystal packing were analyzed. The role of the water molecules hosted within the crystals lattices was also studied.

Keywords: zinc complexes; copper(II) complexes; coordination polymers; chelidonic acid; metallosupramolecular compounds; H-bonding pattern; X-ray diffraction

check for updates

Citation: Carballo, R.; Lago, A.B.; Pino-Cuevas, A.; Gómez-Paz, O.; Fernández-Hermida, N.; Vázquez-López, E.M. Neutral and Cationic Chelidonate Coordination Polymers with *N,N′*-Bridging Ligands. *Chemistry* **2021**, *3*, 256–268. https://doi.org/10.3390/chemistry3010019

Received: 9 January 2021
Accepted: 7 February 2021
Published: 11 February 2021

Publisher's Note: MDPI stays neutral with regard to jurisdictional claims in published maps and institutional affiliations.

1. Introduction

Chelidonic acid (H_2chel, 4-oxo-4H-pyran-2,6-dicarboxylic acid) is a biomolecule found in several plants and is considered to be an active component of some medicinal herbs [1]. The compound exhibits some pharmacological effects such as mild analgesic, antimicrobial, oncostatic and sedative [2] effects and its therapeutic potential in allergic disorders [1], intestinal inflammation [2] and regulation of depression associated with inflammation [3] have been investigated. From the point of view of coordination chemistry, H_2chel after deprotonation is an angular dicarboxylate that can act as a linker between metal ions to yield coordination polymers. From the supramolecular point of view, $chel^{2-}$ presents several oxygen atoms that can act as hydrogen acceptors towards biologically important hydrogen donors such as water molecules. In this way, the dianion chelidonate can be considered a biologically and environmentally interesting building block for the construction of new metallosupramolecular compounds. This kind of building blocks offers advantages such as its natural availability, the presence of various metal-binding sites which allow the display of different coordination modes and consequently a structural diversity and the possibility of establishing different hydrogen bonds patterns in aqueous media.

Several structural studies on metal chelidonates that are coordination polymers of different dimensionality can be found in the literature: 1D compounds with Zn(II) [4], Cu(II) [5,6], Dy(III)-Ba(II) [7] or Lu(III) [8], 2D systems with several lanthanide cations [8] and also with Cu(II) [9] and 3D polymers with Cd(II) [5], Cu(II) and Ag(I) [9], Pr(III), Nd(III), Sm(III), Lu-Ba and Sm-Ba [7] and with Ba(II) [10]. In all these compounds the chelidonate is the only ligand used. However, the structural information available on mixed ligand chelidonate coordination polymers is scarce. To the best of our knowledge only a few examples have been reported: 1D coordination polymers with monodentate ligands such as pyridine [11] or dmso [12] and with the bidentate chelating ligand 2,2′-bipyridine [13], which also stabilised the 2D copper(II) coordination polymer $^2_\infty$[Cu(chel)(bipy)] [13]. In all of these coordination polymers the chelidonate dianion behaves as a bridging ligand between two, three or four metal centres with the coordination modes shown in Scheme 1.

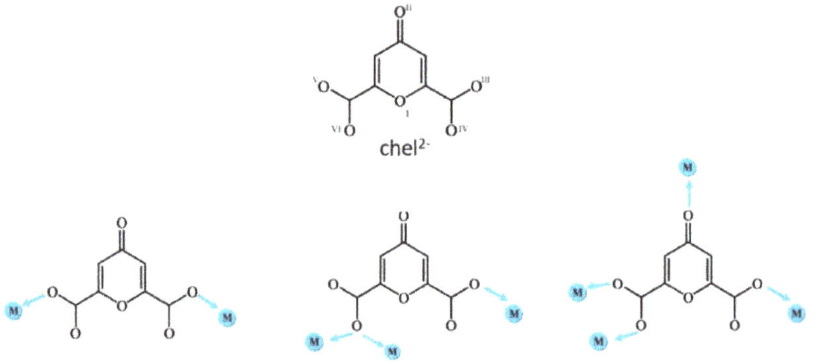

$$\text{chel}^{2-}$$

$$\mu\text{-}1\kappa O^{III}{:}2\kappa O^{V} \qquad \mu_3\text{-}1\kappa O^{III}{;}2\kappa O^{V}{:}3\kappa O^{V} \qquad \mu_4\text{-}1\kappa O^{III}{:}2\kappa O^{V}{:}3\kappa O^{VI}{:}4\kappa O^{II}$$

Scheme 1. Coordination modes of the chelidonate ligand in coordination polymers.

In view of these previous results on the role of the chelidonate anion in coordination polymers, we decided to study its coordinative possibilities and supramolecular role in aqueous media in systems that contain the bridging N,N'-ligands 4,4′-bipyridine (4,4-bipy) and 1,2-bis(4-pyridyl)ethane (bpe), which can facilitate the extension of the metal-ligand interaction in at least one direction. The aim of this work was to prepare crystalline solids of coordination polymers formed with chelidonate anions to analyse their coordinative behaviour, searching for new coordination modes, and to study their participation in the supramolecular arrangements, searching for new hydrogen bond patterns in aqueous media.

We report here the diffusion solvent crystallization, structural characterisation and supramolecular structures of the 2D neutral coordination polymers $^2_\infty$[Zn(chel)(4,4-bipy)(H$_2$O)]·2H$_2$O (**1**) and $^2_\infty$[Zn(chel)(bpe)(H$_2$O)]·H$_2$O (**3**), in which the chelidonate anion acts as a bridging ligand. The cationic 1D coordination polymer $^1_\infty$[Zn(4,4-bipy)(H$_2$O)$_4$]chel·3H$_2$O (**2**) and its analogues $^1_\infty$[Cu(4,4-bipy)(H$_2$O)$_4$]chel·nH$_2$O (n = 3, **4a**; n = 6, **4b**), with the chelidonate anion in the second coordination sphere, were also obtained and structurally characterised.

2. Results and Discussion

2.1. Preparation and Spectroscopic Characterisation of Complexes

The synthetic procedure for the preparation of the target compounds is shown in Scheme 2. The slow diffusion at room temperature of aqueous solutions of the 1D coordination polymer $^1_\infty$[Zn(chel)(H$_2$O)$_2$] [4] and ethanolic solutions of the two N,N' bridging ligands 4,4′-bipyridine (4,4-bipy) and 1,2-bis(4-pyridyl)ethane (bpe) led to the formation of single crystals of the neutral 2D coordination polymers $^2_\infty$[Zn(chel)(4,4-bipy)(H$_2$O)]·2H$_2$O (**1**) and $^2_\infty$[Zn(chel)(bpe)(H$_2$O)]·H$_2$O (**3**). After the separation of crystals of **1** the remaining solution was left to evaporate at room temperature and single crystals of the 1D cationic coordination polymer $^1_\infty$[Zn(4,4-bipy)(H$_2$O)$_4$]chel·3H$_2$O (**2**), with the uncoordinated chelidonate anion, were isolated. Polymer **2** contains the $^1_\infty$[Zn(4,4-bipy)(H$_2$O)$_4$]$^{2+}$ cation, which was also found in several crystalline compounds with other anions such as nitrate [14], biphenyldisulfonate [15], sulfobenzoate [16], sulfamoylbenzoate [17], succinate [18], barbiturate [19] and perchlorate [20] in the second coordination sphere. After a CSD search [21] it was verified that the formation of the polymeric species $^1_\infty$[M(4,4-bipy)(H$_2$O)$_4$]$^{2+}$ is observed for other metal centres in the presence of different anions in the outer coordination sphere. Some examples are Mn(II) or Co(II) with 4-aminobenzenesulfonate [22], Fe(II) or Co(II) with croconate [23], Co(II) with several polycarboxylates (furandicarboxylate, benzenetricarboxylate, benzenetetracarboxylate [24] and 3,3′,4,4′-biphenyltetracarboxylate [25]), Ni(II) with 1,2,4,5-benzenetetracarboxylate, Ni(II) or Cu(II) with 2,6-naphthalenedisulfonate [26], Co(II), Ni(II) or Cu(II) with barbiturate [27] and Cd(II) with 1,3-propanedisulfonate [28].

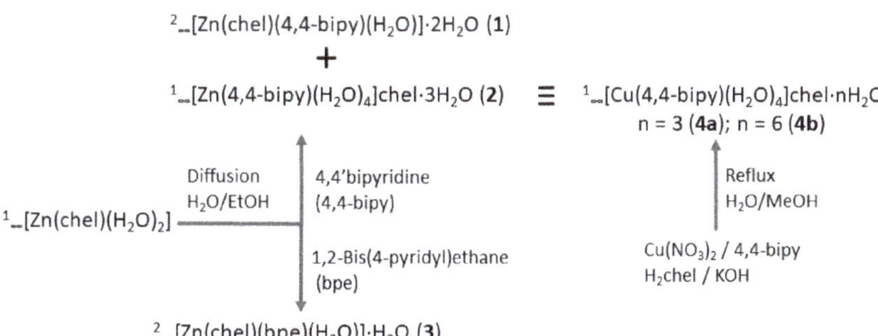

Scheme 2. Synthetic procedure for the preparation of the compounds.

The frequent formation of such polymeric cationic species with different anions that can be considered as weakly coordinating suggests a remarkable stability for this cation. It was decided to explore whether it was possible to obtain the $^1_\infty[Cu(4,4\text{-bipy})(H_2O)_4]^{2+}$ species with the uncoordinated chelidonate anion. After unsuccessfully attempting diffusion methods we carried out a reaction under reflux in $H_2O/MeOH$ of a mixture of $Cu(NO_3)_2/4,4\text{-bipy}$ and H_2chel/KOH. This led to the isolation of single crystals of the pseudopolymorphs $^1_\infty[Cu(4,4\text{-bipy})(H_2O)_4]chel\cdot nH_2O$ ($n = 3$, **4a**; $n = 6$, **4b**).

The IR spectra of all of the compounds exhibit broad bands of medium intensity at around 3400 cm^{-1} corresponding to the OH stretching vibrations of the coordinated and crystallisation water molecules. The strong IR bands between 1600 and 1640 cm^{-1} can be assigned to the $\nu_{asym}(OCO)$ vibration mode and the medium intensity bands at around 1360 cm^{-1} are attributable to the $\nu_{sym}(OCO)$ vibration mode.

2.2. Structural Studies

All of the compounds described here were isolated as single crystals and their structures were elucidated by X-ray diffraction. The significant structural parameters for compounds **1–4** are listed in Tables S1 and S2, and crystal structure and refinement data are listed in Table S5.

Structural analysis revealed that all of the compounds are hydrates and that **2**, **4a** and **4b** are formed by cationic chains that are stabilised by chelidonate anions and water molecules. When the chelidonate ligand is coordinated to the metal centre and forms part of the structural backbone, the resulting compounds are 2D coordination polymers, as revealed by X-ray analysis of **1** and **3**.

2.3. Crystal Structures of the Cationic Polymers 2, 4a and 4b

The coordination environment of the metal ions and the 1D polymeric nature of compounds **2**, **4a** and **4b** are represented in Figure 1. A summary of the bond lengths and angles in the first coordination sphere is provided in Table 1.

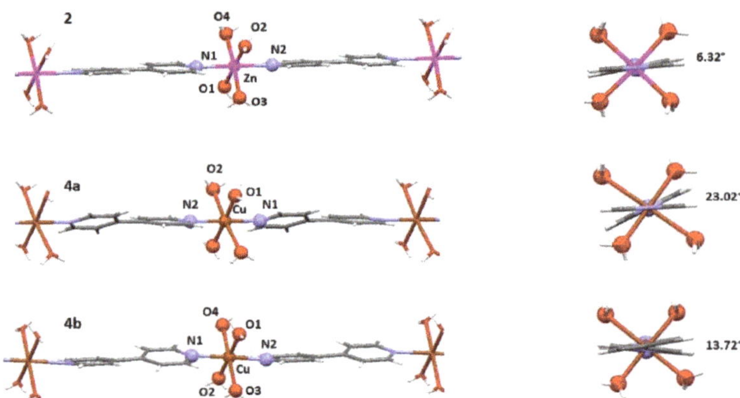

Figure 1. Cationic chains in coordination polymers **2**, **4a** and **4b** showing in detail the coordination environment and the disposition of the 4,4-bipy rings.

Table 1. Summary of bond lengths/Å and angles/° in the first coordination sphere of **2**, **4a** and **4b**.

	2	**4a**	**4b**
M–OH$_2$	2.088(2)–2.105(2)	1.952(2)–2.417(2)	2.028(2)–2.291(2)
M–N	2.174(2) 2.195(2)	2.049(3) 2.031(3)	2.093(3)
cis angles	87.15(9)–93.49(9)	87.30(7)–92.70(7)	88.26(9)–91.70(9)
trans angles	174.83(10)–178.58(8)	174.59(15)–180.0	177.21(9)–179.90(10)

The three structures are polymeric chains based on the cationic complexes [M(4,4-bipy)(OH$_2$)$_4$]$^{2+}$ (M = Cu(II) and Zn(II)), where the N-donor ligands act as a bridge between the metal centres. The second coordination spheres are formed by dianionic chelidonate anions and three or six water molecules. Complexes **4a** and **4b** crystallise in the orthorhombic chiral *P2$_1$2$_1$2* and triclinic *P-1* space groups, respectively, and they can be considered pseudopolymorphs [29] since they have the same crystalline form, albeit with different numbers of water molecules trapped within the crystal networks.

All M^{II} metal centres are {N$_2$O$_4$}-hexacoordinated through the nitrogen bipy atoms in trans positions and four oxygen atoms belonging to four water molecules in the equatorial plane. The copper atoms in **4a** and **4b** are in an elongated octahedral environment and they exhibit the expected Jahn–Teller distortion, with four short metal–ligand bonds (Cu–O$_{water}$ and Cu–N with distances of around 2 Å) and two longer bonds with water molecules (2.417(2) Å in **4a**, 2.291(2) and 2.173(3) Å in **4b**). Clearly the Jahn–Teller distortion is more pronounced in compound **4a**, where the pyridine rings of the 4,4′-bipyridine ligand are further from coplanarity, as shown in Figure 1.

In compound **2** the Zn–O and Zn–N distances are in the range 2.08–2.19 Å (Table 1) and the values for the orthogonal angles in the octahedral polyhedron are between 87.15 and 93.49°. Comparison of the orthogonality of the coordination octahedra in the three compounds shows that the smallest deviation is observed in the copper(II) compound **4b**, with values between 88.26 and 91.70°.

The intermetallic distance through the 4,4-bipy bridging ligand is shorter in copper polymers (11.176 Å in **4a** and 11.290 Å in **4b**) than in **2** (11.487 Å). The aromatic rings in 4,4-bipy are not coplanar and the angles between the planes are 23.02° and 13.72° in **4a** and **4b**, respectively, and the disposition is closer to coplanarity in **2** (around 6°) (Figure 1).

In all structures the second coordination sphere plays a crucial role in the supramolecular organisation. The cationic chains are connected by hydrogen bonds to the chelidonate anions and different numbers of water molecules to form different arrangements in each structure. The main hydrogen-bonding interactions responsible for the 3D crystal packing of compounds **2**, **4a** and **4b** are listed in Table S3. In all cases, the resulting supramolecular networks are formed and propagated through $O_{chelidonate}\cdots O_{water}$ hydrogen bonds, with the formation of different association patterns such as rings, discrete associations or infinite layers.

In compound **2** the three uncoordinated water molecules are not associated with each other, so the notation W_1 has been selected to describe their isolated participation in the supramolecular organisation. One of the water molecules, O2w, and two chelidonate anions form a discrete dimer through H-bonding interactions to yield a four-membered ring, with $O\cdots O$ distances of 2.781 and 2.932 Å (Figure 2, Table S3). Another crystallisation water molecule, O1w, establishes H-bonding interactions with the two carboxylate groups of a chelidonate to form an eight-membered ring and the third water molecule is H-bonded with the ketonic oxygen of the chelidonate. In this way, all of the oxygen atoms of the chelidonate are involved as acceptors in hydrogen-bonding interactions.

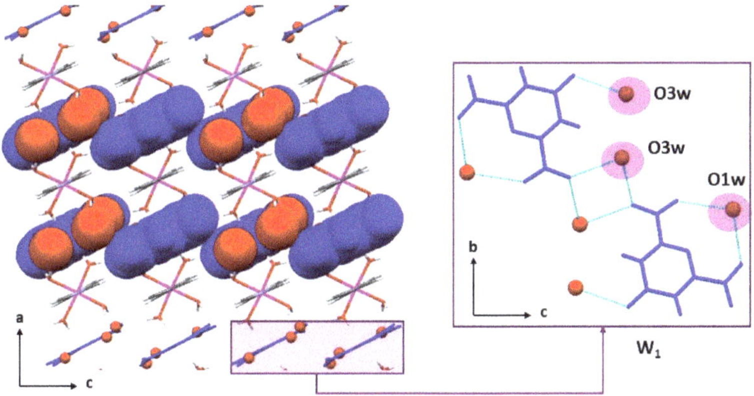

Figure 2. Supramolecular arrangement of **2** showing in detail the interactions established between the crystallisation water molecules and chelidonate anions.

The cationic chains of **2** are anchored to these water-chelidonate dimers through $Ow_{coord}\cdots O_{carboxy}$ and $Ow_{coord}\cdots Ow_{crystallisation}$ interactions (Table S3). Cationic chains and the anionic units are arranged in an alternating manner to form the final supramolecular organisation (Figure 2).

In the structures of **4a** and **4b** chelidonate anions and water molecules are associated through different H-bonding interactions to form different anionic arrangements (Figures 3 and 4) that are extended in two dimensions, thus leading to anionic $[(H_2O)_n chel]^{2-}$ sheets. In this way, the layers act as the 'glue' between the cationic units to form the final metallosupramolecular arrangements [30]. This kind of 2D anion-water association has been observed in other cases, such as $[Ni(4,4\text{-bipy})(H_2O)_4]\cdot0.5(btc)\cdot H_2O$ (H_4btc, 1,2,4,5-benzenetetracarboxylic acid) [25].

In **4a** the crystallisation water molecules are also not interconnected by hydrogen bonds (W1 motif) and only one of them (O2w) is involved in the formation of the supramolecular $[(H_2O)_3 chel)]^{2-}$ layer (Figure 3), with $O\cdots O$ distances of 2.781(2) and 2.913(3) Å. These layers are parallel to one another and this arrangement produces columnar voids in which the cationic chains are placed and stabilised by several hydrogen bonds involving as donors the crystallisation and coordinated water molecules and as acceptors all of the oxygen atoms of chelidonate (Table S3).

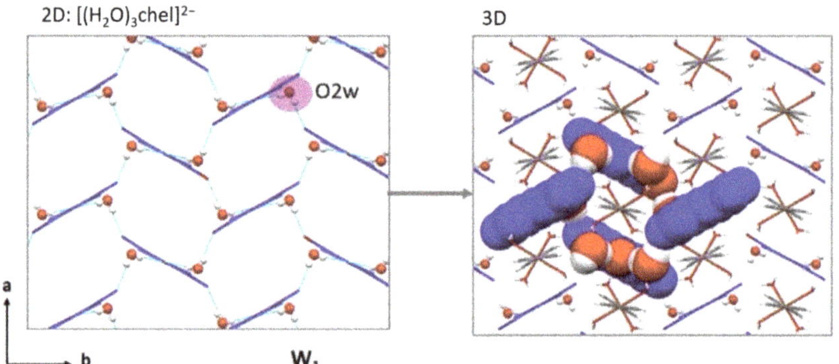

Figure 3. Supramolecular arrangement of **4a** showing in detail the layers formed by the interaction of crystallisation water molecules and chelidonate anions.

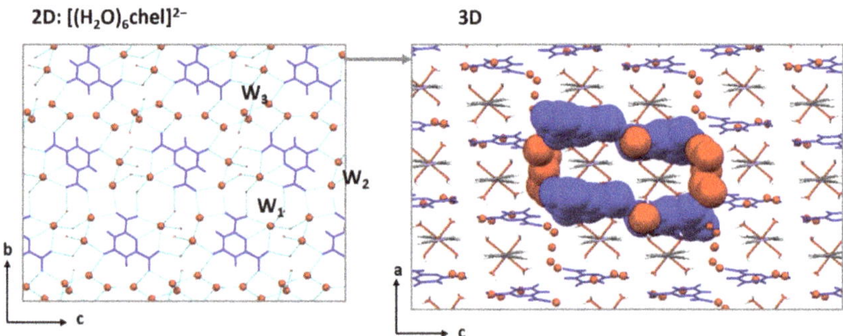

Figure 4. Supramolecular arrangement of **4b** showing in detail the layer formed by the interaction of crystallisation water molecules and chelidonate anions.

In **4b** the six water molecules of crystallisation have three different arrangements, namely one isolated molecule (W1), a dimer (W2) and a trimer (W3) (Figure 4). The isolated molecule, O13w, corresponds to the W1 motif that is also present in all three compounds (**2**, **4a** and **4b**) with the same structural behaviour; as a donor this water molecule is connected with a chelidonate molecule by hydrogen bonds with two oxygen atoms from the two carboxylate groups (O···O distances of 2.721(3) and 2.723(3) Å in **4b**) and as an acceptor with one or two coordinated water molecules of the corresponding cations. The other five water molecules are associated in a dimer (W2) with an O···O distance of 2.855(4) Å and in a trimer (W3) with shorter O···O distances (2.729(5) and 2.737(6) Å). The hydrogen-bonding interactions of these water molecule motifs with the chelidonate anions (Table S3) lead to the organisation of the second coordination sphere being anionic [(H₂O)₆chel]²⁻ ladder-shaped layers (Figure 4) that leave space to accommodate the cationic units anchored through $Ow_{coord} \cdots O_{carboxy}$ and $Ow_{coord} \cdots Ow_{crystallisation}$ interactions.

In the coordination polymers **2**, **4a** and **4b** the different collections of hydrogen bonds modulate the association of the chains and result in a similar packing degree, as indicated by the Kitaigorodskii indexes [31]; 70.1% in **4a**, 70.6% in **4b** and 72.1% in **2**.

2.4. Crystal Structures of Neutral Polymers 1 and 3

The Zn(II) polymeric compounds **1** and **3** crystallise in the *P-1* and *P*2₁/*c* space groups, respectively. The significant structural parameters are listed in Table S2 and a summary of bond lengths/Å and angles/° in the first coordination sphere is provided in Table 2. In both compounds, the two carboxylate groups of the chelidonate ligand are coordinated to the metal centre to produce monodimensional chains. These chains are bridged by the rigid 4,4-bipyridine or the more flexible 1,2-bis(4-pyridyl)ethane *N*-donor ligands to yield the final 2D coordination compounds (Figures 5 and 6).

Table 2. Summary of bond lengths/Å and angles/° in the first coordination sphere of [Zn(chel)(4,4-bipy)(H$_2$O)]·2H$_2$O (**1**) and [Zn(chel)(bpe)(H$_2$O)]·H$_2$O (**3**).

	1	**3**
Zn–O(1W)	2.032(3)	2.158(2)
Zn–Ochel	2.015(3) 2.539(3) 2.130(3)	1.978(2) 2.557(2) 2.085(2)
Zn–N	2.151(4) 2.157(3)	2.147(2) 2.084(2)
cis angles	82.27(12)–116.83(12)°	88.37(8)–97.42(8)
trans angles	149.06(11)–169.87(14)°	138.00(8)–169.69(9)°

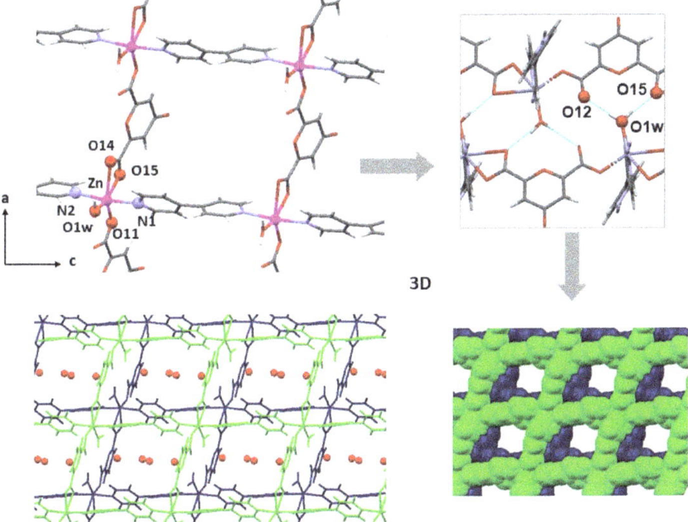

Figure 5. View of the 2D layers of $^2_\infty$[Zn(chel)(4,4-bipy)(H$_2$O)]·2H$_2$O (**1**) and the final 3D supramolecular arrangement, showing the zinc atom environment and the corresponding labelling scheme.

Chelidonate acts in both structures as a bridging ligand in a μ-1κ^{2}O^{III},O^{IV}:2κO^V coordination mode (Scheme 3), which has not been previously observed in other coordination polymers with this ligand. Each *N*-donor ligand acts as a bis-monodentate system to bridge zinc atoms with Zn···Zn distances of 11.346 Å and 13.434 Å in **1** and **3**, respectively. Chelidonate ligands also connect two zinc metal centres to produce similar Zn···Zn distances of 9.928 Å in **1** and 9.611 Å in **3**. In this way rhomboidal grid layer structures are formed.

Each metal centre is in a distorted octahedral geometry, with two nitrogen atoms from *N*-donor molecules in a trans disposition in **1** and a cis arrangement in **3** (Zn–N distances are in the range 2.084–2.157 Å), three oxygen atoms from carboxylate groups of a chelidonate ligand and one oxygen atom from a water molecule (Zn–O$_W$ distances are 2.032(3) Å in **1** and 2.158(2) Å in **3**). One carboxylate group of the chelidonate ligand is anisobidentate, with a bite angle of about 55°, one zinc metal has a short Zn–O distance (2.130(3) Å in **1** and 2.085(2) Å in **3**), the other has a longer distance of around 2.5 Å and the other carboxylate group is monodentate.

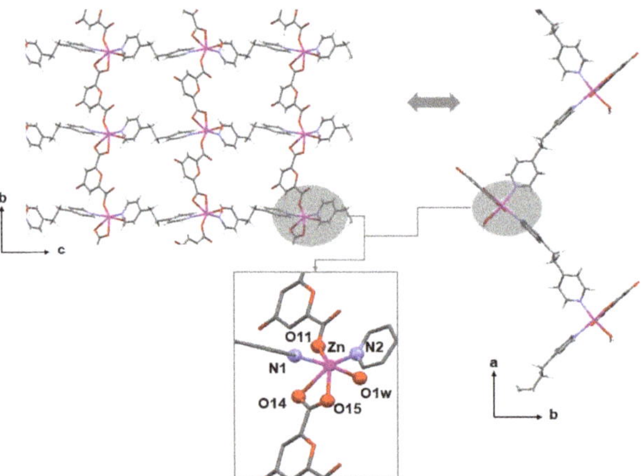

Figure 6. Structure of [Zn(chel)(bpe)(H$_2$O)]·H$_2$O (**3**) showing the zinc atom environment and the corrugated disposition of the layers.

Scheme 3. New coordination mode, μ-1κ²O^{III},O^{IV}:2κO^{V}, of chelidonate in coordination polymers **1** and **3**.

Zinc-chelidonate chains and the bis-monodentate rigid 4,4-bipy bridging ligand form a rhomboidal grid layer in **1** within an sql topology, with rhomboid window dimensions of 11.35 × 9.93 Å and angles of 97° and 82° (defined by Zn···Zn distances and Zn···Zn···Zn angles). Further expansion of the length of the bidentate bipyridine ligand to bpe leads to the formation of a corrugated layer structure in **3** with rhomboidal void dimensions of 13.43 × 9.61 Å and angles of 97° and 82°.

The aromatic rings in 4,4-bipy are not coplanar and show a deviation of 42° in **1**, which is clearly higher than the values observed in the polymeric chains of **4a** and **4b**. The bpe linkers in **3** are twisted with respect to each other through an anticlinal torsion angle of 167.4° (through C–CH$_2$–CH$_2$–C) indicating an anti conformation.

The layers in **1** are stacked along the *b* axis to yield a final 3D arrangement with an ABAB stacking mode (Figure 5). These layers are connected by means of hydrogen bonds involving the coordinated water molecule and oxygen atoms from carboxylate groups of neighbouring layers (d(O1W ··· Ochel) = 2.699(4) and 2.772(5) Å; Table S4, detail shown in Figure 5). The two-crystallisation independent free water molecules (W1) are located in the voids of the rhomboidal grids (Figure 5) and are weakly attached to the 2D layers.

The supramolecular arrangement in the corrugated layers of **3** is achieved by means of the same supramolecular synthon described above for compound **1**: hydrogen bonds involving the coordinated water molecule and oxygen atoms from carboxylate groups of neighbouring layers (d(O1W ··· Ochel) = 2.762(3) and 2.825(3) Å, Table S4 and detail in Figure 7). As in compound **1**, the only crystallisation water molecule (W1) is weakly attached to the metal organic framework through a hydrogen-bonding interaction with the ketonic group of a chelidonate ligand (d(Ow ··· O) = 3.125(5) Å).

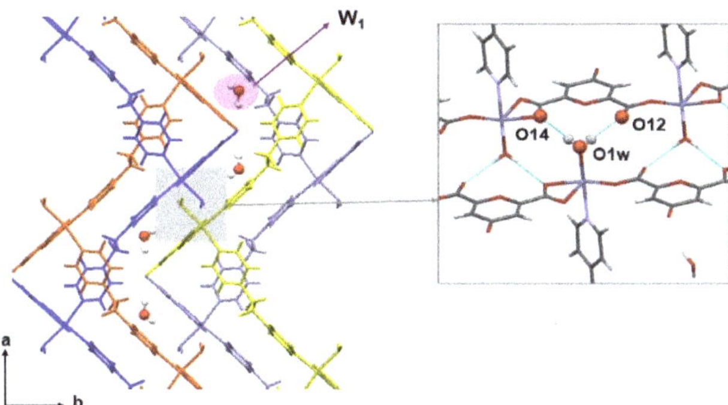

Figure 7. View of the 3D supramolecular arrangement in [Zn(chel)(bpe)(H$_2$O)]·H$_2$O (**3**) showing details of the of O1w–H1w ⋯ O hydrogen-bonding interactions.

In both structures the layers are packed efficiently, as evidenced by the Kitaigorod-skii [31] packing indexes of 64.9% in **1** and 69.7% in **3**.

3. Experimental

3.1. Materials and Physical Measurements

All reagents and solvents were obtained commercially and were used as supplied. Elemental analysis (C, H, N) was carried out on a Fisons EA-1108 microanalyser. IR spectra were recorded from KBr discs (4000–400 cm^{-1}) with a Bruker IFS28FT spectrophotometer.

3.2. Synthesis of the Precursor $^1_\infty$[Zn(chel)(H$_2$O)$_2$]

$^1_\infty$[Zn(chel)(H$_2$O)$_2$] [4] was obtained by reaction of ZnCO$_3$ and chelidonic acid (1:1 molar ratio) in methanol. The resulting suspension was heated under reflux for 2 h and stirred at room temperature for 1 day. The colourless solid was filtered off, washed with MeOH and vacuum dried.

3.3. Synthesis of the Complexes

3.3.1. $^2_\infty$[Zn(chel)(4,4-bipy)(H$_2$O)]·2H$_2$O (**1**)

Colourless single crystals of **1** were obtained by slow diffusion of a solution of $^1_\infty$[Zn(chel)(H$_2$O)$_2$] (0.20 mmol) in water and a solution of 4,4′-bipyridine (4,4-bipy, 0.20 mmol) in ethanol.

Data. Anal. calc. for C$_{17}$H$_{16}$O$_9$N$_2$Zn (457.7): C 44.7, H 3.5, N 6.1%; Found: C 44.5, H 3.4, N 5.7%. IR (KBr, cm^{-1}): 3446 m,b; 1640 vs; 1407 m; 1360 s.

3.3.2. $^1_\infty$[Zn(4,4-bipy)(H$_2$O)$_4$]chel·3H$_2$O (**2**)

Compound **2** was obtained as colourless single crystals by slow evaporation of the resulting solution after filtering off the crystals of **1**.

Data. Anal. calc. for C$_{17}$H$_{24}$O$_{13}$N$_2$Zn (529.8): C 38.6, H 4.6, N 5.3%; Found: C 38.5, H 4.9, N 5.4%. IR (KBr, cm^{-1}): 3444 m,b; 1641 vs; 1416 m; 1361 s.

3.3.3. $^2_\infty$[Zn(chel)(bpe)(H$_2$O)]·H$_2$O (**3**)

Compound **3** was also obtained as colourless single crystals by slow diffusion of a solution of $^1_\infty$[Zn(chel)(H$_2$O)$_2$] (0.20 mmol) in water and a solution of 1,2-bis(4-pyridyl)ethane (bpe, 0.20 mmol) in ethanol.

Data. Anal. calc. for C$_{19}$H$_{18}$O$_8$N$_2$Zn (467.7): C 48.9, H 3.9, N 6.0%; Found: C 48.1, H 4.0, N 5.8%. IR (KBr, cm^{-1}): 3365 m,b; 1635 vs; 1425 m; 1355 s.

3.3.4. $^1_\infty$[Cu(4,4-bipy)(H$_2$O)$_4$]chel·3H$_2$O (4a) and $^1_\infty$[Cu(4,4-bipy)(H$_2$O)$_4$]chel·6H$_2$O (**4b**)

A solution of 4,4$'$-bipyridine (0.156 g, 1 mmol) in methanol (10 mL) was slowly added to a solution of Cu(NO$_3$)$_2$·2.5H$_2$O (0.233 g, 1 mmol) in water (10 mL). The mixture was heated under reflux for 1 h and was left to cool down to room temperature. A solution of chelidonic acid (0.184 g, 1 mmol) and KOH (0.112 g, 2 mmol) in 2:1 MeOH/H$_2$O was added to the mixture, which was then heated under reflux for 1 h and was left to cool down to room temperature. The blue precipitate was filtered off and dissolved in water. Blue crystals of **4a** were obtained by slow evaporation of the aqueous solution. After separation of the crystals of **4a** the slow evaporation of the remaining solution gave rise to some brown crystals of **4b**.

Data for **4a**: Anal. calc. for C$_{17}$H$_{24}$O$_{13}$N$_2$Cu (527.9): C 38.7, H 4.6, N 5.3%; Found: C 39.1, H 4.1, N 5.2%. IR (KBr, cm^{-1}): 3445 s,b; 1637 vs, 1404 m, 1358 m.

Data for **4b**: Anal. calc. for C$_{17}$H$_{30}$O$_{16}$N$_2$Cu (581.1): C 35.1, H 5.2, N 4.8%; Found: C 35.5, H 5.4, N 5.0%. IR (KBr, cm^{-1}): 3420 s,b; 1613 vs, 1403 m, 1354 s.

3.4. Crystallography

Crystallographic data were collected on a Bruker Smart 1000 CCD diffractometer using graphite monochromated Mo-Kα radiation (λ = 0.71073 Å). All data were corrected for Lorentz and polarisation effects. The frames were integrated with the Bruker Saint software package [32] and the data were corrected for absorption using the program SADABS. [33] The structures were solved by direct methods using the program SHELXS97 [34]. All non-hydrogen atoms were refined with anisotropic thermal parameters by full-matrix least-squares calculations on F^2 using the programme SHELXL97 [34] or the programme SHELXL [35] with OLEX2 [36]. Hydrogen atoms were inserted at calculated positions and constrained with isotropic thermal parameters. The hydrogen atoms of the water molecules were located from a difference Fourier map and refined with isotropic parameters but the hydrogen atoms for the crystallisation water molecules in **1** could not be located. Compound **4a** crystallised in the chiral space group $P2_12_12$ and its absolute configuration was confirmed by a value of 0.010(6) for the Flack parameter [37]. Drawings were produced with MERCURY [38]. Crystal data and structure refinement parameters are reported in Table S5.

4. Conclusions

The study described here concerned the crystalline species resulting from chelidonic acid/4,4$'$-bipyridine or 1,2-bis(4-pyridyl)ethane/Zn(II) or Cu(II) systems in aqueous solutions. The process led to the isolation of single crystals of two kinds of coordination polymers; the 1D Zn(II) and Cu(II) cationic polymers (**2**, **4a** and **4b**) with the chelidonate anion in the second coordination sphere and the 2D Zn(II) polymers (**1** and **3**) with the bridging chelidonate ligand showing a new coordination mode, $\mu(\kappa^2O^{III}.O^{IV}{:}\kappa O^V)$. The results highlight the ability of chelidonate to stabilise different coordination compounds through its participation in the first or second coordination sphere. This feature is evidenced by the formation of $^2_\infty$[Zn(chel)(4,4-bipy)(H$_2$O)]·2H$_2$O (**1**) and $^1_\infty$[Zn(4,4-bipy)(H$_2$O)$_4$]chel·3H$_2$O (**2**) from the same reaction by diffusion. The crystallisation of three compounds containing the 1D [M(4,4-bipy)(H$_2$O)$_4$]$^{2+}$ (M = Zn, Cu) indicates the great stability of this cationic polymer, which was confirmed by a CSD search as the one included in Section 2.1. of this paper.

The five coordination polymers crystallise as hydrates but the crystallisation water molecules play different roles in the neutral polymers and in the solids containing the cationic polymers. In the neutral compounds the water molecules are weakly hydrogen bonded to the metal-organic framework but in the compounds based on the cationic polymers, the water molecules play a key role in the formation of the resulting supramolecular networks. These latter compounds are hydrogen-bonded networks through the second coordination sphere, which is organised in [(H$_2$O)$_n$chel]$^{2-}$ (n = 3 or 6) layers.

The presence of non-associated water molecules (W1 motif) is common in the networks for all of the compounds reported here. Furthermore, in all compounds a recurrent supramolecular synthon based on the interaction between a water molecule (coordinated in the neutral compounds and free in the cationic species) and the two carboxylate groups of a chelidonate anion (coordinated or not) is observed. This synthon leads to the formation of an eight-membered ring in all cases.

From the point of view of crystal engineering the biological anion chelidonate, in combination with N,N'-bridging ligands, is an interesting tool to obtain coordination polymers and/or hydrogen-bonded networks.

Supplementary Materials: The following are available online at https://www.mdpi.com/2624-8 549/3/1/19/s1, Figure S1: Infrared spectra, Table S1: Selected bond lengths/Å and angles/° in the cationic polymers $[Zn(4,4\text{-bipy})(H_2O)_4](chel)\cdot3H_2O$ (2), $[Cu(4,4\text{-bipy})(H_2O)_4](chel)\cdot3H_2O$ (4a) and $[Cu(4,4\text{-bipy})(H_2O)_4](chel)\cdot6H_2O$ (4b), Table S2: Selected bond lengths/Å and angles/° in the neutral polymers $^2_\infty[Zn(chel)(4,4\text{-bipy})(H_2O)]\cdot2H_2O$ (1) and $^2_\infty[Zn(chel)(bpe)(H_2O)]\cdot H_2O$ (3), Table S3: Main hydrogen bonds in the cationic polymers $[Cu(4,4\text{-bipy})(H_2O)_4](chel)\cdot3H_2O$ (4a) $[Cu(4,4\text{-bipy})(H_2O)_4](chel)\cdot6H_2O$ (4b) and $[Zn(4,4\text{-bipy})(H_2O)_4](chel)\cdot3H_2O$ (2), Table S4: Main hydrogen bonds in the neutral polymers $^2_\infty[Zn(chel)(4,4\text{-bipy})(H_2O)]\cdot2H_2O$ (1) and $^2_\infty[Zn(chel)(bpe)(H_2O)]\cdot H_2O$ (3), Table S5: Crystal data and structure refinement. CCDC entries 2051289–2051293 contain the supplementary crystallographic data for 1, 2, 3, 4a and 4b.

Author Contributions: Conceptualization, R.C. and A.B.L.; methodology, R.C., A.P.-C. and O.G.-P.; synthesis, characterisation, thermal analysis A.P.-C. and N.F.-H.; formal analysis, R.C., A.P.-C. and O.G.-P.; investigation, O.G.-P.; resources, R.C. and O.G.-P.; writing—original draft preparation, R.C. and A.B.L.; writing—review and editing, R.C., A.B.L. and E.M.V.-L.; visualization, R.C. and A.B.L.; supervision, E.M.V.-L.; project administration, R.C. and E.M.V.-L.; funding acquisition, R.C. and E.M.V.-L. All authors have read and agreed to the published version of the manuscript.

Funding: This research was funded by the Ministerio de Ciencia e Innovación (Spain) (research project PID2019-110218RB-I00).

Institutional Review Board Statement: Not applicable.

Informed Consent Statement: Not applicable.

Data Availability Statement: The data presented in this study are available in the article and in the Supplementary Materials.

Acknowledgments: SC-XRD measurements were performed at the Unidade de Difracción de Raios X de Monocristal (CACTI-Universidade de Vigo), Spain. O.G.P. thanks the Xunta de Galicia for a predoctoral contract.

Conflicts of Interest: The authors declare no conflict of interest. The funders had no role in the design of the study; in the collection, analyses, or interpretation of data; in the writing of the manuscript, or in the decision to publish the results.

Sample Availability: Samples of the compounds are available from the authors.

References

1. Kumar Singh, D.; Gulati, K.; Ray, A. Effects of chelidonic acid, a secondary plant metabolite, on mast cell degranulation and adaptive immunity in rats. *Int. Immunopharmacol.* **2016**, *40*, 229–234. [CrossRef] [PubMed]
2. Kim, D.-S.; Kim, S.-J.; Kim, M.-C.; Jeon, Y.-D.; Um, J.-Y.; Hong, S.-H. The Therapeutic Effect of Chelidonic Acid on Ulcerative Colitis. *Biol. Pharm. Bull.* **2012**, *35*, 666–671. [CrossRef]
3. Jeong, H.-J.; Yang, S.-Y.; Kim, H.-Y.; Kim, N.-R.; Jang, J.-B.; Kim, H.-M. Chelidonic acid evokes antidepressant-like effect through the up-regulation of BDNF in forced swimming test. *Exp. Biol. Med.* **2016**, *241*, 1559–1567. [CrossRef]
4. Zhou, X.-X.; Liu, M.-S.; Lin, X.-M.; Fang, H.-C.; Chen, J.-Q.; Yang, D.-Q.; Cai, Y.-P. Construction of three low dimensional Zn(II) complexes based on different organic-carboxylic acids. *Inorg. Chim. Acta* **2009**, *362*, 1441–1447. [CrossRef]
5. Yasodha, V.; Govindarajan, S.; Lowb, J.N.; Glidewell, C. Cationic, neutral and anionic metal(II) complexes derived from 4-oxo-4*H*-pyran-2,6-dicarboxylic acid (chelidonic acid). *Acta Cryst.* **2007**, *63*, m207–m215.

6. Fainerman-Melnikova, M.; Clegg, J.K.; Pakchung, A.A.H.; Jensen, P.; Codd, R. Structural diversity of complexes between Cu(II) or Ni(II) and endocyclic oxygen- or nitrogen-containing ligands: Synthesis, X-ray structure determinations and circular dichroism spectra. *Cryst. Eng. Comm.* **2010**, *12*, 4217–4225. [CrossRef]

7. Fang, M.; Chang, L.; Liu, X.; Zhao, B.; Zuo, Y.; Chen, Z. Fabrication and Properties of Eight Novel Lanthanide−Organic Frameworks Based on 4-Hydroxypyran-2,6-dicarboxylate and 4-Hydroxypyridine-2,6-dicarboxylate. *Cryst. Growth Des.* **2009**, *9*, 4006–4016. [CrossRef]

8. Zhang, Z.-J.; Zhang, S.-Y.; Li, Y.; Niu, Z.; Shi, W.; Cheng, P. Systematic investigation of the lanthanide coordination polymers with γ-pyrone-2,6-dicarboxylic acid. *Cryst. Eng. Comm.* **2010**, *12*, 1809–1815. [CrossRef]

9. Qu, B.-T.; Lai, J.-C.; Liu, S.; Liu, F.; Gao, Y.-D.; You, X.-Z. Cu- and Ag-Based Metal−Organic Frameworks with 4-Pyranone-2,6-dicarboxylic Acid: Syntheses, Crystal Structures, and Dielectric Properties. *Cryst. Growth Des.* **2015**, *15*, 1707–1713. [CrossRef]

10. Yasodha, V.; Govindarajan, S.; Starosta, W.; Leciejewicz, J. New Metal–Organic Framework Solids Built from Barium and Isoelectronic Chelidamic and Chelidonic Acids. *J. Chem. Crystallogr.* **2011**, *41*, 1988–1997. [CrossRef]

11. Eubank, J.F.; Kravtsov, V.C.; Eddaoudi, M. Synthesis of Organic Photodimeric Cage Molecules Based on Cycloaddition via Metal−Ligand Directed Assembly. *J. Am. Chem. Soc.* **2007**, *129*, 5820–5821. [CrossRef] [PubMed]

12. Lago, A.B.; Carballo, R.; Fernández-Hermida, N.; Rodríguez-Hermida, S.; Vázquez-López, E.M. Coordination polymers with chelidonate (4-oxo-4H-pyran-2,6-dicarboxylate) anions and dmso: [Zn(chel)(dmso)$_2$] and linkage isomers of [Co(chel)(dmso)(OH$_2$)$_3$]·H$_2$O. *J. Mol. Struct.* **2011**, *1003*, 121–128. [CrossRef]

13. Lago, A.B.; Carballo, R.; Fernández-Hermida, N.; Vázquez-López, E.M. Mononuclear discrete complexes and coordination polymers based on metal(II) chelidonate complexes with aromatic *N,N*-chelating ligands. *CrystEngComm.* **2011**, *13*, 941–951. [CrossRef]

14. Carlucci, L.; Ciani, G.; Proserpio, D.M.; Sironi, A. Extended networks via hydrogen bond cross-linkages of [M(bipy)] (M = Zn^{2+} or Fe^{2+}; bipy = 4,4′-bipyridyl) linear co-ordination polymers. *J. Chem. Soc. Dalton Trans.* **1997**, 1801–1804. [CrossRef]

15. Lian, Z.-X.; Cai, J.; Chen, C.-H. A series of metal-organic frameworks constructed with arenesulfonates and 4,4′-bipy ligands. *Polyhedron* **2007**, *26*, 2647–2654. [CrossRef]

16. Zhang, J.; Zhu, L.-G. Syntheses, structures, and supra-molecular assembles of zinc 4-sulfobenzoate complexes with chelating and/or bridging ligands. *J. Mol. Struct.* **2009**, *931*, 87–93. [CrossRef]

17. Xiong, W.; Su, Y.; Zhang, Z.; Chen, Z.; Liang, F.; Wang, L. Synthesis and Crystal Structures of Two Metal Complexes Formed in the Solvothermal Decomposition Reactions of *N*-Carboxyphenylenesulfonyl-*S*-Carboxymethyl-L-Cysteine. *J. Chem. Cryst.* **2011**, *41*, 1510–1514. [CrossRef]

18. Mao, H.; Zhang, C.; Li, G.; Zhang, H.; Hou, H.; Li, L.; Wu, Q.; Zhu, Y.; Wang, E. New types of the flexible self-assembled metal–organic coordination polymers constructed by aliphatic dicarboxylates and rigid bidentate nitrogen ligands. *Dalton Trans.* **2004**, *22*, 3918–3925. [CrossRef]

19. García, H.C.; Diniz, R.; Yoshida, M.I.; de Oliveira, L.F.C. Synthesis, structural studies and vibrational spectroscopy of Fe^{2+} and Zn^{2+} complexes containing 4,4′-bipyridine and barbiturate anion. *J. Mol. Struct.* **2010**, *978*, 79–85. [CrossRef]

20. Colinas, R.; Rojas-Andrade, M.D.; Chakraborty, I.; Oliver, S.R.J. Two structurally diverse Zn-based coordination polymers with excellent antibacterial activity. *CrystEngComm.* **2018**, *20*, 3353–3362. [CrossRef]

21. *CSD Web Interface–Intuitive, Cross-Platform, Web-Based access to CSD Data*; Cambridge Crystallographic Data Centre: Cambridge, UK, 2017.

22. Wang, Y.; Feng, L.; Li, Y.; Hu, C.; Wang, E.; Hu, N.; Jia, H. Novel hydrogen-bonded three-dimensional networks encapsulating one-dimensional covalent chains: [M(4,4′-bipy)(H$_2$O)$_4$](4-abs)2.nH$_2$O (4,4′-bipy = 4,4′-bipyridine; 4-abs = 4-aminobenzenesulfonate) (M = Co, n = 1; M = Mn, n = 2). *Inorg. Chem.* **2002**, *41*, 6351–6357. [CrossRef] [PubMed]

23. Manna, S.C.; Ghosh, A.K.; Zangrando, E.; Ray Chaudhuri, N. 3D supramolecular networks of Co(II)/Fe(II) using the croconate dianion and a bipyridyl spacer: Synthesis, crystal structure and thermal study. *Polyhedron* **2007**, *26*, 1105–1112. [CrossRef]

24. Elahi, S.M.; Chand, M.; Deng, W.-H.; Pal, A.; Das, M.C. Polycarboxylate-Templated Coordination Polymers: Role of Templates for Superprotonic Conductivities of up to 10^{-1} S cm^{-1}. *Angew. Chem. Int. Ed. Engl.* **2018**, *57*, 6662–6666. [CrossRef]

25. Wang, X.-L.; Qin, C.; Wang, E.-B. Polythreading of Infinite 1D Chains into Different Structural Motifs: Two Poly(pseudo-rotaxane) Architectures Constructed by Concomitant Coordinative and Hydrogen Bonds. *Cryst. Growth Des.* **2006**, *6*, 439–443. [CrossRef]

26. Kanoo, P.; Gurunatha, K.L.; Maji, T.K. Temperature-Controlled Synthesis of Metal-Organic Coordination Polymers: Crystal Structure, Supramolecular Isomerism, and Porous Property. *Cryst. Growth Des.* **2009**, *9*, 4147–4156. [CrossRef]

27. García, H.C.; Diniz, R.; Yoshidab, M.I.; de Oliveira, L.F.C. An intriguing hydrogen bond arrangement of polymeric 1D chains of 4,4′-bipyridine coordinated to Co^{2+}, Ni^{2+}, Cu^{2+} and Zn^{2+} ions having barbiturate as counterions in a 3D network. *CrystEng Comm.* **2009**, *11*, 881–888. [CrossRef]

28. Fei, H.; Oliver, S.R.J. Two cationic metal–organic frameworks based on cadmium and α,ω-alkanedisulfonate anions and their photoluminescent properties. *Dalton Trans.* **2010**, *39*, 11193–11200. [CrossRef]

29. Robin, A.Y.; Fromm, K.M. Coordination polymer networks with O- and N-donors: What they are, why and how they are made. *Coord. Chem. Rev.* **2006**, *250*, 2127–2157. [CrossRef]

30. Carballo, R.; Covelo, B.; García-Martínez, E.; Lago, A.B.; Vázquez-López, E.M. Role of inorganic and organic anions in the formation of metallosupramolecular assemblies of silver(I) coordination polymers with the twisted ligand bis (4-pyridylthio) methane. *Polyhedron* **2009**, *5*, 923–932. [CrossRef]

31. Kitaigorodskii, A.I. *Molecular Crystals and Molecules*; Academic Press: New York, NY, USA, 1973.
32. Siemens SAINT. *Version 4, Software Reference Manual*; Siemens Analytical X-Ray Systems, Inc.: Madison, WI, USA, 1996.
33. Sheldrick, G.M. *SADABS, Program for Empirical Absorption Correction of Area Detector Data*; University of Göttingen: Göttingen, Germany, 1997.
34. Sheldrick, G.M. *SHELXS-97 and SHELXL-97, Program for Crystal Structure Solution and Refinement*; University of Göttingen: Göttingen, Germany, 1997.
35. Sheldrick, G.M. Crystal Structure Refinement with SHELXL. *Acta Cryst.* **2015**, *C71*, 3–8.
36. Dolomanov, V.; Bourhis, L.J.; Gildea, R.J.; Howard, J.A.K.; Puschmann, H. OLEX2: A complete structure solution, refinement and analysis program. *J. Appl. Cryst.* **2009**, *42*, 339–341. [CrossRef]
37. Flack, H.D. On enantiomorph-polarity estimation. *Acta Cryst.* **1983**, *39*, 876–881. [CrossRef]
38. Bruno, J.; Cole, J.C.; Edgington, P.R.; Kessler, M.K.; Macrae, C.F.; McCabe, P.; Pearson, J.; Taylor, R. MERCURY. New Software for Visualizing Crystal Structures. *Acta Cryst.* **2002**, *B58*, 389. [CrossRef] [PubMed]

Article

Dimensionality Control in Crystalline Zinc(II) and Silver(I) Complexes with Ditopic Benzothiadiazole-Dipyridine Ligands [†]

Teodora Mocanu [1,2], Nataliya Plyuta [3], Thomas Cauchy [3], Marius Andruh [1,*] and Narcis Avarvari [3,*]

[1] Inorganic Chemistry Laboratory, Faculty of Chemistry, University of Bucharest, Str. Dumbrava Rosie nr. 23, 020464 Bucharest, Romania; teo_mocanu_sl@yahoo.com

[2] Coordination and Supramolecular Chemistry Laboratory, "Ilie Murgulescu" Institute of Physical Chemistry, Romanian Academy, Splaiul Independentei 202, 060021 Bucharest, Romania

[3] MOLTECH-Anjou, UMR 6200, CNRS, UNIV Angers, 2 bd Lavoisier, CEDEX 49045 Angers, France; plyutanataliya@gmail.com (N.P.); thomas.cauchy@univ-angers.fr (T.C.)

* Correspondence: marius.andruh@dnt.ro (M.A.); narcis.avarvari@univ-angers.fr (N.A.); Tel.: +40-74-487-0656 (M.A.); +33-024-173-5084 (N.A.)

† Dedicated to Professor Christoph Janiak on the occasion of his 60th anniversary.

Abstract: Three 2,1,3-benzothiadiazole-based ligands decorated with two pyridyl groups, 4,7-di(2-pyridyl)-2,1,3-benzothiadiazol (2-PyBTD), 4,7-di(3-pyridyl)-2,1,3-benzothiadiazol (3-PyBTD) and 4,7-di(4-pyridyl)-2,1,3 benzothiadiazol (4-PyBTD), generate Zn^{II} and Ag^{I} complexes with a rich structural variety: $[Zn(hfac)_2(2\text{-}PyBTD)]$ **1**, $[Zn_2(hfac)_4(2\text{-}PyBTD)]$ **2**, $[Ag(CF_3SO_3)(2\text{-}PyBTD)]_2$ **3**, $[Ag(2\text{-}PyBTD)]_2(SbF_6)_2$ **4**, $[Ag_2(NO_3)_2(2\text{-}PyBTD)(CH_3CN)]$ **5**, $[Zn(hfac)_2(3\text{-}PyBTD)]$ **6**, $[Zn(hfac)_2(4\text{-}PyBTD)]$ **7**, $[ZnCl_2(4\text{-}PyBTD)_2]$ **8** and $[ZnCl_2(4\text{-}PyBTD)]$ **9** (hfac = hexafluoroacetylacetonato). The nature of the resulting complexes (discrete species or coordination polymers) is influenced by the relative position of the pyridyl nitrogen atoms, the nature of the starting metal precursors, as well as by the synthetic conditions. Compounds **1** and **8** are mononuclear and **2**, **3** and **4** are binuclear species. Compounds **6**, **7** and **9** are 1D coordination polymers, while compound **5** is a 2D coordination polymer, the metal ions being bridged by 2-PyBTD and nitrato ligands. The solid-state architectures are sustained by intermolecular π–π stacking interactions established between the pyridyl group and the benzene ring from the benzothiadiazol moiety. Compounds **1**, **2**, **7–9** show luminescence in the visible range. Density Functional Theory (DFT) and Time Dependent Density Functional Theory (TD-DFT) calculations have been performed on the Zn^{II} complexes **1** and **2** in order to disclose the nature of the electronic transitions and to have an insight on the modulation of the photophysical properties upon complexation.

Keywords: benzothiadiazole; pyridine; nitrogen ligands; coordination polymers; crystal structure determination; photophysical properties

Citation: Mocanu, T.; Plyuta, N.; Cauchy, T.; Andruh, M.; Avarvari, N. Dimensionality Control in Crystalline Zinc(II) and Silver(I) Complexes with Ditopic Benzothiadiazole-Dipyridine Ligands. *Chemistry* **2021**, *3*, 269–287. https://doi.org/10.3390/chemistry3010020

Received: 19 January 2021
Accepted: 8 February 2021
Published: 12 February 2021

Publisher's Note: MDPI stays neutral with regard to jurisdictional claims in published maps and institutional affiliations.

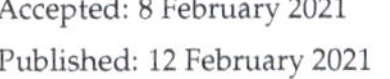

1. Introduction

The 2,1,3-benzothiadiazole (BTD) unit is an electron acceptor fluorophore [1], which has been extensively used in the structure of derivatives and materials for various applications such as organic field-effect transistors [2–4], light emitting diodes [5,6], photovoltaics [7–11], redox switchable donor-acceptor systems [12,13], fluorescent probes [14,15] and so forth. In the solid state, the BTD fragment generally engages in a variety of supramolecular interactions such as hydrogen bonding, chalcogen N···S bonding and π–π stacking interactions [14,16,17]. Moreover, the presence of two sp^2 nitrogen atoms on the thiadiazole ring confers coordinating properties to the BTD unit, which have been exploited, for example, in several transition metal complexes where BTD acted as monotopic [18] or ditopic ligand [19–23]. On the other hand, introduction of coordinating groups on the benzene ring affords neutral or anionic ligands allowing the formation of luminescent discrete complexes or coordination polymers in which the thiadiazole nitrogen atoms can

65

take part, or not, to the coordination of metal centers [24]. In this respect, 4,7-substituted BTD derivatives with carboxylate or carboxylate containing fragments have been used as divergent anionic linkers towards various metal-organic frameworks (MOFs) which showed, for example, luminescence sensing properties in the presence of water [25], organic amines [26], organic dyes [27] or metal ions [28], while neutral 4,7-di(4-pyridyl-vinyl)-BTD ligands, with vinyl spacers between the pyridine rings and the BTD unit, provided luminescent zinc(II) coordination polymers in the presence of dicarboxylate linkers [29]. Particularly interesting is the direct connection of pyridines on the 4,7-positions of BTD, as it allows the access to the series of ditopic ligands 4,7-di(2-pyridyl)-2,1,3-benzothiadiazol (2-PyBTD), 4,7-di(3-pyridyl)-2,1,3-benzothiadiazol (3-PyBTD) and 4,7-di(4-pyridyl)-2,1,3-benzothiadiazol (4-PyBTD) (Scheme 1), the first one having the possibility to act as ditopic bis(chelating) ligand if the BTD nitrogen atoms are involved in coordination.

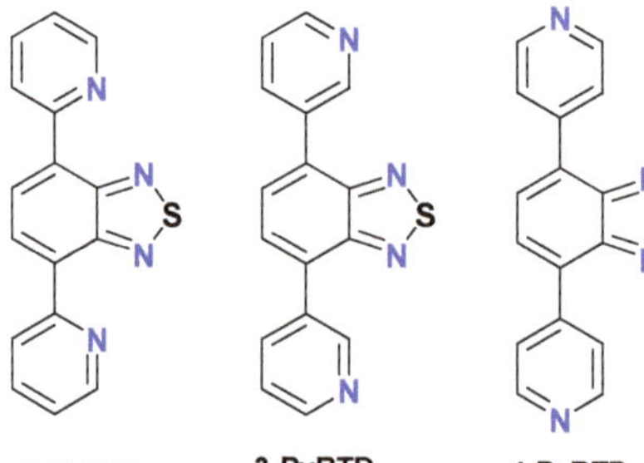

Scheme 1. 4,7-dipyridyl-BTD ligands used in the present study.

The synthesis of the three ligands is straightforward, consisting of the Stille coupling of 4,7-dibromo-2,1,3-benzothiadiazole with the series of 2-, 3- and 4-tributylstanyl-pyridines, respectively [30]. The three ligands are highly fluorescent, with quantum yields in acetonitrile solutions ranging between 68 and 90%. Rather surprisingly, when considering their promising coordination abilities, only the 4-PyBTD ligand has been used so far to prepare 3D coordination polymers of Zn^{II}, Cd^{II}, Co^{II} and Ni^{II}, also containing benzenedicarboxylate linkers, showing modulation of emission properties in the presence of metal ions or picric acid [31,32].

We report herein a first combined coordination chemistry study of the three ditopic ligands 2-PyBTD, 3-PyBTD and 4-PyBTD (Scheme 1) towards zinc(II) and silver(I) based fragments with the objective to explore the topicity of the ligands and the dimensionality of the resulting complexes, for which the crystal structures are described in detail, with a special focus on the supramolecular interactions in the solid state. Moreover, the photophysical properties of the Zn^{II} complexes are reported and compared to those of the ligands.

2. Materials and Methods

The reagents employed were purchased from commercial sources and used without further purification.

2-PyBTD, 3-PyBTD and 4-PyBTD were synthesized according to the literature procedure [30].

[Zn(hfac)$_2$(2-PyBTD)] **1**: 0.0011 g (0.0038 mmol) 2-PyBTD in 4 mL CHCl$_3$ and 2 mL methanol were added to a solution of 0.0039 g (0.0075 mmol) of [Zn(hfac)$_2$(H$_2$O)$_2$]·H$_2$O in 2 mL methanol and stirred for one hour. After filtration, the solution was left to slowly evaporate. Yellow needle like single crystals were formed in a few days (0.0023 g, yield 79%). Selected IR data (KBr, cm^{-1}): 3066(w), 2955(m), 2921(m), 2851(m), 1655(s), 1617(m), 1573(m), 1553(m), 1528(m), 1486(m), 1465(m), 1348(w), 1260(s), 1203(s), 1143(vs), 1099(s), 1067(s), 909(m), 880(m), 842(m), 795(w), 779(w), 743(m), 666(m), 584(m) (Figure S1).

Elemental analysis calcd for $C_{26}H_{12}ZnF_{12}N_4O_4S$: C, 40.56; H, 1.57; N, 7.28; S, 4.17%. Found: C, 40.51; H, 1.26; N, 7.24; S, 4.11%.

[$Zn_2(hfac)_4(2\text{-PyBTD})$] **2**: 0.0035 g (0.0067 mmol) of [$Zn(hfac)_2(H_2O)_2$]·H_2O and 0.001g (0.0034 mmol) 2-PyBTD were solubilized in a mixture of 4 mL n-heptane and 2 mL CH_2Cl_2 and then heated to the n-heptane boiling point. Upon cooling to room temperature, 2 mL of CH_2Cl_2 were added and the solution was left to slowly evaporate. Yellow single crystals were formed within a few days (0.0034 g, yield 81%). Selected IR data (KBr, cm^{-1}): 3123(w), 3045(w), 1655(s), 1600(m), 1558(m), 1531(m), 1479(s), 1457(m), 1348(w), 1253(vs), 1216(vs), 1144(vs), 1097(m), 1064(m), 1020(w), 949(m), 907(m), 866(w), 796(m), 782(m), 741(m), 665(m), 585(m) (Figure S2). Elemental analysis calcd for $C_{36}H_{14}Zn_2F_{24}N_4O_8S$: C, 34.61; H, 1.13; N, 4.48; S, 2.57%. Found: C, 33.99; H, 0.94; N, 4.38; S, 2.49%.

[$Ag(CF_3SO_3)(2\text{-PyBTD})$]$_2$ **3**: An ethylacetate solution (2 mL) of $AgCF_3SO_3$ (0.0021 g, 0.0081 mmol) was left to slowly diffuse through a 4 mL layer of ethylacetate and dichloromethane (1:1 *v/v*) into a solution of 0.0012g (0.0041 mmol) 2-PyBTD solubilized in 2 mL CH_2Cl_2. Yellow single crystals were formed within a few days (0.0013 g, yield 59%). Selected IR data (KBr, cm^{-1}): 3313(w), 3053(w), 2920(w), 1720(w), 1582(m), 1549(m), 1462(m), 1434(m), 1384(m), 1262(vs), 1175(s), 1035(vs), 948(w), 890(w), 775(m), 737(m), 646(s), 578(w) (Figure S3). Elemental analysis calcd for $C_{17}H_{10}AgF_3N_4O_3S_2$: C, 37.31; H, 1.84; N, 10.24; S, 11.72%. Found: C, 37.22; H, 2.04; N, 9.86; S, 11.62%.

[$Ag(2\text{-PyBTD})$]$_2$(SbF_6)$_2$ **4**: Crystalline **4** have been prepared in the same manner as **3**, except that an ethylacetate solution of $AgSbF_6$ (0.0026g, 0.0075 mmol) and dichloromethane solution of 2-PyBTD (0.001g, 0.0034 mmol) were used. Yellow single crystals of **4** crystalized in a week (0.0018 g, yield 84%). Selected IR data (KBr, cm^{-1}): 3109(m), 3050(m), 2991(m), 2934(m), 1721(w), 1596(s), 1553(s), 1527(s), 1464(s), 1433(s), 1373(s), 1350(w), 1284(w), 1256(m), 1159(m), 1094(m), 1054(m), 1018(m), 994(m), 948(m), 892(w), 858(m), 772(s), 747(m), 657(vs), 550(m) (Figure S4). Elemental analysis calcd for $C_{32}H_{20}Ag_2F_{12}N_8S_2Sb_2$: C, 30.31; H, 1.59; N, 8.84; S, 5.06%. Found: C, 30.69; H, 1.72; N, 8.45; S, 4.77%.

[$Ag_2(NO_3)_2(2\text{-PyBTD})(CH_3CN)$] **5**: A solution of $AgNO_3$ (0.005 g, 0.029 mmol) in 2 mL ethylacetate and 2 mL acetonitrile was left to slowly diffuse through a 4 mL layer of ethylacetate and dichloromethane (1:1 *v/v*) into a dichloromethane solution (2 mL) of 2-PyBTD (0.0043 g, 0.014 mmol). Yellow single crystals appeared in three weeks (0.0014 g, yield 15.5%).

[$Zn(hfac)_2(3\text{-PyBTD})$] **6**: A solution of 0.0041 mmol (0.0012g) 3-PyBTD in 4 mL $CHCl_3$ is added to a solution containing 0.0081 mmol [$Zn(hfac)_2(H_2O)_2$]·H_2O (0.0042 g) in 2 mL of methanol. The mixture is stirred for 2 h and then filtered. Yellow single crystals of **6** are formed in few days by slow evaporation of the solvent (0.0013 g, yield 42%). Selected IR data (KBr, cm^{-1}): 3064(w), 2920(w), 1682(m), 1652(m), 1550(m), 1525(m), 1510(m), 1485(m), 1438(m), 1254(s), 1194(s), 1142(s), 1122(s), 1095(s), 847(w), 790(m), 746(m), 725(m), 694(m) (Figure S5).

[$Zn(hfac)_2(4\text{-PyBTD})$] **7**: A solution of 0.0034 mmol (0.001 g) 4-PyBTD in a mixture of 4 mL $CHCl_3$ and 2 mL methanol was mixed with a solution containing 0.0067 mmol [$Zn(hfac)_2(H_2O)_2$]·H_2O (0.0035 g) in 2 mL methanol and then stirred for 2 h. After filtration, the solution was left to slowly evaporate. Yellow needle like single crystals were formed in few days (0.0020 g, yield 77%). Selected IR data (KBr, cm^{-1}): 1650(s), 1613(m), 1558(s), 1533(m), 1495(s), 1460(s), 1430(s), 1375(m), 1259(vs), 1219(vs), 1144(vs), 1019(m), 850(m), 821(m), 808(m), 727(w), 669(s) (Figure S6). Elemental analysis calcd for $C_{26}H_{12}ZnF_{12}N_4O_4S$: C, 40.56; H, 1.57; N, 7.28; S, 4.17%. Found: C, 40.13; H, 1.42; N, 6.96; S, 4.53%.

[$ZnCl_2(4\text{-PyBTD})_2$] **8**: 0.0024g (0.0082 mmol) 4-PyBTD in 4 mL CH_2Cl_2 and 2 mL methanol were added to a solution of 0.0012 g (0.0088 mmol) of $ZnCl_2$ in 2 mL methanol and stirred for 10 min, when the mixture became opaque. After adding 2 mL DMF and heating at 60 °C for 30 min, the precipitate was solubilized and the light-yellow solution was left to slowly evaporate. Dark-yellow single crystals were formed within a few days (0.0025 g, yield 85%). Selected IR data (KBr, cm^{-1}): 3460(m), 1638(s), 1616(s), 1594(m),

1548(m), 1478(w), 1429(m), 1409(m), 1220(m), 1071(m), 1029(m), 883(w), 829(m), 815(m), 721(m), 629(w) (Figure S7).

[ZnCl$_2$(4-PyBTD)] **9**: Light yellow single crystals of **9** were synthesized in the same manner as **8**, except that the opaque solution was further stirred for one hour before 2 mL DMF were added, followed by two hours stirring at 60 °C (0.0019 g, yield 54%). Selected IR data (KBr, cm^{-1}): 3288(m), 2964(m), 2918(m), 1668(m), 1552(m), 1392(m), 1261(s), 1094(vs), 1027(vs), 805(vs), 700(w) (Figure S8).

IR spectra (KBr pellets) were recorded on a Bruker Tensor 37 spectrophotometer in 4000 to 400 cm^{-1} frequencies range. Diffuse reflectance spectra were performed on a JASCO V-670 spectrophotometer. Elemental analysis was carried out on a EuroEa Elemental Analyzer. The fluorescence spectra were carried out on a JASCO FP-6500 spectrofluorimeter.

X-ray data for compounds **2–6** and **8** were collected on an Agilent Supernova diffractometer with CuKα (λ = 1.54184 Å) and for compounds **1**, **7** and **9** on a Rigaku XtaLAB Synergy, Single source at offset/far, HyPix diffractometer using a graphite-monochromated Mo K$_\alpha$ radiation source (λ = 0.71073 Å). The structures were solved by direct methods and refined by full-matrix least squares techniques based on F^2. The non-H atoms were refined with anisotropic displacement parameters. Calculations were performed using SHELXT and SHELXL-2015 crystallographic software packages [33,34]. A summary of the crystallographic data and the structure refinement for crystals **1–9** is given in Tables S1 and S2.

Crystallographic data for the nine structures have been deposited with the Cambridge Crystallographic Data Centre, deposition numbers CCDC 2056006 (**1**), CCDC 2056007 (**2**), CCDC 2056008 (**3**), CCDC 2056009 (**4**), CCDC 2056010 (**5**), CCDC 2056011 (**6**), CCDC 2056012 (**7**), CCDC 2056013 (**8**), CCDC 2056014 (**9**). These data can be obtained free of charge from CCDC, 12 Union road, Cambridge CB2 1EZ, UK (e-mail: deposit@ccdc.cam.ac.uk or http://www.ccdc.cam.ac.uk (accessed on 12 February 2021)).

DFT and TD-DFT calculations have been performed with the Gaussian 09 program [35] using the DFT method with the PBE1PBE functional and the augmented and polarized Ahlrichs triple-zeta basis set TZVP [36]. The full molecular reports, molecular orbitals, electron density differences pictures and calculated spectra have been automatically generated by a homemade Python program, quchemreport [37], based on cclib [38].

3. Results and Discussion

The ligands 2-PyBTD, 3-PyBTD and 4-PyBTD (Scheme 1) have been synthesized according to the literature procedure described by Yamashita et al. [30]. To compare their coordination modes, the precursor {Zn(hfac)$_2$} (hfac = hexafluoroacetylacetonate) has been used throughout the whole series, thus ensuring, in principle, the preparation of neutral complexes and the coordination of two additional ligands in either *cis* or *trans* positions. The coordination ability of the metal center is enhanced thanks to the electron withdrawing effect of the fluorinated hfac ligands [39,40]. Furthermore, while the Zn^{II} ion provides a more rigid system upon coordination, it does not present any d-d transition which could interfere with the ligand based emission. Then, the scope of the study has been extended towards the use of ZnCl$_2$, as it provides tetrahedral ZnCl$_2$L$_2$ complexes with pyridine based ligands [41], at the difference with the {Zn(hfac)$_2$} fragment which favors the formation of octahedral Zn(hfac)$_2$L$_2$ complexes. Finally, silver(I) precursors have been evaluated in coordination with 2-PyBTD in order to take advantage of the propensity of this metal center to adopt various coordination modes as a consequence of the interplay between the coordination flexibility of the 2-PyBTD ligand, that is, chelating vs. divergent ditopic and the coordination behavior of the anionic ligand of the Ag^{I} precursor. In the following discussion we describe in detail the crystal structures of the complexes obtained with each of the three ligands, together with the photophysical properties of the zinc(II) complexes with 2-PyBTD and 4-PyBTD.

3.1. Discrete Complexes and Coordination Polymers with the 2-PyBTD Ligand

3.1.1. Mononuclear [Zn(hfac)₂(2-PyBTD)] 1 and Binuclear [Zn₂(hfac)₄(2-PyBTD)] 2 Complexes

The reaction between 2-PyBTD and the $[Zn(hfac)_2(H_2O)_2]\cdot H_2O$ precursor in a ratio of 1:2 afforded either the mononuclear complex **1** or the binuclear complex **2** depending on the reaction solvent mixture, that is, methanol/chloroform and heptane/methylene chloride, respectively. Complex **1** crystallized in the monoclinic space group $P2_1/c$, with one independent complex molecule in the asymmetric unit. The metal center shows octahedral coordination stereochemistry, with the coordination sphere generated by four oxygen atoms (O1–O4) from the hfac⁻ anionic ligands and two nitrogen atoms from a pyridine (N1) and the BTD unit (N2) (Figure 1a, Table 1 and Table S3).

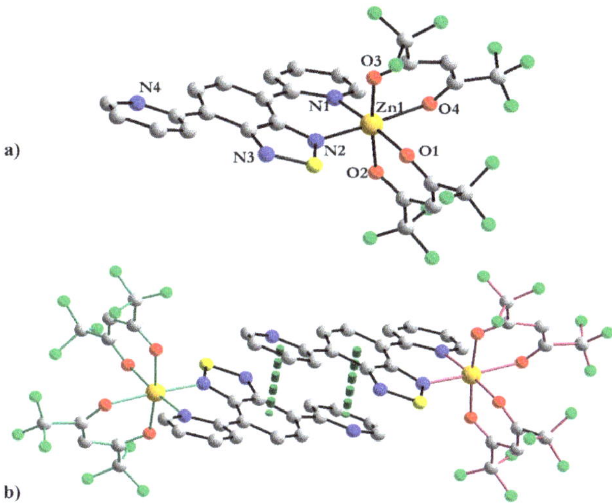

Figure 1. (**a**) Crystal structure of $[Zn(hfac)_2(2\text{-PyBTD})]$ (**1**) with a focus on the coordination sphere of Zn^{II} and partial atom numbering scheme; (**b**) Supramolecular dimer built out of mononuclear units of opposite chirality (green—Λ configuration, violet—Δ configuration) through π–π stacking interactions.

Table 1. Selected bond lengths (Å) for compounds **1** and **2**.

	1		2
Zn1–N1	2.105(6)	Zn1–N1	2.104(2)
Zn1–N2	2.101(7)	Zn1–N2	2.171(2)
Zn1–O1	2.092(5)	Zn1–O1	2.115(2)
Zn1–O2	2.092(6)	Zn1–O2	2.111(2)
Zn1–O3	2.135(6)	Zn1–O3	2.088(2)
Zn1–O4	2.087(6)	Zn1–O4	2.064(2)

The pyridine units of the ligand are *cis-trans* oriented and are slightly twisted with respect to the BTD plane, with values of 14.1° (PyN1) and 15.3° (PyN4) for the corresponding dihedral angles. The mononuclear units arrange in supramolecular dimers, with opposite chiralities Δ and Λ of the octahedral configuration of Zn^{II}, thanks to π stacking interactions between the uncoordinated pyridine and BTD phenyl rings (Figure 1b).

The binuclear complex **2** crystallized in the monoclinic space group $C2/c$ with half a molecule of complex in the asymmetric unit, the other half being generated by the C_2 axis crossing the BTD unit through the S atom located on a special position (Table 1 and Figure 2). Zn–O and Zn–N bond lengths are comparable in the two complexes. However, the Py rings in **2**, having now a *cis-cis* arrangement due to the bis-chelation, are more strongly twisted when compared to **1**, as attested by the value of ±26.9° for the Py–BTD dihedral angle.

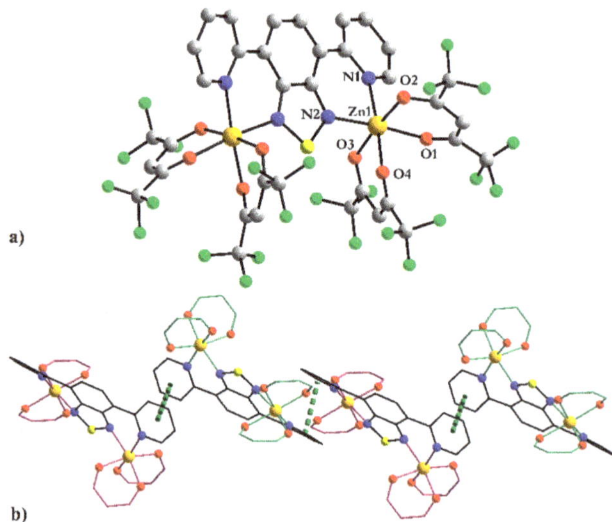

Figure 2. (**a**) Crystal structure of [Zn$_2$(hfac)$_4$(2-PyBTD)] (**2**) with a partial atom numbering scheme; (**b**) View of the packing diagram illustrating the opposite chirality of the metal centers from neighboring binuclear complexes (green—Λ configuration, violet—Δ configuration) (CF$_3$ groups have been omitted for clarity).

Whereas both metal centers of the binuclear unit show the same Δ or Λ configuration, the complexes form supramolecular chains through offset π–π stacking interactions between the pyridyl rings (centroid–centroid distances = 3.64 Å), with an alternation of Δ and Λ configurations (Figure 2b).

3.1.2. Binuclear [Ag(CF$_3$SO$_3$)(2-PyBTD)]$_2$ **3** and [Ag(2-PyBTD)]$_2$(SbF$_6$)$_2$ **4** Complexes

Reaction of 2-PyBTD with the silver(I) salts AgCF$_3$SO$_3$ and AgSbF$_6$ provided the yellow crystalline binuclear complexes [Ag(CF$_3$SO$_3$)(2-PyBTD)]$_2$ **3** and [Ag(2-PyBTD)]$_2$(SbF$_6$)$_2$ **4**, respectively. The isostructural compounds crystallized in the monoclinic space group $P2_1$/c with one centrosymmetric dimeric complex in the unit cell (Figure 3). In the ligand 2-PyBTD the pyridine rings are in *cis-cis* configuration to allow chelation of silver(I), yet the twist between Py and BTD rings is much larger than in complex **2**, with dihedral angles of 39.0° (PyN1) and 38.9° (PyN4) in complex **3** and 44.1° (PyN1) and 31.1° (PyN4) in complex **4**. As a consequence, while the pyridine nitrogen lone pairs point towards the metal center, the coordination with the BTD nitrogen atoms seems reminiscent to Ag(I)-π interactions [42]. In complex **3** the silver ions are pentacoordinated within a slightly distorted square pyramid, with the basal plane formed by three nitrogen atoms from two molecules of BTD (Ag1–N1 = 2.274(3), Ag1–N2 = 2.361(3), Ag1–N4^a = 2.216(3) Å, ^{a} = 1 − x, 1 − y, 1 − z) and an oxygen atom from a triflate anion (Ag1–O1 = 2.658(3) Å), while in the apical position a semi-coordinated nitrogen atom of a thiadiazole ring (Ag1–N3^a = 2.770 Å, ^{a} = 1 − x, 1 − y, 1 − z) is located (Figure 3a, Table 4 and Table S4). The value of τ_5 parameter defined as [(θ−φ)/60] [43], estimating the degree of trigonal distortion from the square pyramid geometry, amounts to 0.08 in complex **3**. In complex **4**, the SbF$_6^-$ anion is not coordinated to the metal center, therefore the coordination stereochemistry can be described as see-saw, according to the value of τ_4 parameter [44] (τ_{4Ag1} = 0.58), with four nitrogen atoms from pyridine and thiadiazole rings of two ligands (Ag1–N1 = 2.192(5), Ag1–N2 = 2.287(5),

Ag1–N3^a = 2.415(5) Ag1–N4^a = 2.221(5) Å, ^{a} = 1 − x, −y, 1 − z) (Figure 3b, Table 2 and Table S4).

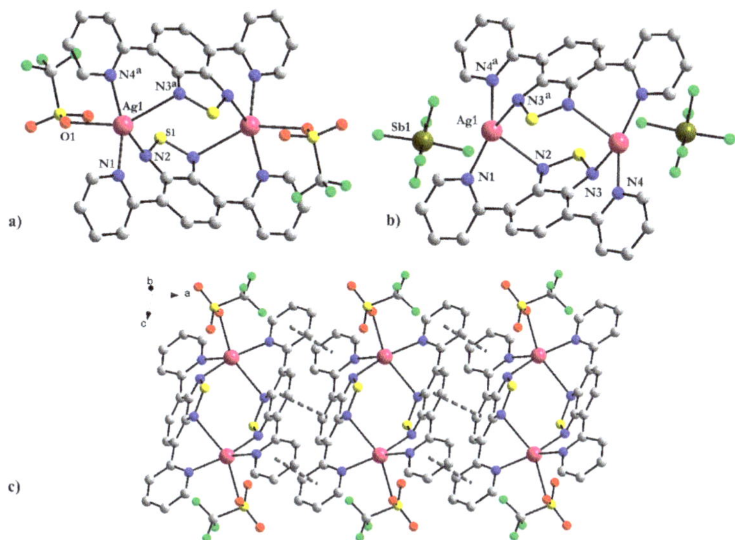

Figure 3. (**a**) Crystal structure of complex [Ag(CF$_3$SO$_3$)(2-PyBTD)]$_2$ **3**. Symmetry operation ^{a} = 1 − x, 1 − y, 1 − z; (**b**) Crystal structure of complex **4**. Symmetry operation ^{a} = 1 − x, − y, 1 − z; (**c**) View of a one-dimensional assembly in compound **3** highlighting the π–π stacking interactions between neighboring binuclear species.

Table 2. Selected bond lengths (Å) for compounds **3** and **4**.

3		4	
Ag1–N1	2.274(3)	Ag1–N1	2.192(5)
Ag1–N2	2.361(3)	Ag1–N2	2.587(5)
Ag1–N3^a	2.770(3)	Ag1–N3^a	2.415(5)
Ag1–N4^a	2.216(3)	Ag1–N4^a	2.221(5)
Ag1–O1	2.657(3)		
^{a} = 1 − x, 1 − y, 1 − z		^{a} = 1 − x, 1 − y, 1 − z	

The intermolecular Ag⋯Ag distances are 5.91 Å in **3** and 5.95 Å in **4**, thus precluding the existence of any argentophilic interaction. The packing diagram reveals the formation of one-dimensional stair-like supramolecular motifs upon π–π stacking offset interactions between pyridine (centroid–centroid distance 3.59 Å (**3**) and 3.72 Å (**4**)) and phenyl (centroid–centroid distance 3.62 Å (**3**) and 3.66 Å (**4**)) rings belonging to neighboring binuclear species (Figure 3c and Figure S9).

3.1.3. Coordination Polymer [Ag$_2$(NO$_3$)$_2$(2-PyBTD)(CH$_3$CN)] 5

Reaction between 2-PyBTD and AgNO$_3$ provided a 2D coordination polymer with a 2:1 metal to ligand ratio. The compound crystallized in the monoclinic space group $P2_1/c$, with one ligand, two silver(I), two nitrate ions and one acetonitrile molecule in the asymmetric unit. The structure of the coordination polymer can be described as follows: the nitrate anions link the silver(I) ions in double chains (Figure 4a) and 2-PyBTD acts as tridentate ligand, thus connecting the inorganic chains in bidimensional layers (Figure 4b). The two metal centers are crystallographically inequivalent and present different coordination geometry. Accordingly, Ag1 ions are hexacoordinated and their coordination stereochemistry can be at best described as pentagonal pyramid following the SHAPE analysis [45,46] (Table S5). Five oxygen atoms, with three shorter distances (Ag1–O1 = 2.471(5), Ag1–O4 = 2.575(3), Ag1–O6^a = 2.441(3) Å) and two longer ones Ag1–O3 = 2.795(3), Ag1–O5^a = 2.768(3) Å, ^{a} = 1 − x, 0.5 + y, 0.5 − z), indicative of semi-coordination, together with a pyridine nitrogen atom (Ag1–N1 = 2.304(3) Å) form the coordination environment of Ag1 (Table 3 and Table S6). Ag2 ions are pentacoordinated and adopt a distorted square pyramid stereochemistry, with τ_{5Ag2} = 0.49. The basal plane is formed by one pyridine nitrogen atom (Ag2–N4^b = 2.370(5) Å, ^{b} = 1 − x, 1 − y, 1 − z), one thiadiazole nitrogen atom (Ag2–N2 = 2.299(4) Å), one acetonitrile nitrogen atom (Ag1–N7 = 2.264(6) Å) and one semi-coordinated nitrato oxygen atom (Ag1–O6 = 2.737(3) Å), while the apical position is occupied by a semi-coordinated nitrato oxygen atom (Ag2–O5^c = 2.653(2) Å, ^{c} = 1 − x, −0.5 + y, 0.5 − z). The BTD ligand adopts a *cis-trans* conformation and shows a tridentate coordination mode, connecting two metal centers of the same double chain through a pyridine (N1) and a thiadiazolyl (N2) nitrogen atom and a third silver ion from a neighboring chain through the second pyridine nitrogen atom (N4) (Figure 4c). The two nitrate anions adopt as well different coordination modes, with N5 being bidentate chelate to Ag1 and N6 linking four metal centers. Within a chain π–π stacking contacts establish between pyridine rings (centroid–centroid distance of 3.54 Å) and also short Ag···Ag distances of 3.55 Å for Ag1···Ag2 are observed.

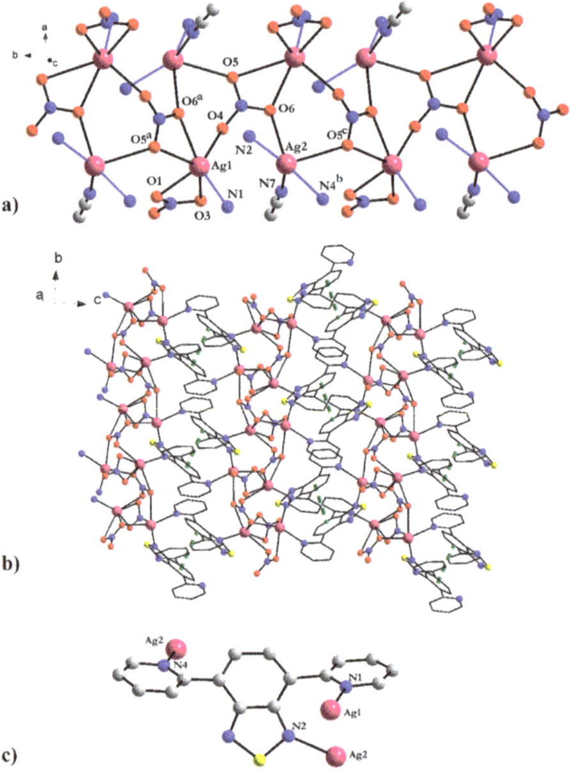

Figure 4. (**a**) Structural detail of a double chain formed by Ag(I) and nitrate ions in the crystal structure of compound **5**. Symmetry operations ^{a} = 1 − x, 0.5 + y, 0.5 − z; ^{b} = 1 − x, 1 − y, 1 − z; ^{c} = 1 − x, −0.5 + y, 0.5 − z; (**b**) 2D layer resulting from the cross-linking of double chains with 2-PyBTD ligands; (**c**) Tridentate coordination mode adopted by 2-PyBTD in **5**.

Table 3. Selected bond lengths (Å) for compound **5**.

	5		
Ag1–N1	2.302(6)	Ag2–N2	2.294(4)
Ag1–O1	2.474(5)	Ag2–N4[b]	2.374(6)
Ag1–O3	2.793(6)	Ag2–N7	2.275(6)
Ag1–O4	2.571(6)	Ag2–O6	2.736(6)
Ag1–O5[a]	2.767(5)	Ag2–O5[c]	2.654(5)
Ag1–O6[a]	2.436(5)		

[a] $= 1 - x, 0.5 + y, 0.5 - z$; [b] $= 1 - x, 1 - y, 1 - z$; [c] $= 1 - x, -0.5 + y, 0.5 - z$

3.2. Coordination Polymer with the 3-PyBTD Ligand

Reaction between 3-Py-BTD and [Zn(hfac)$_2$(H$_2$O)$_2$]·2H$_2$O afforded yellow crystals of the coordination polymer formulated as [Zn(hfac)$_2$(3-PyBTD)] (**6**). The compound crystallized in the monoclinic space group $P2_1/n$ with one ligand and one Zn(hfac)$_2$ fragment in the asymmetric unit. The ligand adopts a ditopic coordination mode through the pyridine nitrogen atoms, connecting the Zn(hfac)$_2$ nodes to form wavy chains in spite of the *cis* arrangement of the pyridine ligands in the coordination sphere of Zn(II) (Figure 5). The coordination geometry of the metal ions is octahedral, with four hfac oxygen atoms (Zn1–O1 = 2.120(3), Zn1–O2 = 2.124(3), Zn1–O3 = 2.112(3) Å, Zn1–O4 = 2.119(3) Å) and two pyridine nitrogen atoms from two different 3-PyBTD ligands (Zn1–N1 = 2.124(4) Å, Zn1–N4^a = 2.120(4) Å, ^{a} = $-0.5 + x, 1.5 - y, -0.5 + z$) (Table 4 and Table S7), disposed, as already mentioned, in a *cis* configuration. The two pyridine rings of the ligand are arranged in *trans-trans* with respect to the BTD unit, with dihedral angles of 40.5° (N1) and 39.9° (N4).

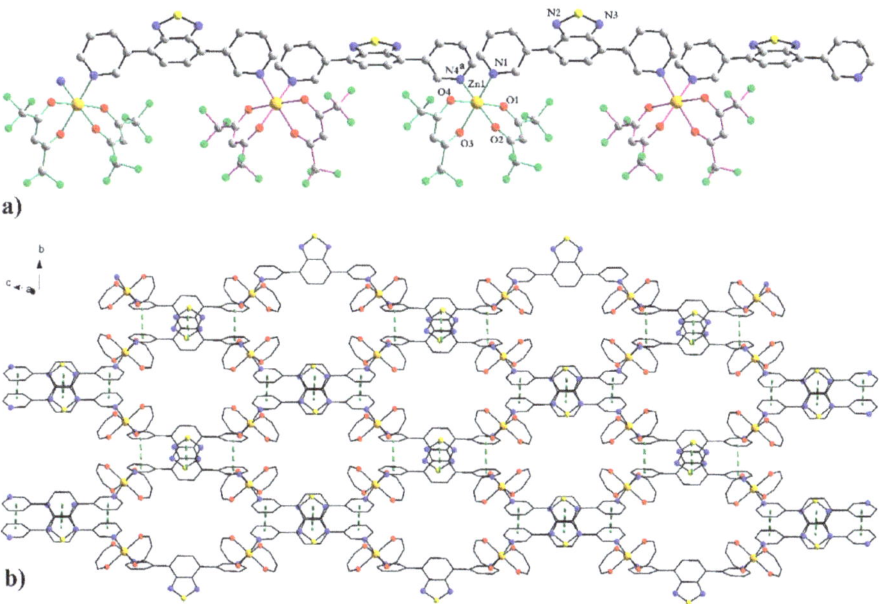

Figure 5. (**a**) Structural detail of a chain built from Zn(hfac)$_2$ nodes and 3-PyBTD spacers in the crystalline structure of compound **6** illustrating the opposite chirality of the alternated metal centers (green—Λ configuration, violet—Δ configuration). Symmetry operations ^{a} = $-0.5 + x, 1.5 - y, -0.5 + z$; (**b**) View along the *a* direction of the 2D layers assembled through π–π stacking between adjacent chains (CF$_3$ groups have been omitted for clarity).

Table 4. Selected bond lengths (Å) for compound **6**.

	6
Zn1–O1	2.120(3)
Zn1–O2	2.124(3)
Zn1–O3	2.112(3)
Zn1–O4	2.119(3)
Zn1–N1	2.124(4)
Zn1–N4[a]	2.120(4)

$$^{a} = -0.5 + x, 1.5 - y, -0.5 + z$$

The metal nodes within the chains show opposite alternated chirality and the chains are associated in supramolecular layers thanks to π–π stacking between the pyridine (3.50 Å) and benzene (3.69 Å) rings (Figure 5b and Figure S10).

3.3. Discrete Complexes and Coordination Polymers with the 4-PyBTD Ligand

3.3.1. Coordination Polymer [Zn(hfac)$_2$(4-PyBTD)] 7

Compound **7** has been obtained as yellow crystals in a similar manner as compound **6**. It crystallized in the triclinic space group *P*–1, with one independent ligand 4-PyBTD in general position, two Zn(II) ions located on inversion centers and one hfac ligand on each metal center, the other hfac ligands being generated through the inversion centers. At the difference with complex **6**, in **7** the pyridine ligands are located in *trans* positions (Figure 6). As expected, the coordination geometry around the metal centers is octahedral, with the equatorial positions occupied by four hfac oxygen atoms (Zn1–O1 = 2.041(3), Zn1–O2 = 2.035(3), Zn1–O3 = 2.057(3) Å, Zn1–O4 = 2.053(3) Å) and the axial ones by two pyridine nitrogen atoms (Zn1–N1 = 2.084(3) Å, Zn1–N2 = 2.042(3) Å) (Table 5 and Table S8). Again, the pyridine rings are twisted with respect to the BTD unit, the corresponding dihedral angles amounting at 39.0° (N1) and 41.6° (N4), values which are slightly superior to those observed in the free ligand (36.8°) [30].

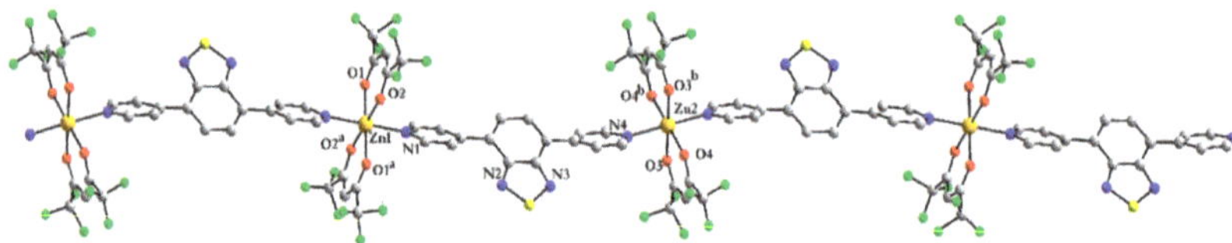

Figure 6. Linear chain in the crystal structure of compound **7**. Symmetry operation $^{a} = -x, 2 - y, 2 - z;$ $^{b} = 2 - x, -y, 1 - z$.

Table 5. Selected bond lengths (Å) for compound **7**.

	7
Zn1–O1	2.076(3)
Zn1–O2	2.075(2)
Zn1–N1	2.161(2)
Zn2–O3	2.073(2)
Zn2–O4	2.096(2)
Zn2–N4	2.136(2)

3.3.2. Coordination Complexes of 4-PyBTD with ZnCl$_2$

The reaction of 4-PyBTD with ZnCl$_2$ afforded either the mononuclear complex [ZnCl$_2$(4-PyBTD)$_2$] **8** or the coordination polymer [ZnCl$_2$(4-PyBTD)] **9** by slightly varying the experimental conditions. The former crystallized in the monoclinic space group *P*2$_1$/c with one complex in general position in the asymmetric unit, while the latter crystallized in the non-centrosymmetric orthorhombic space group *P*2$_1$2$_1$2$_1$, with one 4-PyBTD ligand and one ZnCl$_2$ fragment in the asymmetric unit. In both complexes the coordination geometry around the metal center is tetrahedral, with values of the bond lengths Zn–Cl and Zn–N in the usual range (Table 6 and Table S9).

Table 6. Selected bond lengths (Å) for compounds **8** and **9**.

8		9	
Zn1–N1	2.049(2)	Zn1–N1[a]	2.064(5)
Zn1–N5	2.060(2)	Zn1–N4	2.068(5)
Zn1–Cl1	2.217(2)	Zn1–Cl1	2.202(2)
Zn1–Cl2	2.214(1)	Zn1–Cl2	2.232(2)
[a] $= 1.5 - x, 1 - y, 0.5 + z$			

The pyridine rings of the two 4-PyBTD ligands show large distortions with respect to the BTD units in complex **8**, with dihedral angles of 46.3° (N1), 39.4° (N4), 38.6° (N5) and 39.8° (N8) (Figure S11a), and, comparable, much weaker distortions in the coordination polymer **9** (23.2° (N1) and 16.9° (N4)) (Figure 7). In the structure of **8** a 2D network is established thanks to π–π stacking interactions between pyridine rings, with centroid–centroid distances of 3.79 and 3.78 Å (Figure S11b). In the coordination polymer **9**, the bidentate tecton connects the metal ions in zig-zag chains (Figure 7a).

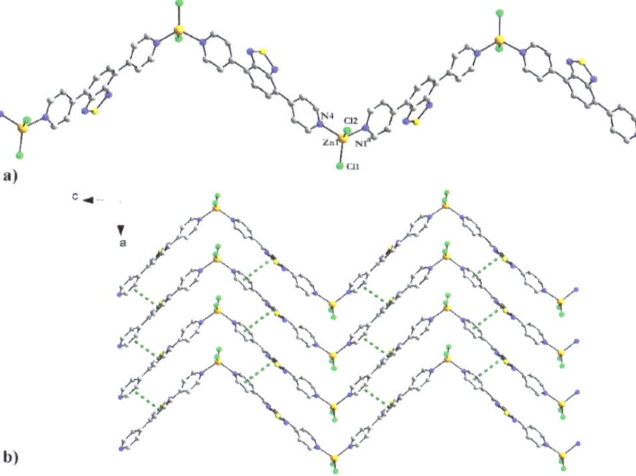

Figure 7. (**a**) Structural detail of a zig-zag chain in compound **9**; (**b**) Supramolecular layer from π–π stacking interactions. Symmetry operation [a] $= 1.5 - x, 1 - y, 0.5 + z$.

The chains arrange in layers upon π–π stacking interactions between pyridine and benzene rings, with an interplanar distance of 3.89 Å (Figure 7b).

The purity of the different complexes have been checked, when the amounts were sufficient, by elemental analysis and powder X-ray diffraction (PXRD) analysis, by comparing the experimental PXRD diffractograms with those simulated from the single crystal X-ray data (Figures S12–S16).

3.4. Photophysical Properties

As mentioned in the Introduction, the Py-BTD ligands are luminescent. Their coordination to metal ions with d^{10} configurations such as Zn(II) should provide luminescent complexes with, in principle, an increase of the emission efficiency as a consequence of a more rigid structure [47–50]. Moreover, emission wavelength could vary because of the different dihedral angles between the pyridine and BTD units, influencing the extension of the π-conjugated system. We have therefore investigated the photophysical properties of the Zn(II) complexes of ligands 2-PyBTD and 4-Py-BTD and compared them with those of the ligands, previously reported [30]. In order to have reliable comparisons between ligands, molecular complexes and coordination polymers, all the measurements have been performed in the solid state.

3.4.1. Complexes 1 and 2

The ligand 2-PyBTD presents in the UV-Vis absorption spectrum a very broad intense band with a maximum at 420 nm and another distinct weaker band at 258 nm encompassing several π–π* transitions [30] (Figure S17). Upon excitation at λ_{ex} = 400 nm, an intense emission band centered at λ_{em} = 523 nm is observed, which, interestingly, is strongly bathochromically shifted when compared to the value λ_{em} = 469 nm (λ_{ex} = 299 nm) measured in acetonitrile solutions [30] (see Figure S18 for spectra measured in CH_2Cl_2 solutions). Clearly, packing effects have a strong influence on the absorption and emission wavelengths. However, a direct comparison between the photophysical properties of 2-Py-BTD and its zinc(II) complexes **1** and **2** is not straightforward since in the former the pyridine rings adopt a *trans-trans* conformation [30], while in the complexes the respective conformations are *cis-trans* and *cis-cis*.

The Zn(II) ion does not possess any d-d transition, therefore, following its coordination to 2-PyBTD, ligand based π–π* transitions centered at 247, 343 and 440 nm are observed in the UV-Vis spectrum of complex **1** (Figure 8). In its emission spectrum, the band observed at λ_{em} = 526 nm upon excitation at λ_{ex} = 440 nm is much more intense than the one of the free ligand.

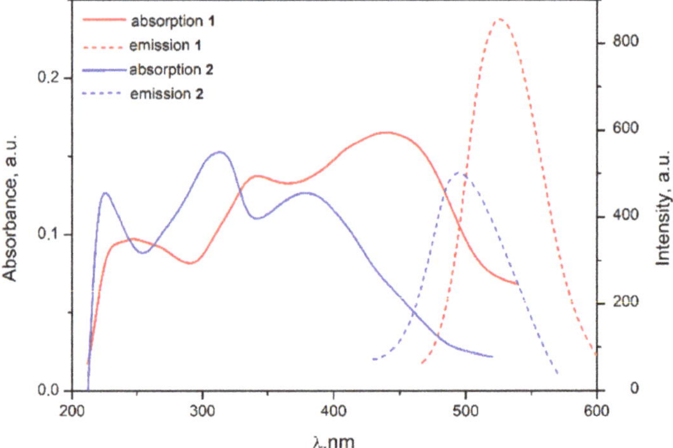

Figure 8. Solid state UV-Vis absorption (solid lines) and emission (dashed lines) spectra of complexes **1** (red) and **2** (blue).

Coordination of two Zn(hfac)$_2$ fragments induces massive changes in the photophysical properties of complex **2** compared to the free ligand and complex **1**. In the diffused reflexion UV-Vis absorption spectrum of **2**, three bands at 379, 313 and 226 nm are observed, while the emission spectrum shows an intense band centered at λ_{em} = 495 nm upon excitation at λ_{ex} = 380 nm (Figure 8). The blue-shift of the absorption and emission bands of complex **2** compared to complex **1** can be tentatively explained, on the one hand, by the different conformations adopted by the pyridine rings, that is, *cis-cis* in **2** and *cis-trans* in **1** and also by the decrease of planarity in the former system upon coordination of 2-PyBTD to two metal centers. Indeed, in complex **2** the twist between the two pyridine rings amounts to 52.9° compared to 7.3° in **1** and the distortion with respect to the BTD chromophore are 26.9° in **2** and 14.1° and 15.3° in **1** (see Figures 1 and 2). On the other hand, photophysical properties measured in the solid state can be strongly influenced by the intermolecular interactions in the packing. As discussed above, in complex **1** there is formation of supramolecular dimers through π–π stacking interactions between the uncoordinated pyridine and BTD, while in **2** the complexes form supramolecular chains through π–π stacking between the pyridine units (*vide supra*). The influence of all these factors on the absorption/emission properties is difficult to be disclosed.

In order to have an insight on the nature of the electronic transitions involved in the absorption and emission bands, we have performed DFT and TD-DFT calculations on both complexes **1** and **2** (see the details in the SI). The optimized geometries are in agreement with the experimental ones obtained by X-ray structure analysis. The simulated UV-Vis absorption spectra of complexes **1** and **2** are shown in Figure 9.

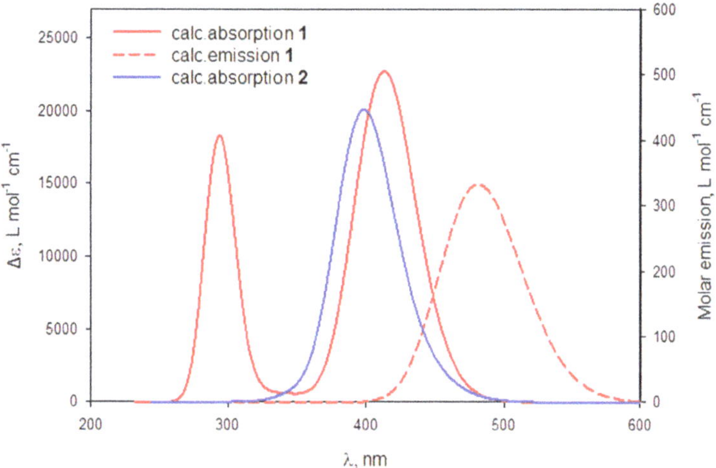

Figure 9. Calculated UV-Vis absorption (solid lines) and emission (dashed line) spectra of complexes **1** (red) and **2** (blue) with a gaussian broadening (FWHM = 3000 cm^{-1}); TD-DFT/PBE1PBE/TZVP.

The first calculated allowed transition in complex **1**, occurring at 413 nm (SI), corresponds to a HOMO → LUMO excitation of π–π* type (Figure 10 left), HOMO being based mainly on the pyridine and benzene rings, while LUMO is mainly delocalized on the BTD unit, with a large participation of the thiadiazole ring. The corresponding calculated emission wavelength values amounts to 478 nm, hence blue-shifted compared to the experimental value measured in the solid state but in agreement with the value of 467 nm for the emission in CH_2Cl_2 solution (Figure S19). One can thus hypothesize that the red-shift for the absorption/emission bands observed in the solid state is the consequence of intermolecular π–π stacking interactions, all the more since HOMO and LUMO are localized on the π system involved in the corresponding electronic transition. In complex **2** the highest four occupied orbitals, that is, from HOMO–3 to HOMO, are in-phase and out-of-phase combinations of hfac based orbitals, while the LUMO has a similar distribution as in complex **1**, namely on the BTD unit, with a large participation of the thiadiazole ring. Consequently, the first four singlet excited states, which are HOMO–n → LUMO excitations ($n = 0$–3), present very weak oscillator strengths. The first intense absorption band is described by a HOMO–4 → LUMO excitation, as shown by the electron density difference (EDD) between the 5th excited state and the ground state (Figure 10 right) and corresponds, as the S_1-S_0 transition in complex **1**, to a charge transfer from the benzene-dipyridine skeleton to thiadiazole. The calculated wavelength value for this absorption is 397 nm, thus slightly blue-shifted compared to **1**, yet remaining consistent with the experimental absorption band measured in CH_2Cl_2 solution occurring at $\lambda_{max} = 387$ nm

(Figure S20). Unfortunately, the optimization of the excited emissive state for complex **2** could not be carried out due to interstate mixing. It should be emphasized once again, that it is not straightforward to directly compare the absorption/emission wavelengths values between ligand 2-PyBTD, complex **1** and complex **2**, as they show the three possible conformations *trans-trans*, *cis-trans* and *cis-cis*, respectively. However, as stated above, the blue-shift observed in the solid state for the absorption/emission bands of **2** compared to those of **1** is very likely due to the different π–π stacking interactions involving the aromatic rings responsible for the corresponding electronic transitions.

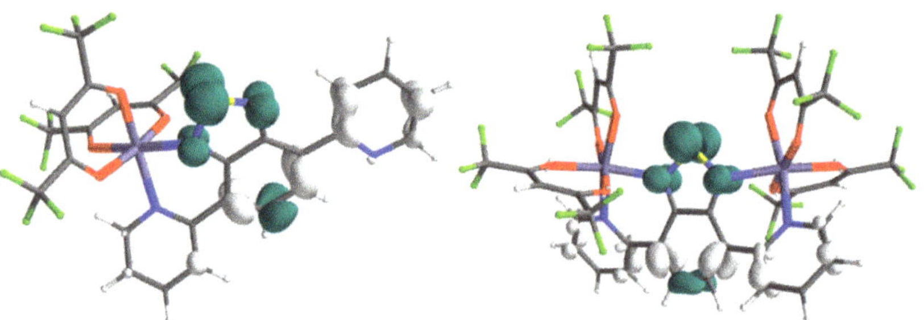

Figure 10. Representation of the Electron Density Differences (EDD) for the S_1-S_0 transition of complex **1** (**left**) and the S_5-S_0 transition of complex **2** (**right**). The excited electron and hole regions are indicated by, respectively, blue and white surfaces.

3.4.2. Complexes **7–9**

The UV-Vis spectrum of 4-PyBTD ligand, which provided complexes **7–9**, shows the presence of one very broad band with a maximum at 397 nm and an additional weaker band at 260 nm. In the emission spectrum an intense band centered at λ_{em} = 472 nm (λ_{ex} = 370 nm) is observed (Figure S21), in agreement with a previous report [31]. At the difference with ligand 2-PyBTD (*vide supra*), a much weaker bathochromic shift is observed now for the solid state emission of 4-PyBTD compared to the emission band in acetonitrile solution occurring at λ_{em} = 448 nm (λ_{ex} = 299 nm) [30].

Coordination of Zn(hfac)$_2$ fragments to provide the coordination polymer **7** is accompanied by a strong bathochromic shift of the luminescence, as witnessed by the emission band centered at 527 nm (λ_{ex} = 460 nm) and an increase of the emission intensity compared to the free ligand (Figure 11). A similar bathochromic shift has been observed for a mixed coordination polymer of Zn(II), 4-PyBTD and isophthalic acid [31].

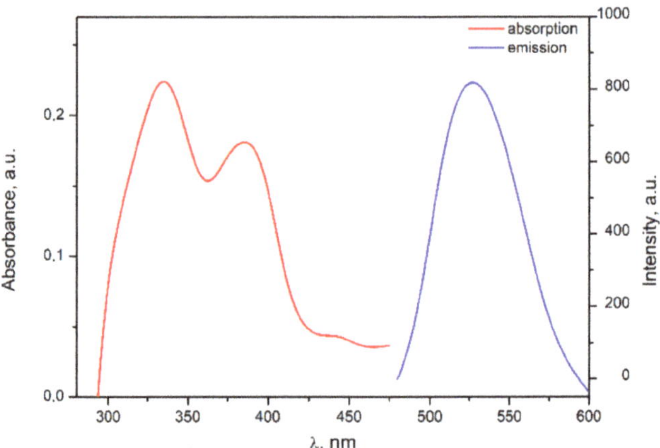

Figure 11. Solid state UV-Vis absorption and emission spectra of complex **7**.

Complexes **8** and **9**, containing $ZnCl_2$ fragments instead of $Zn(hfac)_2$, show strong blue luminescence at emission wavelengths of λ_{em} = 476 nm (λ_{ex} = 410 nm) and λ_{em} = 471 nm (λ_{ex} = 400 nm) (Figure 12, see Figure S22 for the absorption spectra), respectively, comparable to the emission wavelength of the free ligand (*vide supra*). The excitation spectra of both complexes overlap with the corresponding wavelength regions of the absorption spectra.

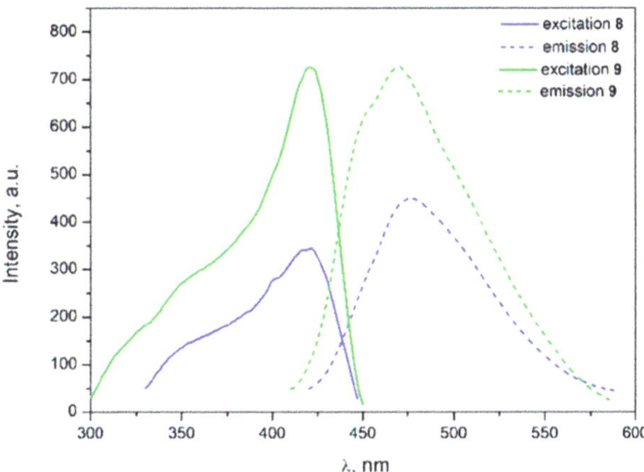

Figure 12. Emission (dotted curves, λ_{ex} = 410 nm for **8** and λ_{ex} = 400 nm for **9**) and excitation (solid curves, λ_{em} = 475 nm for **8** and λ_{em} = 470 nm for **9**) for complexes **8** (blue) and **9** (green).

The absence of bathochromic shift of the luminescence in the case of complexes **8** and **9** when compared to complex **7**, might have its origin, besides the different molecular packing in the solid state, in the difference of molecular orbital levels involved in the emission, as a consequence of the strong electron withdrawing effect exerted by the hfac ligands with respect to the chloride ligands which have a π-donating effect.

4. Conclusions

4,7-dipyridyl-2,1,3-benzothiadiazol ligands 2-PyBTD, 3-PyBTD and 4-PyBTD have been used to prepare coordination complexes with d^{10} transition metals group in order to take advantage of their flexible topicity and luminescence properties. Compounds **1–5** with 2-PyBTD and **6** with 3-PyBTD represent the first reported complexes based on these ligands. While ligand 2-PyBTD showed chelating behavior towards zinc(II) and silver(I) ions, involving coordination by the pyridine and thiadiazole nitrogen atoms, in the zinc(II) complexes based on 3-PyBTD and 4-PyBTD the ligands act as bridges through the pyridine nitrogen atoms to provide the coordination polymers **6**, **7** and **9**. Moreover, the mononuclear complex $[ZnCl_2(4\text{-PyBTD})_2]$ **8** has been prepared as well, as a brick model for the coordination polymer **9**. In the crystal structures of all the complexes the occurrence of supramolecular interactions directing the overall architecture has been thoroughly discussed. A very peculiar coordination behavior of 2-PyBTD has been revealed in the coordination polymer **5** with silver nitrate, where the ligand shows a tridentate coordination mode leading to the formation of an original 2D structure. A striking difference has been observed between 3-PyBTD and 4-PyBTD towards the $Zn(hfac)_2$ fragments in complexes **6** and **7**, since in the former the pyridine nitrogen atoms are in *cis* configuration, while in the latter they coordinate the metal ion in axial positions. Solid state photophysical properties of the zinc(II) complexes of ligands 2-PyBTD and 4-PyBTD indicate an enhancement of the emission intensity when compared to the free ligands. The emission wavelength shows modulation with the nuclearity of the complexes, that is, blue-shift in **2** compared to **1** very likely because of the different intermolecular π–π stacking interactions, and with the coordinated fragment, that is, red-shift in **7** compared to **8** and **9**. DFT calculations on complexes **1** and **2** shed light on the nature of the electronic transitions involved in the low energy absorption band responsible for the emissive properties. This first coordination chemistry study of PyBTD with d^{10} metals clearly highlight the interest of these ligands for the access to multifunctional ligands and open the way towards the preparation of whole series of complexes. Particularly attractive, for example, is the use of 2-PyBTD ligand as bridge between paramagnetic metal centers where the magnetic interaction can be mediated by the oxidation state of the ligand [51], when considering the possibility to reduce the BTD unit into radical anion species. These directions are currently explored in our groups.

Supplementary Materials: The following are available online at https://www.mdpi.com/2624-854 9/3/1/20/s1. Table S1. Crystallographic data, details of data collection and structure refinement parameters for compounds **1–5**. Table S2. Crystallographic data, details of data collection and structure refinement parameters for compounds **6–9**. Table S3. Selected bond angles (°) for compounds **1** and **2**. Table S4. Selected bond angles (°) for compounds **3** and **4**. Table S5. Values of the SHAPE parameter for the Ag1 ion in compound **5** for hexacoordination. Table S6. Selected bond angles (°) for compound **5**. Table S7. Selected bond angles (°) for compound **6**. Table S8. Selected bond angles (°) for compound **7**. Table S9 Selected bond angles (°) for compounds **8** and **9**. Figure S1. IR spectrum of compound **1**. Figure S2. IR spectrum of compound **2**. Figure S3. IR spectrum of compound **3**. Figure S4. IR spectrum of compound **4**. Figure S5. IR Spectrum of compound **6**. Figure S6. IR spectrum of compound **7**. Figure S7. IR spectrum of compound **8**. Figure S8. IR spectrum of compound **9**. Figure S9. Perspective of the 1D supramolecular assembly in **4**. Figure S10. Schematic representation of the supramolecular layer in crystal structure of **6**. Figure S11. (**a**) Molecular structure of complex [ZnCl$_2$(4-PyBTD)$_2$] **8**; (**b**) Perspective view of the 2D layers resulting from π–π stacking between pyridine rings from adjacent mononuclear species. Figure S12. Simulated (black) and experimental (red) powder X-ray diffractograms for compound **1**. Figure S13. Simulated (black) and experimental (red) powder X-ray diffractograms for compound **2**. Figure S14. Simulated (black) and experimental (red) powder X-ray diffractograms for compound **4**. Figure S15. Simulated (black) and experimental (red) powder X-ray diffractograms for compound **7**. Figure S16. Simulated (black) and experimental (red) powder X-ray diffractograms for compound **8**. Figure S17. UV-Vis absorption spectrum (red curve), emission spectrum (blue curve, λ_{ex} = 400 nm) and excitation spectrum (green curve, λ_{em} = 525 nm) of ligand 2-PyBTD. Figure S18. Absorption (red), emission (blue) and excitation (green) spectra of 2-PyBTD recorded in CH$_2$Cl$_2$. (Absorption: λ = 236, 290, 382 nm; Emission: λ_{ex} = 380 nm, λ_{max} = 466 nm; Excitation: λ_{em} = 465 nm, λ = 292, 385 nm). Figure S19. Absorption (red), emission (blue) and excitation (green) spectra of compound **1** recorded in CH$_2$Cl$_2$. (Absorption: λ = 265, 314, 381 nm; Emission: λ_{ex} = 390 nm, λ_{max} = 467 nm; Excitation: λ_{em} = 465 nm, λ = 282, 305, 329, 398 nm). Figure S20. Absorption (red), emission (blue) and excitation (green) spectra of compound **2** recorded in CH$_2$Cl$_2$. (Absorption: λ = 272, 321, 387 nm; Emission: λ_{ex} = 390 nm, λ_{max} = 470 nm; Excitation: λ_{em} = 470 nm, λ = 273, 330, 389 nm). Figure S21. UV-Vis absorption spectrum (red curve), emission spectrum (blue curve, λ_{ex} = 370 nm) and excitation spectrum (green curve, λ_{em} = 470 nm) of ligand 4-PyBTD. Figure S22. (**a**) UV-Vis absorption spectra of complex **8**; (**b**) UV-Vis absorption spectra of complex **9**. Figure S23. Two views of the optimized geometry of **1** together with the atom numbering scheme. Figure S24. HOMO, LUMO, HOMO-1 and LUMO+1 (from top to bottom, two views each) of complex **1**. Figure S25. Representation of the F+ function (two views) for complex **1**. The blue color indicates the most electrophilic regions. Figure S26. Representation of the F- function (two views) for complex **1**. The blue color indicates the most nucleophilic regions. Figure S27. Representation of the Electron Density Difference (S1-S0 left) and (S2-S0 right) for complex **1**. The excited electron and the hole regions are indicated by, respectively, blue and white surfaces. Figure S28. Calculated UV-visible absorption spectrum of complex **1** with a gaussian broadening (FWHM = 3000 cm^{-1}). Figure S29. Calculated UV-visible emission spectrum of complex **1** with a gaussian broadening (FWHM = 3000 cm^{-1}). Figure S30. Two views of the optimized geometry of **2** together with the atom numbering scheme. Figure S31. HOMO, LUMO, HOMO-1 and LUMO+1 (from top to bottom, two views each) of complex **2**. Figure S32. Representation of the Electron Density Difference (S1-S0 left) and (S2-S0 right) for complex **2**. The excited electron and the hole regions are indicated by, respectively, blue and white surfaces. Figure S33. Calculated UV-visible absorption spectrum of complex **2** with a gaussian broadening (FWHM = 3000 cm^{-1}).

Author Contributions: N.A. conceived and designed the experiments; T.M. synthesized and characterized the ligands and the complexes; N.P. and T.C. performed the DFT calculations; N.A. and M.A. wrote and/or reviewed the manuscript with contributions from all the authors. All authors have read and agreed to the published version of the manuscript.

Funding: Financial support in France from the CNRS and University of Angers is gratefully acknowledged. The Erasmus exchange program between the University of Angers, France and University of Bucharest, Romania, is acknowledged (internship grant to T.M.).

Acknowledgments: Magali Allain (Plateau CRISTAL, SFR Matrix, University of Angers) is gratefully acknowledged for help with the X-ray structure of complex **6**.

Conflicts of Interest: The authors declare no conflict of interest. The funders had no role in the design of the study; in the collection, analyses, or interpretation of data; in the writing of the manuscript, or in the decision to publish the results.

References

1. Thomas, K.R.J.; Lin, J.T.; Velusamy, M.; Tao, Y.-T.; Chuen, C.-H. Color Tuning in Benzo[1,2,5]thiadiazole-Based Small Molecules by Amino Conjugation/Deconjugation: Bright Red-Light-Emitting Diodes. *Adv. Funct. Mater.* **2004**, *14*, 83–90. [CrossRef]
2. Kono, T.; Kumaki, D.; Nishida, J.; Sakanoue, T.; Kakita, M.; Tada, H.; Tokito, S.; Yamashita, Y. High-Performance and Light-Emitting n-Type Organic Field-Effect Transistors Based on Dithienylbenzothiadiazole and Related Heterocycles. *Chem. Mater.* **2007**, *19*, 1218–1220. [CrossRef]
3. Zhang, M.; Tsao, H.N.; Pisula, W.; Yang, C.; Mishra, A.K.; Müllen, K. Field-Effect Transistors Based on a Benzothiadiazole–Cyclopentadithiophene Copolymer. *J. Am. Chem. Soc.* **2007**, *129*, 3472–3473. [CrossRef]
4. Geng, Y.; Pfattner, R.; Campos, A.; Hauser, J.; Laukhin, V.; Puigdollers, J.; Veciana, J.; Mas-Torrent, M.; Rovira, C.; Decurtins, S.; et al. A Compact Tetrathiafulvalene–Benzothiadiazole Dyad and Its Highly Symmetrical Charge-Transfer Salt: Ordered Donor π-Stacks Closely Bound to Their Acceptors. *Chem. Eur. J.* **2014**, *20*, 7136–7143. [CrossRef] [PubMed]
5. Chen, C.T. Evolution of Red Organic Light-Emitting Diodes: Materials and Devices. *Chem. Mater.* **2004**, *16*, 4389–4400. [CrossRef]
6. Neto, B.A.D.; Lapis, A.A.M.; da Silva Júnior, E.N.; Dupont, J. 2,1,3-Benzothiadiazole and Derivatives: Synthesis, Properties, Reactions and Applications in Light Technology of Small Molecules. *Eur. J. Org. Chem.* **2013**, 228–255. [CrossRef]
7. Velusamy, M.; Thomas, K.R.J.; Lin, J.T.; Hsu, Y.; Ho, K. Organic Dyes Incorporating Low-Band-Gap Chromophores for Dye-Sensitized Solar Cells. *Org. Lett.* **2005**, *7*, 1899–1902. [CrossRef]
8. Lin, L.-Y.; Lu, C.-W.; Huang, W.-C.; Chen, Y.-H.; Lin, H.-W.; Wong, K.-T. New A-A-D-A-A-Type Electron Donors for Small Molecule Organic Solar Cells. *Org. Lett.* **2011**, *13*, 4962–4965. [CrossRef]
9. Lin, L.-Y.; Chen, Y.-H.; Huang, Z.-Y.; Lin, H.-W.; Chou, S.-H.; Lin, F.; Chen, C.-W.; Liu, Y.-H.; Wong, K.-T. A Low-Energy-Gap Organic Dye for High-Performance Small-Molecule Organic Solar Cells. *J. Am. Chem. Soc.* **2011**, *133*, 15822–15825. [CrossRef]
10. Haid, S.; Marszalek, M.; Mishra, A.; Wielopolski, M.; Teuscher, J.; Moser, J.-E.; Humphry-Baker, R.; Zakeeruddin, S.M.; Grätzel, M.; Bäuerle, P. Significant Improvement of Dye-Sensitized Solar Cell Performance by Small Structural Modification in π-Conjugated Donor–Acceptor Dyes. *Adv. Funct. Mater.* **2012**, *22*, 1291–1302. [CrossRef]
11. Geng, Y.; Pop, F.; Yi, C.; Avarvari, N.; Grätzel, M.; Decurtins, S.; Liu, S.-X. Electronic tuning effects via π-linkers in tetrathiafulvalene-based dyes. *New J. Chem.* **2014**, *38*, 3269–3274. [CrossRef]
12. Pop, F.; Amacher, A.; Avarvari, N.; Ding, J.; Lawson Daku, L.M.; Hauser, A.; Koch, M.; Hauser, J.; Liu, S.-X.; Decurtins, S. Tetrathiafulvalene-Benzothiadiazoles as Redox-Tunable Donor-Acceptor Systems: Synthesis and Photophysical Study. *Chem. Eur. J.* **2013**, *19*, 2504–2514. [CrossRef]
13. Pop, F.; Seifert, S.; Hankache, J.; Ding, J.; Hauser, A.; Avarvari, N. Modulation of the charge transfer and photophysical properties in non-fused tetrathiafulvalene-benzothiadiazole derivatives. *Org. Biomol. Chem.* **2015**, *13*, 1040–1047. [CrossRef] [PubMed]
14. Neto, B.A.D.; Carvalho, P.H.P.R.; Correa, J.R. Benzothiadiazole Derivatives as Fluorescence Imaging Probes: Beyond Classical Scaffolds. *Acc. Chem. Res.* **2015**, *48*, 1560–1569. [CrossRef] [PubMed]
15. Aldakov, D.; Palacios, M.A.; Anzenbacher, P. Benzothiadiazoles and Dipyrrolyl Quinoxalines with Extended Conjugated Chromophores–Fluorophores and Anion Sensors. *Chem. Mater.* **2005**, *17*, 5238–5241. [CrossRef]
16. Cozzolino, A.F.; Vargas-Baca, I.; Mansour, S.; Mahmoudkhani, A.H. The Nature of the Supramolecular Association of 1,2,5-Chalcogenadiazoles. *J. Am. Chem. Soc.* **2005**, *127*, 3184–3190. [CrossRef] [PubMed]
17. Ams, M.R.; Trapp, N.; Schwab, A.; Milić, J.V.; Diederich, F. Chalcogen Bonding "2S–2N Squares" versus Competing Interactions: Exploring the Recognition Properties of Sulfur. *Chem. Eur. J.* **2019**, *25*, 323–333. [CrossRef] [PubMed]
18. Alcock, N.W.; Hill, A.F.; Roe, M.S. Hydrido(benzochalcogenadiazole) complexes of ruthenium: Crystal structure of [RuCl(H)(CO)(PPh$_3$)(SN$_2$C$_6$H$_4$)]. *J. Chem. Soc. Dalton Trans.* **1990**, *5*, 1737–1740. [CrossRef]
19. Munakata, M.; Kuroda-Sowa, T.; Maekawa, M.; Nakamura, M.; Akiyama, S.; Kitagawa, S. Architecture of 2D Sheets with Six-Membered Rings of Coppers Interconnected by 2,1,3-Benzothiadiazoles and a Layered Structure Composed of the 2D Sheets. *Inorg. Chem.* **1994**, *33*, 1284–1291. [CrossRef]

20. Renner, M.W.; Barkigia, K.M.; Melamed, D.; Smith, K.M.; Fajer, J. Ligand-Bridged Heterobimetallic Polymers: Silver(I)-Benzothiadiazole-Nickel Porphyrin Cation-Benzothiadiazole Arrays. *Inorg. Chem.* **1996**, *35*, 5120–5121. [CrossRef]

21. Papaefstathiou, G.S.; Perlepes, S.P.; Escuer, A.; Vicente, R.; Gantis, A.; Raptopoulou, C.P.; Tsohos, A.; Psycharis, V.; Terzis, A.; Bakalbassis, E.G. Topological Control in Two-Dimensional Cobalt(II) Coordination Polymers by π–π Stacking Interactions: Synthesis, Spectroscopic Characterization, Crystal Structure and Magnetic Properties. *J. Solid State Chem.* **2001**, *159*, 371–378. [CrossRef]

22. Papaefstathiou, G.S.; Tsohos, A.; Raptopoulou, C.P.; Terzis, A.; Psycharis, V.; Gatteschi, D.; Perlepes, S.P. Crystal Engineering: Stacking Interactions Control the Crystal Structures of Benzothiadiazole (btd) and Its Complexes with Copper(II) and Copper(I) Chlorides. *Cryst. Growth Des.* **2001**, *1*, 191–194. [CrossRef]

23. Bashirov, D.A.; Sukhikh, T.S.; Kuratieva, N.V.; Naumov, D.Y.; Konchenko, S.N.; Semenov, N.A.; Zibarev, A.V. Iridium complexes with 2,1,3-benzothiadiazole and related ligands. *Polyhedron* **2012**, *42*, 168–174. [CrossRef]

24. Sukhikh, T.S.; Ogienko, D.S.; Bashirov, D.A.; Konchenkoa, S.N. Luminescent complexes of 2,1,3-benzothiadiazole derivatives. *Russ. Chem. Bull.* **2019**, *68*, 651–661. [CrossRef]

25. Marshall, R.J.; Kalinovskyy, Y.; Griffin, S.L.; Wilson, C.; Blight, B.A.; Forgan, R.S. Functional Versatility of a Series of Zr Metal–Organic Frameworks Probed by Solid-State Photoluminescence Spectroscopy. *J. Am. Chem. Soc.* **2017**, *139*, 6253–6260. [CrossRef]

26. Zhang, W.-Q.; Li, Q.-Y.; Cheng, J.-Y.; Cheng, K.; Yang, X.; Li, Y.; Zhao, X.; Wang, X.-J. Ratiometric Luminescent Detection of Organic Amines Due to the Induced Lactam–Lactim Tautomerization of Organic Linker in a Metal–Organic Framework. *ACS Appl. Mater. Interfaces* **2017**, *9*, 31352–31356. [CrossRef] [PubMed]

27. Zhao, D.; Yue, D.; Jiang, K.; Cui, Y.; Zhang, Q.; Yang, Y.; Qian, G. Ratiometric dual-emitting MOF⊃∩dye thermometers with a tunable operating range and sensitivity. *J. Mater. Chem. C* **2017**, *5*, 1607–1613. [CrossRef]

28. Cheng, Q.; Han, X.; Tong, Y.; Huang, C.; Ding, J.; Hou, H. Two 3D Cd(II) Metal–Organic Frameworks Linked by Benzothiadiazole Dicarboxylates: Fantastic S@Cd$_6$ Cage, Benzothiadiazole Antidimmer and Dual Emission. *Inorg. Chem.* **2017**, *56*, 1696–1705. [CrossRef] [PubMed]

29. Song, W.-C.; Liang, L.; Cui, X.-Z.; Wang, X.-G.; Yang, E.-C.; Zhao, X.-J. Assembly of Zn^{II}-coordination polymers constructed from benzothiadiazole functionalized bipyridines and V-shaped dicarboxylic acids: Topology variety, photochemical and visible-light-driven photocatalytic properties. *CrystEngComm* **2018**, *20*, 668–678. [CrossRef]

30. Akhtaruzzaman, M.; Tomura, M.; Nishida, J.; Yamashita, Y. Synthesis and Characterization of Novel Dipyridylbenzothiadiazole and Bisbenzothiadiazole Derivatives. *J. Org. Chem.* **2004**, *69*, 2953–2958. [CrossRef]

31. Ju, Z.; Yan, W.; Gao, X.; Shi, Z.; Wang, T.; Zheng, H. Syntheses, Characterization and Luminescence Properties of Four Metal–Organic Frameworks Based on a Linear-Shaped Rigid Pyridine Ligand. *Cryst. Growth Des.* **2016**, *16*, 2496–2503. [CrossRef]

32. Shen, K.; Ju, Z.; Qin, L.; Wang, T.; Zheng, H. Two sTable 3D porous metal-organic frameworks with high selectivity for detection of PA and metal ions. *Dyes Pigm.* **2017**, *136*, 515–521. [CrossRef]

33. Sheldrick, G.M. Crystal structure refinement with SHELX. *Acta Cryst.* **2015**, *C71*, 3–8.

34. Sheldrick, G.M. SHELXT–Integrated space-group and crystal-structure determination. *Acta Cryst.* **2015**, *A71*, 3–8. [CrossRef] [PubMed]

35. Frisch, M.J.; et al. *Gaussian∼09 Revision D.01*.

36. Schäfer, A.; Huber, C.; Ahlrichs, R. Fully optimized contracted Gaussian basis sets of triple zeta valence quality for atoms Li to Kr. *J. Chem. Phys.* **1994**, *100*, 5829–5835. [CrossRef]

37. Cauchy, T.; Da Mota, B. Quchemreport. A Python Program for Control Quality and Automatic Generation of Quantum Chemistry Results. University of Angers, Angers, France, 2020.

38. O'boyle, N.M.; Tenderholt, A.L.; Langner, K.M. Cclib: A Library for Package-Independent Computational Chemistry Algorithms. *J. Comput. Chem.* **2008**, *29*, 839–845. [CrossRef] [PubMed]

39. Lee, C.-J.; Wei, H.-H. Synthesis, crystal structure and magnetic properties of zinc(II)-hexafluoro-acetylacetonate complexes with pyridyl-substituted nitronyl and imino nitroxide radicals. *Inorg. Chim. Acta* **2000**, *310*, 89–95. [CrossRef]

40. Zaman, M.B.; Udachin, K.A.; Ripmeester, J.A. One-dimensional coordination polymers generated from 1,2-bis(x-pyridyl)butadiyne (x = 3 and 4) and bis(hexafluoroacetylacetonato)M(II) (M = Cu, Mn, Zn). *CrystEngComm* **2002**, *4*, 613–617. [CrossRef]

41. Steffen, W.L.; Palenik, G.J. Infrared and Crystal Structure Study of σ vs. π Bonding in Tetrahedral Zinc(II) Complexes. Crystal and Molecular Structures of Dichlorobis(4-substituted pyridine)zinc(II) Complexes. *Inorg. Chem.* **1977**, *16*, 1119–1127. [CrossRef]

42. Baudron, S.A. Ag(I)-π interactions with pyrrolic derivatives. *Coord. Chem. Rev.* **2019**, *380*, 318–329. [CrossRef]

43. Addition, A.W.; Nageswara, R.T.; Reedjik, J.; van Rijn, J.; Verschoor, G.C. Synthesis, structure and spectroscopic properties of copper(II) compounds containing nitrogen–sulphur donor ligands; the crystal and molecular structure of aqua[1,7-bis(*N*-methylbenzimidazol-2-yl)-2,6-dithiaheptane]copper(II) perchlorate. *J. Chem. Soc. Dalton Trans.* **1984**, 1349–1356. [CrossRef]

44. Yang, L.; Powell, D.R.; Houser, R.P. Structural variation in copper(I) complexes with pyridylmethylamide ligands: Structural analysis with a new four-coordinate geometry index, τ_4. *Dalton Trans.* **2007**, 955–964. [CrossRef] [PubMed]

45. Casanova, D.; Llunell, M.; Alemany, P.; Alvarez, S. The Rich Stereochemistry of Eight-Vertex Polyhedra: A Continuous Shape Measures Study. *Chem. Eur. J.* **2005**, *11*, 1479–1494. [CrossRef]

46. Llunell, M.; Casanova, D.; Cirera, J.; Bofill, J.M.; Alemany, P.; Alvarez, S.; Pinsky, M.; Avnir, D. *SHAPE*; Version 2.3; University of Barcelona: Barcelona, Spain; Hebrew University of Jerusalem: Jerusalem, Israel, 2013.

47. Li, M.; Lu, H.-Y.; Liu, R.-L.; Chen, J.-D.; Chen, C.-F. Turn-On Fluorescent Sensor for Selective Detection of Zn^{2+}, Cd^{2+} and Hg^{2+} in Water. *J. Org. Chem.* **2012**, *77*, 3670–3673. [CrossRef]

48. Ducloiset, C.; Jouin, P.; Paredes, E.; Guillot, R.; Sircoglou, M.; Orio, M.; Leibl, W.; Aukauloo, A. Monoanionic Dipyrrin–Pyridine Ligands: Synthesis, Structure and Photophysical Properties. *Eur. J. Inorg. Chem.* **2015**, 5405–5410. [CrossRef]

49. Kumar, R.S.; Kumar, S.K.A.; Vijayakrishna, K.; Sivaramakrishna, A.; Paira, P.; Rao, C.V.S.B.; Sivaraman, N.; Sahoo, S.K. Bipyridine bisphosphonate-based fluorescent optical sensor and optode for selective detection of Zn^{2+} ions and its applications. *New J. Chem.* **2018**, *42*, 8494–8502. [CrossRef]

50. Diana, R.; Panunzi, B. The Role of Zinc(II) Ion in Fluorescence Tuning of Tridentate Pincers: A Review. *Molecules* **2020**, *25*, 4984. [CrossRef] [PubMed]

51. Stetsiuk, O.; Abhervé, A.; Avarvari, N. 1,2,4,5-Tetrazine based ligands and complexes. *Dalton Trans.* **2020**, *49*, 5759–5777. [CrossRef]

chemistry

MDPI

Article

Functionalised Terpyridines and Their Metal Complexes—Solid-State Interactions [†]

Young Hoon Lee [1], Jee Young Kim [2], Sotaro Kusumoto [3], Hitomi Ohmagari [4], Miki Hasegawa [4], Pierre Thuéry [5], Jack Harrowfield [6,*], Shinya Hayami [3] and Yang Kim [3]

[1] Department of Chemistry, University of Ulsan, Ulsan 44610, Korea; dasis75@daum.net
[2] Department of Chemistry & Advanced Materials, Kosin University, 194 Wachiro, Yongdo-Gu, Busan 49104, Korea; pinkjy0117@hanmail.net
[3] Department of Chemistry, Graduate School of Science and Technology, Kumamoto University, 2-39-1 Kurokami, Chuo-ku, Kumamoto 860-8555, Japan; 185d9042@st.kumamoto-u.ac.jp (S.K.); hayami@kumamoto-u.ac.jp (S.H.); ykim@kumamoto-u.ac.jp (Y.K.)
[4] Department of Chemistry & Biological Science, College of Science & Engineering, Aoyama Gakuin University, Sagamihara, Kanagawa 252-5258, Japan; ohmagari@chem.aoyama.ac.jp (H.O.); hasemiki@chem.aoyama.ac.jp (M.H.)
[5] Université Paris-Saclay, CEA, CNRS, NIMBE, 91191 Gif-sur-Yvette, France; pierre.thuery@cea.fr
[6] ISIS, Université de Strasbourg, 8 allée Gaspard Monge, 67083 Strasbourg, France
[*] Correspondence: harrowfield@unistra.fr; Tel.: +33-03-6885-5165
[†] Dedicated to Professor Christoph Janiak in recognition of his seminal contributions to the study of weak interactions in crystalline coordination complexes.

check for updates

Citation: Lee, Y.H.; Kim, J.Y.; Kusumoto, S.; Ohmagari, H.; Hasegawa, M.; Thuéry, P.; Harrowfield, J.; Hayami, S.; Kim, Y. Functionalised Terpyridines and Their Metal Complexes—Solid-State Interactions. *Chemistry* **2021**, *3*, 199–227. https://doi.org/10.3390/chemistry3010016

Received: 12 January 2021
Accepted: 27 January 2021
Published: 5 February 2021

Publisher's Note: MDPI stays neutral with regard to jurisdictional claims in published maps and institutional affiliations.

Abstract: Analysis of the weak interactions within the crystal structures of 33 complexes of various 4′-aromatic derivatives of 2,2′:6′,2″-terpyridine (**tpy**) shows that interactions that exceed dispersion are dominated, as expected, by cation···anion contacts but are associated with both ligand–ligand and ligand–solvent contacts, sometimes multicentred, in generally complicated arrays, probably largely determined by dispersion interactions between stacked aromatic units. With V(V) as the coordinating cation, there is evidence that the polarisation of the ligand results in an interaction exceeding dispersion at a carbon bound to nitrogen with oxygen or fluorine, an interaction unseen in the structures of M(II) (M = Fe, Co, Ni, Cu, Zn, Ru and Cd) complexes, except when 1,2,3-trimethoxyphenyl substituents are present in the 4′-**tpy**.

Keywords: terpyridines; metal complexes; crystal structures; Hirshfeld surfaces; weak interactions

1. Introduction

Ligands based on the 2,2′:6′,2″-terpyridine (**tpy**) substructure have been employed in an extraordinary variety of studies concerning their metal-ion coordination chemistry [1–7]. Much of this chemistry has concerned condensed phases—crystals, liquid crystals, monolayers on surfaces, intercalation complexes, and films—where interactions between complex ion units and between them and their environment can be particularly important, as is well-known for 2,2′:6′,2″-terpyridine itself [8]. The postulate of the "terpyridine embrace" [9–12] as a means of rationalising the forms of the lattices of many crystalline terpyridine complexes provides a long-known example of the recognition of this fact. This form of supposed aromatic–aromatic interaction, however, is known not to be universally operative for such complexes [11,13,14], and more importantly, it is well recognised that terms, such as "π-stacking" and "π–π interactions" (as well as "OFF" = offset face-to-face and "EF" = edge-to-face aromatic interactions), widely applied to the description of putative interactions between aromatic entities [15–17] can be rather eclectically interpreted and that there may be infrequent reason to attribute special significance to the association of small aromatic systems beyond the recognition of dispersion interactions [18–24]. An even broader issue is that of whether intermolecular interactions should be considered parallel to molecular

85

bonding in being the result of the combination of proximal two-centre interactions [25–27]. Thus, the dissection of a crystal lattice in terms of atom–atom "contacts", which may or may not be indicative of attractive, labile interactions, is not a straightforward task [25–32], so in the work now reported, we crystallographically characterised a series of complexes of related 4′ derivatives of 2,2′:6′,2″-terpyridine (Figure 1) in the hope that systematic structural variations might facilitate the identification of major interactions. To assess the interactions within the lattices, we employed the calculation of Hirshfeld surfaces [29] using the program CrystalExplorer [30]. This work is a complement to other structural studies of functionalised terpyridines and their complexes that we have reported [14,33–37] and provides results to be integrated with the present and with the extant literature in general. The ultimate objective, in general, is, of course, to use such information in the design and synthesis of functional materials [38].

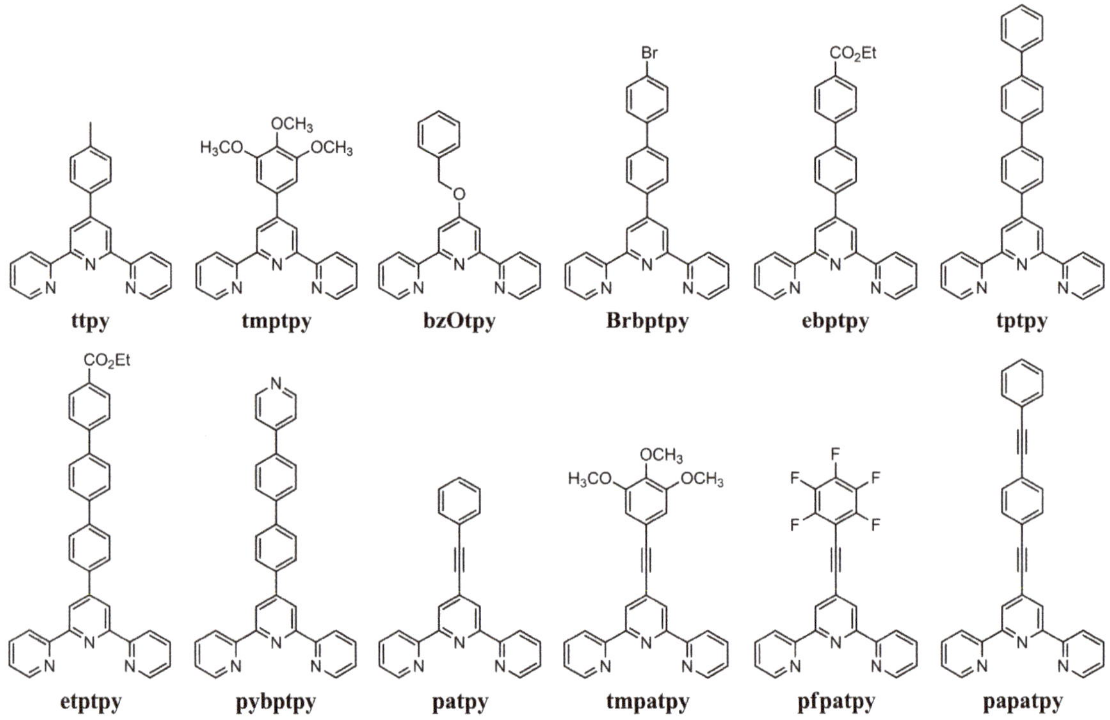

Figure 1. 4′-substituted derivatives of 2,2′:6′,2″-terpyridine studied in the present work.

2. Experimental Section

2.1. General

Dichloromethane and acetonitrile (spectrophotometric grade) for the measurement of electronic absorption and emission spectra were purchased from Merck (Seoul, S. Korea). Other organic solvents were purchased from Aldrich (Basel, Switzerland). Reagents were purchased from Wako (Osaka, Japan) (pyridin-4-yl-4-boronic acid and 4-(ethoxycarbonyl) phenylboronic acid), Tokyo Chemical Industry (Tokyo, Japan) (benzylbromide, 2,2′:6′,2″-terpyridine (**tpy**), 1-bromo-4-iodobenzene, Pd(PPh$_3$)$_4$ and 1,2,3,4,5-pentafluoro-6- iodobenzene), Alfa Aesar (Sulzbach, Germany) ((trimethylsilyl)acetylene and 1-ethynylbenzene) and Aldrich (Basel, Switzerland) (diisopropylamine, CuI, ZnCl$_2$, CdCl$_2$, RuCl$_3$·xH$_2$O, Co(BF$_4$)$_2$·6H$_2$O, Fe(ClO$_4$)$_2$·H$_2$O, Ni(ClO$_4$)$_2$·6H$_2$O, Cu(ClO$_4$)$_2$·6H$_2$O, Zn(ClO$_4$)$_2$·6H$_2$O, Cd(ClO$_4$)$_2$·H$_2$O and K$_2$CO$_3$) and used without further purification. Elemental analyses (C, H and N) were carried out at the Instrumental Analysis Centre of Kumamoto University, Japan. ^{1}H-NMR spectra were recorded with JEOL 500-ECX (500 MHz) and Bruker 300 AM spectrometers (300.13 MHz) in deuterated solvents using TMS as an internal reference. Electronic absorption spectra were recorded on SCINCO S-2100 and Varian Cary 100 spectrophotometers. Fluorescence spectra were recorded on Perkin Elmer LS55 and HORIBA FluoroMax-4P instruments. 1-bromo-4-(2-phenylethynyl)benzene [39], 2,2′:6′,2″-terpyridine-4′-ol [40], 2,6-bis(pyridin-2-yl)pyridin-4-yl trifluoromethanesulfonate [41], 2-(4-ethynyl-6-(pyridin-2-yl)pyridin-2-yl)pyridine (4′-ethynyl-2,2′:6′,2″-terpyridine) [42], 4′-(biphenyl-4-yl)-2,2′:6′,2″-terpyridine (**bptpy**) [43], 4′-(tolyl)-2,2′:6′,2″-terpyridine (**ttpy**) [44], 4′-(3,4,5-trimethoxy-phenyl)-2,2′:6′,2″-terpyridine (**tmptpy**) [45], 2-(4-(2-phenylethynyl)-6-(pyridin-2-yl)pyridin-2-yl)pyridine (**patpy**) [46–48], 4′-((ethoxycarbonyl)biphenyl-4-yl)-2,2′:6′,2″-terpyridine (**ebptpy**) [49], 4′-((ethoxycarbonyl)terphenyl-4-yl)-2,2′:6′,2″-terpyridine (**etptpyb**) [33], 4′-(benzyloxy)-2,2′:6′,2″-terpyridine, (**bzOtpy**) [33], 4′-(pentafluorophenyl)-2,2′:6′,2″-terpyridine (**pfpatpy**) [33], 4′-terphenyl-2,2′:6′,2″-terpyridine (**tptpy**) [34], 4′-(4′-pyridin-4-yl-biphenyl-4-yl)-2,2′:6′,2″-terpyridine (**pybptpy**) [35], 4′-(4-bromobiphenyl-4-yl)-2,2′:6′,2″-terpyridine (**Brbptpy**) [36], 2-(4-(2-(3,4,5-trimethoxyphenyl)ethynyl)-6-(pyridin-2-yl)pyridin-2-yl)pyridine (**tmpatpy**) [37] and [Ru(**tpy**)Cl$_3$] [50] were prepared by literature methods. The synthesis of the ligand 4′-(4‴-(ethynylphenyl)phenylethynyl)-2,2′:6′,2″-terpyridine (**papatpy**) is described below along with that of its complexes.

2.2. Synthesis

CAUTION! Most of the complexes described were isolated as perchlorate salts and are thus potential explosives. Although no difficulties were encountered, such materials should always be prepared in the minimum possible quantities.

2.2.1. Vanadium(V) Complex of 4′-(Biphenyl)-2,2′:6′,2″-terpyridine, Bptpy

(**1**) **[VO$_2$(bptpy)]ClO$_4$: Bptpy** (120 mg, 0.3 mmol) was dissolved in CHCl$_3$/CH$_3$OH (20 mL, 1:1, *v/v*) and VOSO$_4$·3H$_2$O (65 mg, 0.3 mmol) in methanol (10 mL), added with stirring. Under exposure to a normal atmosphere, the initially blue solution slowly changed colour to yellow–green over several hours, after which NaClO$_4$ (0.4 g) was dissolved in it, and the solution was allowed to slowly evaporate under ambient temperature to produce yellow crystals suitable for an X-ray diffraction study (yield: 90 mg, 53%). Anal. calc. for C$_{27}$H$_{19}$ClN$_3$O$_6$V: C, 57.11; H, 3.37; N, 7.40. Found: C, 56.88; H, 3.50; N, 7.26%. CCDC number, 2055133.

2.2.2. Complexes of 4′-Terphenyl-2,2′:6′,2″-terpyridine, Tptpy (and Mixed Complexes with 2,2′:6′,2″-Terpyridine)

(**2**) **[Fe(tptpy)$_2$](ClO$_4$)$_2$·CH$_3$OH**: Fe(ClO$_4$)$_2$·6H$_2$O (9.0 mg, 0.025 mmol) in DMF (5 mL) was added to a solution of **tptpy** (30 mg, 0.065 mmol) in hot DMF (5 mL). After stirring the solution for 2 h at 100 °C, the solvent was removed under reduced pressure. The deep purple residue was washed with methanol and chloroform, and dried at ambient temperature (yield: 25 mg, quantitative). Anal. calc. for C$_{67}$H$_{49}$Cl$_2$N$_6$FeO$_9$: C, 66.57; H, 4.09; N, 6.95. Found: C, 66.52; H, 4.07; N, 6.98%. ^{1}H-NMR (300 MHz) in CD$_3$CN: δ 9.26 (s, 4H), 8.66 (d, *J* = 8.0 Hz, 4H), 8.46 (d, *J* = 8.0 Hz, 4H), 8.17 (d, *J* = 8.0 Hz, 4H), 7.99 (d, *J* = 8.0 Hz, 4H), 7.94 (t, *J* = 8.5 Hz, 4H), 7.88 (d, *J* = 8.0 Hz, 4H), 7.78 (d, *J* = 7.5 Hz, 4H), 7.55 (t, *J* = 7.5 Hz, 4H), 7.45 (t, *J* = 7.5 Hz, 2H), 7.22 (d, *J* = 5.5 Hz, 4H), 7.11 (t, *J* = 7.0 Hz, 4H). The vapour diffusion of methanol into a solution of the complex in a minimum volume of DMF produced purple crystals suitable for X-ray diffraction studies. CCDC number, 2055134.

(**3**) **[Ni(tptpy)$_2$](ClO$_4$)$_2$·CH$_3$OH**: Ni(ClO$_4$)$_2$·6H$_2$O (8.0 mg, 0.022 mmol) in methanol (5 mL) was added to a solution of **tptpy** (20 mg, 0.043 mmol) in CHCl$_3$/CH$_3$OH (6 mL, 1:1, *v/v*). After stirring the solution for 1 h at 60 °C, the solvent was removed under reduced pressure. The yellow residue was washed with methanol and chloroform, and then dried at ambient temperature (yield: 20 mg, 91%). Anal. calc. for C$_{67}$H$_{49}$Cl$_2$N$_6$NiO$_9$: C, 66.41; H, 4.08; N, 6.94. Found: C, 66.50; H, 4.02; N, 7.01%. The vapour diffusion of methanol into a solution of the complex in a minimum volume of DMF produced yellow crystals suitable for X-ray diffraction studies. CCDC number, 2055135.

(4) [Ru(tptpy)$_2$](PF$_6$)$_2$·3CH$_3$CN: A mixture of **tptpy** (100 mg, 0.22 mmol) and RuCl$_3$·xH$_2$O (50 mg, 0.22 mmol) in ethanol (30 mL) was heated at 70 °C for 6 h to produce a turbid, reddish solution. After cooling to ambient temperature, the brown precipitate formed was collected by filtration, washed with cold ethanol and then dried under vacuum. The elemental analyses were consistent with its formulation as [Ru(tptpy)Cl$_3$] (yield: 80 mg). [Ru(**tptpy**)Cl$_3$] (30 mg, 0.045 mmol), **tptpy** (30 mg, 0.065 mmol) and *N*-ethylmorpholine (five drops) were added to methanol/chloroform (30 mL, 1:1, *v/v*), and the mixture was heated at 70 °C for 2 h. The resulting deep red solution was filtered through Celite to remove a small amount of purple precipitate. The filtrate was loaded on a short silica column, and the column eluted with acetonitrile/saturated aqueous KNO$_3$/water at 7:1:0.5, *v/v*. The major red component was collected, and methanolic NH$_4$PF$_6$ was added to the eluate to produce a red powder within 1 h. This was collected by filtration, washed with methanol and then dried under vacuum (yield: 30 mg, 50%). Anal. calc. for C$_{72}$H$_{55}$F$_{12}$N$_9$P$_2$Ru: C, 60.17; H, 3.86; N, 8.77. Found: C, 60.22; H, 3.83; N, 8.80%. Electronic absorption spectrum (in CH$_3$CN): λ_{max} (ε_{max}/M^{-1} cm^{-1}) = 285 nm (58,900), 312 nm (75,500), 330 nm (72,300), 494 nm (32,100). ^{1}H-NMR (500 MHz) in CD$_3$CN: δ 9.08 (s, 4H), 8.69 (d, *J* = 7.5 Hz, 4H), 8.35 (d, *J* = 8.0 Hz, 4H), 8.13 (d, *J* = 8.0 Hz, 4H), 7.98 (m, 8H), 7.87 (d, *J* = 7.5 Hz, 4H), 7.54 (t, *J* = 8.0 Hz, 4H), 7.46 (m, 6H), 7.21 (t, *J* = 6.5 Hz, 4H). The slow evaporation of an acetonitrile solution of the complex at ambient temperature produced red crystals suitable for X-ray structure determination. CCDC number, 2055219.

(5) [Ru(tptpy)(tpy)](PF$_6$)$_2$·2DMF: Tptpy (60 mg, 0.13 mmol), [Ru(tpy)Cl$_3$] (50 mg, 0.11 mmol) and N-ethylmorpholine (five drops) were added to methanol/chloroform (30 mL, 1:1, *v/v*), and the mixture was heated at 70 °C for 2 h. The resulting deep red solution was filtered through Celite to remove a small amount of brown precipitate, and the filtrate was subjected to chromatography on silica using acetonitrile/saturated aqueous KNO$_3$/water at 7:1:0.5, *v/v*, as the eluant. The major red component was collected, and methanolic NH$_4$PF$_6$ was added to the eluate to produce a red precipitate, which was filtered and washed with methanol, and then dried in a vacuum (yield: 23 mg, 20%). Anal. calc. for C$_{54}$H$_{48}$F$_{12}$N$_8$O$_2$P$_2$Ru: C, 52.64; H, 3.93; N, 9.10. Found: C, 52.55; H, 3.85; N, 9.05%. IR spectrum (KBr disc, cm^{-1}): 3106, 3038, 3030 (Ar C−H), 863 (PF$_6$). Electronic absorption spectrum (in CH$_3$CN): λ_{max} (ε_{max}, M^{-1} cm^{-1}) = 275 nm (43,700), 282 nm (43,900), 309 nm (73,900), 485 nm (24,800). ^{1}H-NMR (500 MHz) in CD$_3$CN: δ 9.06 (s, 2H), 8.75 (d, *J* = 8.0 Hz, 2H), 8.67 (d, *J* = 8.0 Hz, 2H), 8.50 (d, *J* = 8.5 Hz, 2H), 8.42 (t, *J* = 8.5 Hz, 1H), 8.33 (d, *J* = 8.0 Hz, 2H), 8.12 (d, *J* = 8.0 Hz, 2H), 7.96 (m, 7H), 7.86 (d, *J* = 8.0 Hz, 2H), 7.77 (d, *J* = 7.5 Hz, 2H), 7.54 (t, *J* = 7.5 Hz, 2H), 7.44 (d, *J* = 5.0 Hz, 2H), 7.35 (d, *J* = 5.5 Hz, 2H), 7.18 (m, 4H). Upon dissolving the complex in hot DMF and then allowing the solution to stand for one week at ambient temperature, red crystals suitable for X-ray structure determination were obtained. CCDC number, 2055220.

(6) [Ru(tptpy)(tpy)](ClO$_4$)$_2$·2DMF: [Ru(**tptpy**)(tpy)](PF$_6$)$_2$ (20 mg, 0.018 mmol) was dissolved in DMF (3 mL), and LiClO$_4$ (0.01 g) was added. The slow deposition of red crystals suitable for X-ray structure determination occurred upon leaving the solution to stand at ambient temperature. Anal. calc. for C$_{54}$H$_{48}$Cl$_2$N$_8$O$_{10}$Ru: C, 56.84; H, 4.24; N, 9.82. Found: C, 56.78; H, 4.27; N, 9.78%. Electronic absorption spectrum in CH$_3$CN: λ_{max} (ε_{max}, M^{-1} cm^{-1}) = 274 nm (44,000), 281 nm (43,300), 308 nm (71,400), 484 nm (23,600). ^{1}H-NMR (500 MHz) in CD$_3$CN: δ 9.06 (s, 2H), 8.75 (d, *J* = 8.5 Hz, 2H), 8.67 (d, *J* = 8.0 Hz, 2H), 8.50 (d, *J* = 8.0 Hz, 2H), 8.42 (t, *J* = 8.0 Hz, 1H), 8.33 (d, *J* = 8.5 Hz, 2H), 8.12 (d, *J* = 8.0 Hz, 2H), 7.96 (m, 7H), 7.86 (d, *J* = 8.0 Hz, 2H), 7.77 (d, *J* = 8.5 Hz, 2H), 7.54 (t, *J* = 7.5 Hz, 2H), 7.44 (d, *J* = 6.0 Hz, 2H), 7.35 (d, *J* = 5.0 Hz, 2H), 7.18 (m, 4H). CCDC number, 2055221.

2.2.3. Zinc(II) Complex of 2-(4-(2-Phenylethynyl)-6-(pyridin-2-yl)yyridin-2-yl) pyridine, Patpy

(7) [Zn(patpy)$_2$](ClO$_4$)$_2$ (YK758): Zn(ClO$_4$)$_2$·6H$_2$O (30 mg, 0.08 mmol) and **patpy** (50 mg, 0.15 mmol) were dissolved in acetonitrile (10 mL) and heated at 70 °C for 2 h. After cooling to ambient temperature, the solvent was removed by evaporation under reduced pressure. The colourless residue was washed with methanol and dichloromethane and dried at ambient temperature (yield: 39 mg, 53%). Anal. calc. for $C_{46}H_{30}Cl_2N_6ZnO_8$: C, 59.34; H, 3.25; N, 9.03. Found: C, 59.41; H, 3.32; N, 9.20%. ^{1}H-NMR (300 MHz) in CD_3CN: δ 8.90 (s, 4H), 8.61 (d, J = 8.0 Hz, 4H), 8.22 (td, J = 7.9, 1.6 Hz, 4H), 7.86 (m, 8H), 7.62 (m, 6H), 7.46 (dd, J = 5.1, 0.9 Hz, 2H), 7.43 (dd, J = 5.1, 0.9 Hz, 2H). The slow evaporation of an acetonitrile solution of the complex at ambient temperature provided colourless crystals suitable for X-ray diffraction studies. CCDC number, 2055222.

2.2.4. Complexes of 4′-(4‴-(Ethynylphenyl)phenylethynyl)-2,2′:6′,2″-terpyridine, Papatpy

(8) papatpy: 2-(4-ethynyl-6-(pyridin-2-yl)pyridin-2-yl)pyridine (0.25 g, 0.97 mmol), 1-bromo-4-(2-phenylethynyl)benzene (0.30 g, 1.12 mmol), Pd(PPh$_3$)$_4$ (0.10 g, 0.086 mmol) and CuI (0.020 g, 0.11 mmol) were placed into a three-necked round-bottom flask under a nitrogen atmosphere. Degassed i-Pr$_2$NH/THF (30 mL, 1:1, v/v) was added to the mixture and heated under reflux for 12 h under nitrogen. The resultant dark brown, turbid solution was filtered, and the filter was washed with THF. The filtrate and wash solution were evaporated to dryness under reduced pressure, and then dissolved in a minimum volume of dichloromethane. Adsorption on a short silica column and elution with ethylacetate/CH$_2$Cl$_2$ at 1:4, v/v, produced a colourless eluate, which was evaporated under reduced pressure to produce a white powder (yield: 0.20 g). ^{1}H-NMR (300 MHz) in CDCl$_3$: δ 8.78 (dq, J = 4.1, 0.8 Hz, 2H), 8.69 (d, J = 8.0 Hz, 2H), 8.12 (s, 2H), 7.96 (td, J = 7.8, 1.8 Hz, 2H), 7.59 (m, 6H), 7.43 (m, 5H). Anal. calc. for $C_{31}H_{19}N_3 \cdot 0.5H_2O$: C, 84.14; H, 4.56; N, 9.50. Found: C, 83.70; H, 4.48; N, 9.33%.

(9) [Co(papatpy)$_2$](BF$_4$)$_3 \cdot$2CH$_3$CN: Co(BF$_4$)$_2 \cdot$6H$_2$O (0.0080 g, 0.023 mmol) in DMF (5 mL) was mixed with **papatpy** (20 mg, 0.046 mmol) in DMF (10 mL). After heating this solution for 2 h at 100 °C, the solvent was removed under reduced pressure to produce a brown residue, which was washed with methanol and chloroform, and then dried at ambient temperature (yield: 20 mg, 69%). Anal. calc. for $C_{66}H_{44}B_3CoF_{12}N_8$: C, 62.49; H, 3.50; N, 8.83. Found: C, 62.55; H, 3.42; N, 8.90%. The vapour diffusion of diethyl ether into an acetonitrile solution of the complex provided yellow, rod-like crystals suitable for X-ray diffraction studies. The structure determination and colour of the complex showed that the oxidation of Co(II) to Co(III) had occurred during the synthesis. CCDC number, 2055223.

(10) [Ni(papatpy)$_2$](ClO$_4$)$_2 \cdot$3DMF: Ni(ClO$_4$)$_2 \cdot$6H$_2$O (9.0 mg, 0.025 mmol) in DMF (5 mL) was mixed with **papatpy** (20 mg, 0.046 mmol) in DMF (10 mL). After heating this solution for 2 h at 100 °C, the solvent was removed under reduced pressure to produce a yellow residue, which was washed with methanol and chloroform, and then dried at ambient temperature (yield: 20 mg, 63%). Anal. calc. for $C_{71}H_{59}Cl_2N_9NiO_{11}$: C, 63.45; H, 4.43; N, 9.38. Found: C, 63.56; H, 4.42; N, 9.42%. The slow evaporation of a DMF solution of the complex at ambient temperature led to the deposition of yellow, rod-like crystals suitable for X-ray diffraction studies. CCDC number, 2055137.

(11) [Cu(papatpy)$_2$](ClO$_4$)$_2 \cdot$2CH$_3$CN$\cdot$0.5H$_2$O: Cu(ClO$_4$)$_2 \cdot$6H$_2$O (9.0 mg, 0.024 mmol) in DMF (5 mL) was mixed with **papatpy** (20 mg, 0.046 mmol) in DMF (10 mL). After heating this solution for 2 h at 100 °C, the solvent was removed under reduced pressure. The green residue was washed with methanol and chloroform, and then dried at ambient temperature (yield: 19 mg, 66%). Anal. calc. for $C_{132}H_{90}Cl_4Cu_2N_{16}O_{17}$: C, 64.95; H, 3.72; N, 9.18. Found: C, 65.06; H, 3.52; N, 9.22%. The slow evaporation of an acetonitrile solution of the complex at ambient temperature led to the deposition of green, block-like crystals suitable for X-ray diffraction studies. CCDC number, 2055138.

(12) [Zn(papatpy)$_2$](ClO$_4$)$_2 \cdot$3DMF: Zn(ClO$_4$)$_2 \cdot$6H$_2$O (9.0 mg, 0.023 mmol) in DMF (5 mL) was mixed with **papatpy** (20 mg, 0.046 mmol) in DMF (10 mL). After heating this solution for 2 h at 100 °C, the solvent was removed under reduced pressure to produce a colourless residue, which was washed with methanol and chloroform, and then dried at ambient temperature (yield: 21 mg, 68%). Anal. calc. for $C_{71}H_{59}Cl_2N_9ZnO_{11}$: C, 63.14; H, 4.40; N, 9.33. Found: C, 63.10; H, 4.45; N, 9.30%. ^{1}H-NMR (300 MHz; CD$_3$CN): δ = 8.90 (s,

4H), 8.60 (d, *J* = 9.0 Hz, 4H), 8.23 (td, *J* = 7.8, 1.5 Hz, 4H), 7.86 (m, 8H), 7.76 (d, *J* = 8.5 Hz, 4H), 7.65 (m, 4H), 7.50 (m, 6H), 7.47 (dd, *J* = 5.1, 0.9 Hz, 2H), 7.43 (dd, *J* = 5.1, 1.0 Hz, 2H). The slow evaporation of a DMF solution of the complex at ambient temperature led to the deposition of colourless, rod-like crystals suitable for X-ray diffraction studies. CCDC number, 2055139.

(13) [Ru(papatpy)$_2$](PF$_6$)$_2$: A mixture of **papatpy** (40 mg, 0.092 mmol) and RuCl$_3$·xH$_2$O (0.020 g) in ethanol (30 mL) was heated at 70 °C for 5 h to produce a turbid, reddish-purple solution. The precipitate formed after cooling to ambient temperature was collected and washed on a filter with cold ethanol to produce a brown powder, taken to be [Ru(**papatpy**)Cl$_3$]. [Ru(**papatpy**)Cl$_3$] (25 mg, 0.039 mmol), **papatpy** (20 mg, 0.046 mmol) and *N*-ethylmorpholine (five drops) in CH$_3$OH/CHCl$_3$ (40 mL, 3:1, *v/v*) were heated at 70 °C for 4 h. The resultant deep red solution was filtered through Celite to remove some insoluble purple material, and the solvent was evaporated off under reduced pressure. The dark red residue was dissolved in a minimum volume of acetonitrile and subjected to chromatography on a short silica column using acetonitrile/saturated aqueous KNO$_3$/water at 7:1:0.5, *v/v*, as the eluent. The major red component was collected, and methanolic NH$_4$PF$_6$ was added to this eluate. After 1 h, the red precipitate was filtered off, washed with methanol and then dried under vacuum (yield: 23 mg, 47%). Anal. calc. for C$_{62}$H$_{38}$F$_{12}$N$_6$P$_2$Ru: C, 59.19; H, 3.04; N, 6.68. Found: C, 59.10; H, 3.12; N, 6.72%. ^{1}H-NMR (300 MHz) in CD$_3$CN: *δ* 8.92 (s, 4H), 8.55 (dq, *J* = 8.1, 0.4 Hz, 4H), 8.01 (td, *J* = 7.8, 1.5 Hz, 4H), 7.85 (d, *J* = 8.6 Hz, 4H), 7.76 (d, *J* = 8.6 Hz, 4H), 7.66 (m, 4H), 7.50 (m, 6H), 7.45 (dq, *J* = 5.5, 0.6 Hz, 4H), 7.25 (dd, *J* = 5.6, 1.3 Hz, 2H), 7.22 (dd, *J* = 5.6, 1.3 Hz, 2H). The slow evaporation of a solution of the complex in CH$_3$CN at ambient temperature produced red crystals suitable for X-ray structure determination. CCDC number, 2055140.

(14) [Ru(papatpy)(tpy)](PF$_6$)$_2$·3CH$_3$CN: [Ru(**tpy**)Cl$_3$] (50 mg, 0.11 mmol) and **papatpy** (60 mg, 0.14 mmol) were dissolved in methanol/chloroform (30 mL, 1:1, *v/v*), and *N*-ethylmorpholine (five drops) was added. The mixture was heated under reflux for 2 h and filtered through Celite to remove a small amount of brown precipitate. The deep red filtrate was chromatographed with a short silica-gel column using a mixed eluent (acetonitrile/saturated aqueous KNO$_3$/water at 7:1:0.5, *v/v*). The major red band was collected, and excess methanolic NH$_4$PF$_6$ was added. The red precipitate obtained was filtered, washed with methanol and dried under vacuum (yield: 28 mg, 22%). The product was dissolved in CH$_3$CN and left for one week at room temperature to afford red crystals suitable for X-ray diffraction studies. Anal. calc. for C$_{52}$H$_{39}$F$_{12}$N$_9$P$_2$Ru: C, 52.89; H, 3.33; N, 10.67. Found: C, 52.95; H, 3.20; N, 10.81%. ^{1}H-NMR (300 MHz, CD$_3$CN): *δ* = 8.90 (s, 2H), 8.78 (d, *J* = 8.2 Hz, 2H), 8.54–8.50 (m, 4H), 8.45 (t, *J* = 8.2 Hz, 1H), 7.94 (qd, *J* = 7.9, 1.5 Hz, 4H), 7.83 (d, *J* = 8.0 Hz, 2H), 7.74 (d, *J* = 8.5 Hz, 2H), 7.65–7.62 (m, 2H), 7.50–7.47 (m, 3H), 7.42–7.37 (m, 4H), 7.24–7.16 (m, 4H). CCDC number, 2055141.

(15) [Cd(papatpy)$_2$](ClO$_4$)$_2$·2CH$_3$CN·H$_2$O: Cd(ClO$_4$)$_2$·H$_2$O (8.0 mg, 0.026 mmol) in DMF (5 mL) was mixed with **papatpy** (20 mg, 0.046 mmol) in DMF (10 mL). After heating this solution for 2 h at 100 °C, the solvent was removed under reduced pressure. The colourless residue was washed with methanol and chloroform, and then dried at ambient temperature (yield: 18 mg, 55%). Anal. calc. for C$_{66}$H$_{46}$Cl$_2$N$_8$CdO$_9$: C, 62.01; H, 3.63; N, 8.76. Found: C, 61.97; H, 3.55; N, 8.82%. ^{1}H-NMR (300 MHz) in CD$_3$CN: *δ* 8.85 (s, 4H), 8.64 (d, *J* = 8.1 Hz, 4H), 8.26 (td, *J* = 7.8, 1.6 Hz, 4H), 8.13 (s, 4H), 7.82 (d, *J* = 8.4 Hz, 4H), 7.75 (d, *J* = 8.4 Hz, 4H), 7.65 (m, 4H), 7.56 (d, *J* = 5.1 Hz, 2H), 7.53 (d, *J* = 5.1 Hz, 2H), 7.49 (m, 6H). The vapour diffusion of diethyl ether into an acetonitrile solution of the complex provided colourless, rod-like crystals suitable for X-ray diffraction studies. CCDC number, 2055142.

(16) [Cd(papatpy)Cl$_2$]·CH$_2$Cl$_2$: CdCl$_2$ (13 mg, 0.071 mmol) in CH$_3$CN/CHCl$_3$ (10 mL, 1:1, *v/v*) was mixed with **papatpy** (30 mg, 0.069 mmol) in chloroform (5 mL). After heating this solution at 70 °C for 2 h, the solvent was removed under reduced pressure. The colourless residue was washed with methanol and chloroform, and then dried at ambient temperature (yield: 32 mg, 64%). Anal. calc. for C$_{32}$H$_{21}$Cl$_4$N$_3$Cd: C, 54.77; H, 3.02; N, 5.99. Found: C, 54.67; H, 3.11; N, 5.88%. ^{1}H-NMR (300 MHz) in DMSO-d$_6$: *δ* 8.79 (d, *J* = 4.3 Hz,

4H), 7.78 (d, J = 8.1 Hz, 4H), 7.73 (d, J = 8.4 Hz, 4H), 7.62 (m, 3H), 7.48 (m, 4H). The slow evaporation of a solution of the complex in CH_3CN/CH_2Cl_2 (1:1, *v/v*) produced colourless, rod-like crystals suitable for X-ray diffraction studies. CCDC number, 2055224.

2.2.5. Iron(II) Complex of 4′-(Pentafluorophenylethynyl)-2,2′:6′,2″-terpyridine, **pfpatpy**

(17) [Fe(pfpatpy)$_2$](ClO$_4$)$_2$·CH$_3$CN·H$_2$O: Fe(ClO$_4$)$_2$·6H$_2$O (9.0 mg, 0.025 mmol) in DMF (5 mL) was added to a solution of **pfpatpy** (30 mg, 0.065 mmol) in hot DMF (5 mL). After stirring for 2 h at 100 °C, the solvent was removed under reduced pressure, and the deep purple residue was washed with CH_3OH and $CHCl_3$, and then dried at room temperature (yield: 25 mg, 89%). It was dissolved in DMF, and CH_3CN vapour was diffused in to produce purple crystals suitable for X-ray diffraction studies. Anal. calc. for $C_{46}H_{20}Cl_2F_{10}FeN_6O_8$·0.5CH$_3$CN·H$_2$O: C, 49.56; H, 2.08; N, 7.99. Found: C, 49.98; H, 1.91; N, 7.70%. (There was a partial loss of CH_3CN by comparison with the crystal.) CCDC number, 2055225.

2.2.6. Complexes of 4′-(4′-Pyridin-4-yl-biphenyl-4-yl)-2,2′:6′,2″-terpyridine, **pybptpy**

(18) [Ni(pybptpy)$_2$](ClO$_4$)$_2$·2DMF·H$_2$O: Ni(ClO$_4$)$_2$·6H$_2$O (20 mg, 0.055 mmol) in DMF (5 mL) was added to **pybptpy** (50 mg, 0.11 mmol) in DMF (10 mL). After stirring the solution for 2 h at 100 °C, the solvent was removed under reduced pressure, and the yellow residue was washed with methanol and chloroform, and then dried at ambient temperature (yield: 54 mg, 82%). Anal. calc. for $C_{70}H_{60}Cl_2N_{10}NiO_{11}$: C, 62.42; H, 4.49; N, 10.40. Found: C, 62.4; H, 4.45; N, 10.21%. The vapour diffusion of methanol into a DMF solution of the complex produced yellow crystals suitable for X-ray diffraction studies. CCDC number, 2055226.

(19) [Cu(pybptpy)$_2$](ClO$_4$)$_2$·3DMF·H$_2$O: Cu(ClO$_4$)$_2$·6H$_2$O (20 mg, 0.055 mmol) in DMF (5 mL) was added to **pybptpy** (50 mg, 0.11 mmol) in DMF (10 mL). After stirring the solution for 2 h at 100 °C, the solvent was removed under reduced pressure, and the green residue was washed with methanol and chloroform, and then dried at ambient temperature (yield: 58 mg, 74%). Anal. calc. for $C_{73}H_{67}Cl_2CuN_{11}O_{12}$: C, 61.54; H, 4.74; N, 10.81. Found: C, 61.33; H, 4.82; N, 10.88%. The vapour diffusion of methanol into a DMF solution of the complex produced green crystals suitable for X-ray diffraction studies. CCDC number, 2055227.

(20) [Zn(pybptpy)$_2$](ClO$_4$)$_2$·2DMF·H$_2$O: Zn(ClO$_4$)$_2$·6H$_2$O (20 mg, 0.055 mmol) in DMF (5 mL) was added to **pybptpy** (50 mg, 0.11 mmol) in DMF (10 mL), and the solution was stirred for 2 h at 100 °C before the solvent was removed under reduced pressure. The white residue was washed with methanol and chloroform, and then dried at ambient temperature (yield: 49 mg, 63%). Anal. calc. for $C_{70}H_{60}Cl_2N_{10}O_{11}Zn$: C, 62.11; H, 4.47; N, 10.35. Found: C, 62.07; H, 4.55; N, 10.51%. ^{1}H-NMR (300 MHz) in DMSO-d$_6$: δ 9.50 (s, 4H), 9.23 (d, J = 8.2 Hz, 4H), 8.73 (d, J = 5.6 Hz, 4H), 8.63 (d, J = 7.8 Hz, 4H), 8.35 (t, J = 7.6 Hz, 4H), 8.22 (d, J = 8.2 Hz, 4H), 8.12 (d, J = 8.3 Hz, 4H), 8.05 (m, 8H), 7.86 (d, J = 5.8 Hz, 4H), 7.56 (t, J = 6.5 Hz, 4H). The vapour diffusion of methanol into a DMF solution of the complex produced colourless crystals suitable for X-ray diffraction studies. CCDC number, 2055228.

2.2.7. Complexes of 4′-(4-Bromobiphenyl-4-yl)-2,2′:6′,2″-terpyridine, Brbptpy

(21) [Co(Brbptpy)$_2$](BF$_4$)$_2$·2CH$_3$CN: Co(BF$_4$)$_2$·6H$_2$O (20 mg, 0.060 mmol) in dimethyl-formamide (DMF; 5 mL) was added to a solution of **Brbptpy** (50 mg, 0.11 mmol) in hot DMF (10 mL). After stirring for 2 h at 100 °C, the solvent was removed under reduced pressure. The brown residue was washed with methanol and dried at ambient temperature (yield: 0.048 g, 71%). Anal. calc. for $C_{58}H_{42}B_2Br_2CoF_8N_8$: C, 56.03; H, 3.40; N, 9.0. Found: C, 56.18; H, 3.36; N, 8.95%. The vapour diffusion of diethyl ether into a solution of the complex in a minimum volume of acetonitrile produced brown crystals suitable for X-ray diffraction studies. CCDC number, 2055148.

(22) [Ni(Brbptpy)₂](ClO₄)₂·DMF·H₂O: $Ni(ClO_4)_2 \cdot 6H_2O$ (40 mg, 0.12 mmol) in methanol (10 mL) was added to a solution of **Brbptpy** (100 mg, 0.22 mmol) in $CHCl_3/CH_3OH$ (10 mL, 1:1, *v/v*). After stirring the solution for 1 h at 60 °C, the solvent was removed under reduced pressure, and the yellow residue was washed with cold methanol and chloroform, and then dried at ambient temperature (yield: 100 mg, 68%). Anal. calc. for $C_{111}H_{79}Br_4$ $C_{14}N_{13}Ni_2O_{18}$: C, 54.16; H, 3.2; N, 7.40. Found: C, 54.32; H, 3.30; N, 7.45%. The vapour diffusion of methanol into a solution of the complex in a minimum volume of DMF produced yellow crystals suitable for X-ray diffraction studies. CCDC number, 2055149.

(23) [Cu(Brbptpy)₂](ClO₄)₂·2CH₃OH: $Cu(ClO_4)_2 \cdot 6H_2O$ (45 mg, 0.12 mmol) in methanol (10 mL) was added to a solution of **Brbptpy** (100 mg, 0.22 mmol) in $CHCl_3/CH_3OH$ (10 mL, 1:1, *v/v*). After stirring the solution for 1 h at 60 °C, the solvent was removed under reduced pressure, and the green residue was washed with cold methanol and chloroform, and then dried at ambient temperature (yield: 110 mg, 75%). Anal. calc. for $C_{56}H_{43}Br_2Cl_2CuN_6O_{10}$: C, 53.63; H, 3.46; N, 6.70. Found: C, 56.28; H, 3.38; N, 6.75%. The complex was dissolved in boiling methanol, and the solution was allowed to evaporate slowly at ambient temperature to provide brown crystals suitable for X-ray diffraction studies. CCDC number, 2055150.

(24) [Ru(Brbptpy)₂](PF₆)₂·3CH₃CN: With the substitution of **Brbptpy** for **tptpy**, the same procedure as used with complex 4 provided red crystals (with a 51% yield) of the complex suitable for structure determination. Anal. calc. for $C_{60}H_{45}Br_2F_{12}N_9P_2Ru$: C, 49.95; H, 3.14; N, 8.74. Found: C, 49.35; H, 3.30; N, 8.82%. Due to the loss of the sample, an NMR spectrum was not obtained. CCDC number, 2055151.

2.2.8. Complexes of 4-((Ethoxycarbonyl)biphenyl-4-yl)-2,2′:6′,2″-terpyridine, Ebptpy

(25) [Ni(ebptpy)₂](ClO₄)₂·3DMF: $Ni(ClO_4)_2 \cdot 6H_2O$ (8.0 mg, 0.022 mmol) in DMF (5 mL) was added to **ebptpy** (20 mg, 0.044 mmol) in DMF (10 mL). The resulting solution was heated for 2 h at 100 °C before the solvent was removed under reduced pressure. The yellow residue was washed with methanol and chloroform and then dried at ambient temperature (yield: 21 mg, 69%). Anal. calc. for $C_{69}H_{67}Cl_2N_9NiO_{15}$: C, 59.54; H, 4.85; N, 9.06. Found: C, 59.60; H, 4.53; N, 9.21%. The vapour diffusion of methanol into a DMF solution of the complex produced yellow, rod-like crystals suitable for X-ray diffraction studies. CCDC number, 2055152.

(26) [Cu(ebptpy)₂](ClO₄)₂·CH₃CN: $Cu(ClO_4)_2 \cdot 6H_2O$ (8.1 mg, 0.022 mmol) in DMF (5 mL) was added to **ebptpy** (20 mg, 0.044 mmol) in DMF (10 mL). The resulting solution was heated for 2 h at 100 °C before the solvent was removed under reduced pressure. The green residue was washed with methanol and chloroform, and dried at ambient temperature (yield: 0.018 g, 67%). Anal. calc. for $C_{62}H_{49}Cl_2N_7CuO_{12}$: C, 61.11; H, 4.05; N, 8.05. Found: C, 60.99; H, 4.02; N, 8.10%. The vapour diffusion of diethyl ether into an acetonitrile solution of the complex provided green, rod-like crystals suitable for X-ray diffraction studies. CCDC number, 2055153.

2.2.9. Zinc(II) Complex of 4-((Ethoxycarbonyl)terphenyl-4-yl)-2,2′:6′,2″-terpyridine, **etptpy**

(27) [Zn(etptpy)₂](ClO₄)₂·2DMF·H₂O: $Zn(ClO_4)_2 \cdot 6H_2O$ (7.0 mg, 0.016 mmol) in DMF (5 mL) was added to **etptpy** (20 mg, 0.037 mmol) in DMF (10 mL). After stirring the resulting solution for 2 h at 100 °C, the solvent was removed under reduced pressure. The white residue was washed with methanol and chloroform and then dried at ambient temperature (yield: 19 mg, 84%). Anal. calc. for $C_{75}H_{61}Cl_2N_7O_{13}Zn$: C, 64.13; H, 4.38; N, 6.98. Found: C, 64.08; H, 4.52; N, 6.70%. ^{1}H-NMR (300 MHz) in DMSO-d_6: δ 9.50 (s, 4H), 9.23 (d, J = 7.5 Hz, 4H), 8.62 (d, J = 7.5 Hz, 4H), 8.33 (t, J = 6.8 Hz, 4H), 8.21 (d, J = 7.7 Hz, 4H), 8.13 (t, J = 8.0 Hz, 8H), 7.99 (m, 12H), 7.55 (t, J = 5.5 Hz, 4H), 4.41 (q, J = 6.7 Hz, 4H, –CH_2–), 1.40 (t, J = 7.2 Hz, 6H, –CH_3). The vapour diffusion of methanol into a DMF solution of the complex produced pale-yellow, rod-like crystals suitable for X-ray diffraction studies. CCDC number, 2055154.

2.2.10. Complexes of 4′-(Benzyloxy)-2,2′:6′,2″-terpyridine, **bzOtpy**

(28) [Fe(bzOtpy)₂](ClO₄)₂: A mixture of Fe(ClO₄)₂·6H₂O (12 mg, 0.033 mmol) and **bzOtpy** (30 mg, 0.088 mmol) in CH₂Cl₂/CH₃OH (20 mL, 1:1, *v/v*) was heated under reflux for 2 h. Upon cooling, the solvent was evaporated off under reduced pressure. The purple residue was washed with methanol and dichloromethane and dried at ambient temperature (yield: 30 mg, 97%). Anal. calc. for C₄₄H₃₄Cl₂N₆FeO₁₀: C, 56.61; H, 3.67; N, 9.00. Found: C, 56.48; H, 3.89; N, 9.11%. ¹H-NMR (300 MHz) in CD₃CN: δ 8.60 (s, 4H), 8.47 (d, *J* = 7.8 Hz, 4H), 7.92 (t, *J* = 7.7 Hz, 4H), 7.77 (d, *J* = 7.5 Hz, 4H), 7.64 (m, 6H), 7.20 (d, *J* = 5.4 Hz, 4H), 7.13 (t, *J* = 6.4 Hz, 4H), 5.72 (s, 4H, –*CH₂*–). The slow evaporation of an acetonitrile solution of the complex at ambient temperature produced purple crystals suitable for X-ray diffraction studies. CCDC number, 2055155.

(29) [Ni(bzOtpy)₂](ClO₄)₂: Ni(ClO₄)₂·6H₂O (17 mg, 0.047 mmol) and **bzOtpy** (30 mg, 0.088 mmol) were dissolved in CH₂Cl₂/CH₃OH (20 mL, 1:1, *v/v*) and heated under reflux for 2 h. Upon cooling, the solvent was evaporated off under reduced pressure. The yellow residue was washed with methanol and dichloromethane and then dried at ambient temperature (yield: 36 mg, 82%). Anal. calc. for C₄₄H₃₄Cl₂N₆NiO₁₀: C, 56.44; H, 3.66; N, 8.98. Found: C, 56.70; H, 3.74; N, 9.08%. The slow evaporation of an acetonitrile solution of the complex at ambient temperature produced purple crystals suitable for X-ray diffraction studies. CCDC number, 2055156.

2.2.11. 4′-(3,4,5-Trimethoxy-phenyl)-2,2′:6′,2″-terpyridine, **tmptpy**, and its Zn(II) Complex

(30) Tmptpy [34] in its diprotonated form as the perchlorate salt, hemi-tetrahydrofuran solvate, [H₂tmptpy](ClO₄)₂·0.5THF: tmptpy (100 mg) was dissolved in THF (30 mL), and 1 mL of 1 M HClO₄ was added with stirring over 30 min, after which the solution was allowed to slowly evaporate under ambient temperature to produce colourless crystals suitable for structure determination. Anal. calc. for C₅₂H₅₄Cl₄N₆O₂₃: C, 49.07; H, 4.28; N, 6.60; Found: C, 48.54; H, 4.31; N, 6.48%. CCDC number, 2055157.

(31) [Zn(tmptpy)₂](ClO₄)₂·CH₃CN: Zn(ClO₄)₂·6H₂O (7.0 mg, 0.016 mmol) in DMF (5 mL) was added to **tmptpy** (20 mg, 0.037 mmol) in DMF (10 mL). The solvent was removed under reduced pressure after stirring the solution for 2 h at 100 °C. The white residue was washed with methanol and chloroform and then dried at ambient temperature (yield: 19 mg, quantitative). Anal. calc. for C₅₀H₄₅Cl₂N₇O₁₄Zn: C, 54.39; H, 4.11; N, 8.88. Found: C, 54.44; H, 4.08; N, 8.96%. ¹H-NMR (300 MHz) in CD₃CN: δ 9.00 (s, 4H), 8.82 (dt, *J* = 8.1, 0.8 Hz, 4H), 8.23 (td, *J* = 7.8, 1.7 Hz, 4H), 7.87 (dq, *J* = 5.1, 0.6 Hz, 4H), 7.46 (dd, *J* = 5.1, 1.0 Hz, 2H), 7.44 (s, 4H), 7.43 (dd, *J* = 5.1, 1.0 Hz, 2H), 4.10 (s, 12H, –*CH₃*), 3.92 (s, 6H, –*CH₃*). The vapour diffusion of methanol into a DMF solution of the complex produced pale-yellow, rod-like crystals suitable for X-ray diffraction studies. CCDC number, 2055158.

2.2.12. Complexes of 2-(4-(2-(3,4,5-Trimethoxyphenyl)ethynyl)-6-(pyridin-2-yl)pyridin-2-yl)pyridine, **tmpatpy**

(32) [Ni(tmpatpy)₂](ClO₄)₂·CH₃CN·H₂O: Ni(ClO₄)₂·6H₂O (9.0 mg, 0.025 mmol) in acetonitrile (5 mL) was added to **tmpatpy** (20 mg, 0.047 mmol) in CH₃CN/CHCl₃ (10 mL, 1:1, *v/v*). After stirring the resulting solution for 1 h at 60 °C, the solvent was removed under reduced pressure. The yellow residue was washed with methanol and chloroform and then dried at ambient temperature (yield: 21 mg, 57%). Anal. calc. for C₁₀₆H₈₉Cl₄N₁₃O₂₉Ni₂: C, 56.13; H, 3.96; N, 8.03. Found: C, 56.22; H, 3.98; N, 8.11%. The slow evaporation of an acetonitrile solution of the complex at ambient temperature provided yellow, rod-like crystals suitable for X-ray diffraction studies. CCDC number, 2055159.

(33) [Zn(tmpatpy)₂](ClO₄)₂·CH₃CN·3H₂O: Zn(ClO₄)₂·6H₂O (9.0 mg, 0.024 mmol) in acetonitrile (5 mL) was added to **tmpatpy** (20 mg, 0.047 mmol) in CH₃CN/CHCl₃ (10 mL, 1:1, *v/v*). After stirring the resulting solution for 1 h at 60 °C, the solvent was removed under reduced pressure, and the bright yellow residue was washed with methanol and chloroform, and then dried at ambient temperature (yield: 22 mg, 76%). Anal. calc. for C₅₄H₅₁Cl₂N₇O₁₇Zn: C, 53.77; H, 4.26; N, 8.13. Found: C, 53.57; H, 4.20; N, 8.18%. ¹H-NMR

(300 MHz) in CD$_3$CN: δ 8.86 (s, 4H), 8.60 (dt, J = 8.1, 0.8 Hz, 4H), 8.22 (td, J = 7.7, 1.6 Hz, 4H), 7.86 (dq, J = 5.1, 0.8 Hz, 4H), 7.46 (dd, J = 5.1, 1.0 Hz, 2H), 7.43 (dd, J = 5.1, 0.9 Hz, 2H), 7.10 (s, 4H), 3.95 (s, 12H, –CH$_3$), 3.85 (s, 6H, –CH$_3$). The slow evaporation of an acetonitrile solution of the complex at ambient temperature provided pale yellow, rod-like crystals suitable for X-ray diffraction studies. CCDC number, 2055160.

(34) [Ru(tmpatpy)$_2$](PF$_6$)$_2$·1.5DMF: A mixture of **tmpatpy** (50 mg, 0.12 mmol) and RuCl$_3$·xH$_2$O (25 mg) in ethanol (30 mL) was heated at 70 °C for 4 h to produce a turbid, reddish-purple solution. This was cooled to ambience to produce a precipitate of a brown powder, washed on the filter with ethanol, which was assumed to be [Ru(**tmpatpy**)Cl$_3$]. [Ru(**tmpatpy**)Cl$_3$] (40 mg, 0.063 mmol), **tmpatpy** (30 mg, 0.071 mmol) and N-ethylmorpholine (five drops) were dissolved in CH$_3$OH/CHCl$_3$ (40 mL, 3:1, v/v) and heated at 70 °C for 2 h. The resultant deep red solution was filtered through Celite to remove some insoluble purple material, and the solvent was evaporated under reduced pressure. The dark red residue was dissolved in a minimum volume of acetonitrile and loaded on a short silica column, and the column was eluted with acetonitrile/saturated aqueous KNO$_3$/water (7:1:0.5, v/v). The major red component was collected, methanolic NH$_4$PF$_6$ was added, and 1 h was allowed for the precipitation of a red powder, which was then collected, washed on the filter with methanol and dried under vacuum (yield: 35 mg, 41%). Anal. calc. for C$_{113}$H$_{105}$F$_{24}$N$_{15}$O$_{15}$P$_4$Ru$_2$: C, 50.36; H, 3.93; N, 7.80. Found: C, 50.41; H, 3.80; N, 7.78%. ^{1}H-NMR (300 MHz) in CD$_3$CN: δ 8.88 (s, 4H), 8.54 (dq, J = 8.2, 0.8 Hz, 4H), 8.01 (td, J = 7.7, 1.5 Hz, 4H), 7.45 (dq, J = 5.5, 0.6 Hz, 4H), 7.24 (dd, J = 5.6, 1.3 Hz, 2H), 7.22 (dd, J = 5.6, 1.3 Hz, 2H), 7.09 (s, 4H), 3.96 (s, 12H, –CH$_3$), 3.85 (s, 6H, –CH$_3$). The slow evaporation of an acetonitrile solution of the complex provided red crystals suitable for X-ray structure determination. CCDC number, 2055161.

3. Crystallography

Diffraction data for crystals of complexes **1–3**, **7–18**, **20**, **21**, **23–28** and **30–33** were obtained at 100(2) K using an ADSC Quantum 210 detector at 2D SMC with a silicon (111) double crystal monochromator (DCM) for various synchrotron wavelengths between 0.62988 and 0.75000 Å at the Pohang Accelerator Laboratory, Pohang, South Korea. The ADSC Q210 ADX program [51] was used for data collection (detector distance, 63 mm; omega scan; $\Delta\omega$ = 1°; exposure time, 1 s per frame), and HKL3000sm (Ver. 703r) [52] was used for cell refinement, reduction and absorption correction. The structures were solved by direct methods and refined by full-matrix least-squares fitting on F^2 using SHELXL-2014 [53]. All the non-hydrogen atoms were refined with anisotropic displacement parameters. The carbon-bound hydrogen atoms were introduced at calculated positions, and all the hydrogen atoms were treated as riding atoms with an isotropic displacement parameter equal to 1.2 times that of the parent atom.

Diffraction data for complexes **4–6**, **19**, **22**, **29** and **34** were obtained on a RIGAKU Saturn CCD diffractometer with a confocal mirror, using graphite-monochromated Mo K_α radiation (λ = 0.71073 Å) or, for **19** only, Cu K_α radiation (λ = 1.54187 Å). The structures were solved by direct methods and refined on F^2 by full-matrix least-squares treatment [53].

For complexes **7**, **9**, **13**, **14**, **18**, **20**, **28**, **29**, **32** and **34**, the SQUEEZE option of PLATON [54] was used to take into account the contribution of disordered solvents during refinement. The results (the size of the voids, number of electrons present in the voids and possible solvent molecules present) are summarized in Table S1.

Full details of all the structure determinations have been deposited as cifs in the Cambridge Crystallographic Data Collection under CCDC deposition numbers 2055133-2055135, 2055137-2055142, 2055148-2055161 and 2055219-2055228. Note that for structures **7** and **9**, the solutions were unsatisfactory in that the residual (R_1 factor) exceeded 0.10, but they have been included for comparative purposes.

The nature of the atoms involved in the interactions exceeding dispersion was established using CrystalExplorer [30], and their specific identities (atom numbers and locations) were established using CrystalMaker [55], which was then used to prepare figures with the interactions shown as dashed lines. In the case of the complexes for which SQUEEZE was applied during the structure refinement (see above), the Hirshfeld surfaces calculated with CrystalExplorer may lack indications of some interactions since not all the solvent molecules present were included in the calculation.

4. Results and Discussion

Most of the ligands used in the present work are well-known species for which the coordination chemistry is well-established. We have described in detail, elsewhere [33], the syntheses of three new ligands (**bzOtpy, etptpy** and **pfatpy**) and, here, that of **papatpy** (Figure 1), closely based on methods established for related compounds, as were the syntheses of their metal-ion complexes, so a detailed discussion of these aspects of the present work is not warranted. With the exception of the kinetically inert Ru(III) ions, the metal ions used herein are all labile species, so the synthesis of their complexes is trivial. The crystallisation of the complexes in a form suitable for X-ray diffraction measurements was not necessarily trivial and generally involved the exploration of several methods, although only in the case of complexes of **pfpatpy** were real difficulties encountered, these being such that only a single complex of this ligand was structurally characterised. In regard to the coordination spheres of the metal ions and the stacking tendency of the cations in the 33 compounds structurally characterised, nothing setting them apart from their extensive context is apparent, and our focus was on just the solid-state interactions discernible from analysis of the Hirshfeld surfaces.

As a background to the analysis of the crystal structures of complexes of $4'$-substituted $2,2':6',2''$-terpyridine, it is useful to consider the structures of free ligands that are available, although these are relatively limited in number, and this we have done in a separate work [37]. All free terpyridines show non-chelating, bis-transoid arrays of the three N-donor atoms, except in the case of the $4'$-hydroxy derivative, which exists as a 1:1 mixture of the hydroxypyridine and pyridone tautomers in the solid state [56–58], where the pyridone tautomer adopts a bis-cisoid array, while the hydroxypyridine form has a bis-transoid form. Significantly, although the stacking of the terpyridine units is a common feature of free ligand structures, this appears in most instances other than those where large, polar substituents are present, to be associated simply by dispersion interactions between terpyridine units, and only the pyridone tautomer of $4'$-hydroxy-$2,2':6',2''$-terpyridine is clearly involved in additional, specific C$\cdots$C interactions between facing aromatic planes, with interactions of the peripheral H-atoms being the dominant form of interactions, exceeding dispersion in all cases.

While much of the interest in $2,2':6',2''$-terpyridine chemistry has been focused on metal-ion complexes in low oxidation states [1–7], complexes of high-oxidation-state metal ions are also known, with vanadium(V) [59] and uranium(VI) (as the uranyl ion, UO_2^{2+}) [60–63] providing examples of extremes. In that the formation of a coordinate bond is formally a process where the metal ion gains electrons and the ligand loses them, such extreme cases might be expected to be associated with charge redistribution favouring the appreciable interaction of the bound ligand with nucleophiles, akin to the addition to an imine bond. In the case of crystalline [VO$_2$(**bptpy**)]ClO$_4$, **1**, the Hirshfeld surfaces of the two inequivalent cations provide evidence for exactly this, with the most marked region of interaction exceeding dispersion, that being due to the contact of vanadyl-O (O1) with one of the carbon atoms (C10) adjacent to the central N-atom of the ligand (Figure 2a). Reciprocal, remarkably short contacts (O1$\cdots$C10^i = O1$^i\cdots$C10, 2.81(1) Å; i = $1 - x$, $1 - y$, $2 - z$) of this type result in the presence of pairs of the two inequivalent cations, where the slightly bowed terpyridine units lie close to parallel, with a separation of ~3.5 Å, and where there are only barely perceptible indications of other interactions beyond the dispersion of the C and H atoms of ligand units lying both close to parallel and nearly perpendicular to the ligand of the cation considered (Figure S1). Multiple interactions involving peripheral aromatic H-atoms and vanadyl- or perchlorate-O atoms serve to link the whole structure together, but none involve unusually short contacts, and as is seen in the examples discussed later, they appear to be a common feature of structures regardless of the metal ion present. However, as some interactions involve the 4′-biphenyl substituent on the terpyridine unit, it is unsurprising that there are significant differences in the structure of the "parent" species [VO$_2$(**tpy**)]ClO$_4$ [59] when compared to that of **1** (Figure 2b). Here, the most obvious interaction beyond the dispersion of a given cation again involves the contact of a O-atom with a C adjacent to a N, but the O-atom is from perchlorate, and two adjacent C atoms, one from the central and one from an outer pyridine unit (O4$\cdots$C6, 3.09(1) Å; O4$\cdots$C5, 3.162(8) Å), are involved. Vanadyl-O$\cdots$C contacts are still apparent and are associated with the maintenance of a stacked arrangement of (equivalent-) cation pairs but involve C-atoms separated by another from N (O1$\cdots$C4′, 3.162(8) Å; ′ = 1 − x, 1 − y, 2 − z). Note that while in **1**, the nearer phenyl ring of the biphenyl substituent is close to coplanar with the terpyridine unit, indicating, possibly, a degree of delocalisation as an influence upon the electronic properties of the whole, this is also seen in some M(II) complexes of the ligand [43] and appears to be a consequence of CH interactions with counteranions.

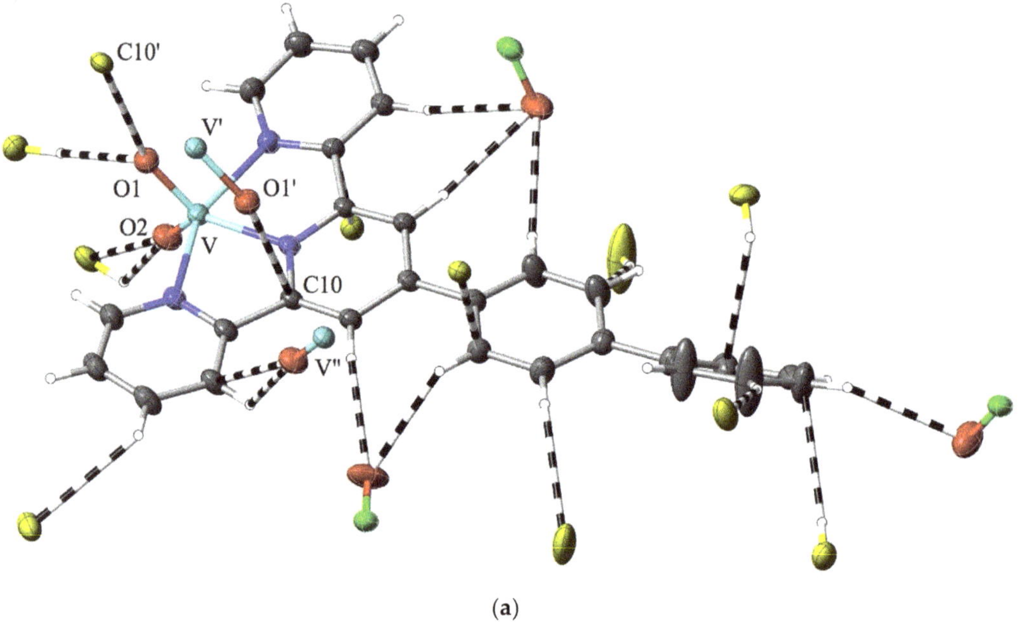

(**a**)

Figure 2. *Cont.*

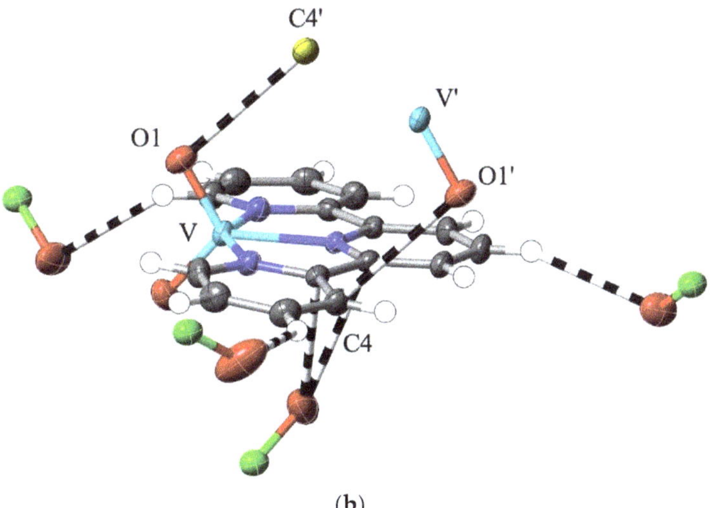

(b)

Figure 2. Interactions beyond dispersion, shown as dashed lines, of the cations present in (**a**) [VO$_2$(**bptpy**)]ClO$_4$, **1** and (**b**) [VO$_2$(**tpy**)]ClO$_4$ [44]. O atoms are identified by showing the atoms to which they are attached. Colour coding: C = grey, N = dark blue, O = red, Cl = green, and V = light blue, except for carbon atoms from adjacent cations, which are shown in yellow. H-atoms are shown as small, colourless spheres of arbitrary 0.2 Å radius.

In known structures of uranyl-ion complexes of 2,2′:6′,2″-terpyridine [60–63] (and, in some cases [60,61], of 4′-chloro-2,2′:6′,2″-terpyridine), the close contact of at least one O-atom with one or two C-atoms bound to the N of another complex forming a stacked array is common but not universal. In [UO$_2$(**tpy**)(**tdc**)]$_2$ (**tdc** = thiophene-1,4-dicarboxylate) and its chloroterpyridine analogue [45], there are such contacts with C6···O2′(uranyl) 3.039(5) Å and C6···O4′(carboxylate) 2.947(8) Å, respectively, although much more obvious in the former are reciprocal uranyl-O2···N2′ contacts of 2.783(4) Å (Figure S2a). In [(UO$_2$)$_2$(OH)(**tpy**)(**dcb**)$_3$] (**dcb** = 3,5-dichlorobenzoate) and its chloroterpyridine analogue [61], there are contacts with C15···O2′(uranyl) 3.014(7) Å and C15···O2′(uranyl) 2.915(6) Å, respectively, while in [(UO$_2$)$_2$(OH)$_2$(**tpy**)$_2$](ClO$_4$)$_2$ [62], where there are two inequivalent dimers within the structure, the C···O contacts C5···O8′(uranyl) 2.949(6) Å, C6···O8′(uranyl) 2.959(7) Å and C41···O2′(uranyl) 2.953(6) Å are apparent (Figure S2b). In [UO$_2$(**tpy**)(**bpdc**)] (**bpdc** = 2,2′-bipyridine-3,3′-dicarboxylate) [63], there is a similarly short contact of C5···O1′(uranyl) 2.937(3) Å, but in the four complexes [UO$_2$(**tpy**)(**tph**)] and [UO$_2$(tpy)(npdc)] (**tph** = terephthalate {benzene-1,4-dicarboxylate}; npdc = naphthalene-1,4-dicarboxylate) and their chloroterpyridine analogues, such interactions are not evident. Here, extended stacking arrays of the aromatic units appear to block the access of other species to C-atoms bound to N in the ligand. This is a timely reminder that any weak interactions beyond dispersion evident in a Hirshfeld surface representation are exactly that, weak, and that the balance between any number of such interactions may be extremely sensitive to compositional differences.

If a positive charge on a bound metal does produce charge depletion on the ligand in the region of the donor atoms, then any effect should diminish with a decreasing charge, and the smallest positive charge is that carried by the proton, H$^+$. The diprotonation of a terpyridine is required to produce the di-cisoid configuration found in metal-ion complexes such as **1**, and the structure of compound **30**, [H$_2$**tmptpy**](ClO$_4$)$_2$·0.5THF, adds to several known for such species. (A summary of relevant literature is given in [33].) Here, as is broadly true [33], there is no evidence on the Hirshfeld surface, except for an indication that the interaction of one perchlorate-O (O17) may be better considered as involving the C25–H25 bond rather than H25 alone, for interactions of nucleophiles with C-atoms attached to any of the N-donor atoms, and this is consistent with what is seen in complex **1** and several other species discussed above, being evidence for the significant polarising effects of highly charged cations. Peripheral CH···O interactions of aromatic CH with perchlorate-O, however, are numerous, and the cations lie in infinite stacks largely involving dispersion interactions only, except for barely discernible C···C interactions between alternate pairs of the two inequivalent cations of the structure, with C11···C44^i 3.31(4) Å and C18···C29^i 3.34(4) Å (i = x + 0.5, y + 0.5, z) (Figure 3). While the atoms C11 and C29 are adjacent to N in the terpyridine unit, C18 and C44 are C-atoms of the substituent to which methoxy groups are attached, showing that purely dispersive stacking of the terpyridine units as seen in unfunctionalised ligand salts such as [H$_2$**tpy**]Cl$_2$·H$_2$O [64] can be disrupted by substituents.

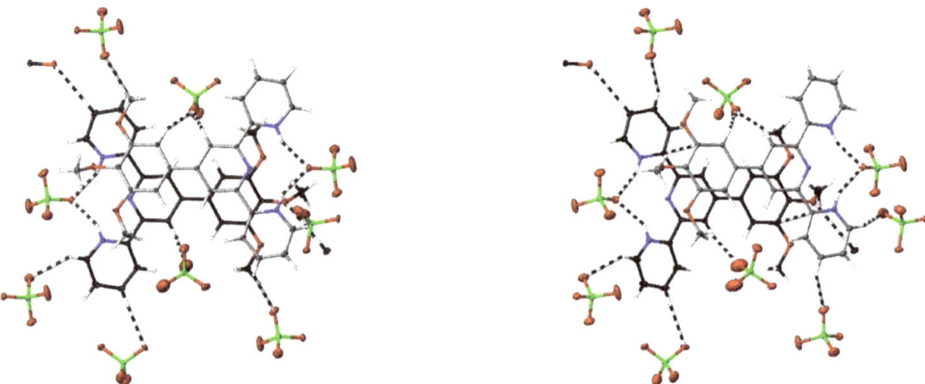

Figure 3. Views perpendicular (**left**) and at 30° (**right**) to the mean plane of one of the stacked pairs of inequivalent cations found in the structure of [H$_2$**tmptpy**](ClO$_4$)$_2$·0.5THF, **30**, showing (dashed lines) the interactions exceeding dispersion of the cations. Carbon atoms of one cation are shown in black; of the other, in grey. C···C interactions are visible in the 30° view.

The vast majority of the structural studies of complexes of 2,2′:6′,2″-terpyridine and its derivatives concern metal(II) and metal(III) species, and the present work largely complements such investigations, which range, for Co(II), for example, from early determinations on a single compound for which H-atom coordinates are not available, e.g., [65], to those of several complexes of various derivatives with full coordinates, e.g., [66], such variations reflecting, in part, major advances in the ease of the synthesis of the ligands [1–7] and in crystallographic equipment. The complexes **2–6** of the **tptpy** ligand add Fe(II), Ni(II) and Ru(II) to the series, with Co(II), Cu(II), Zn(II) and Cd(II) studied earlier [34,37], and form part of a rather large family of structurally characterised complexes of 4′ derivatives of 2,2′:6′,2″-terpyridine, where the substituent is an apolar aromatic unit. An obvious question to ask in relation to the nine structures of 4′-terphenyl-2,2′:6′,2″-terpyridine we studied is whether or not the changes in the electron configurations of the exclusively M(II) cations involved have an effect detectable in the form of the Hirshfeld surfaces of the complexes in the solid state. The answer is that none is clearly evident. A feature of all the structures is that the terphenyl tail is significantly twisted, both within its chain and relative to the terpyridine unit, indicating that there is no extended delocalisation within the whole ligand and, thus, that the addition of the terphenyl tail does not produce a larger, planar aromatic unit that might engage in stronger stacking interactions than in simple 2,2′:6′,2-terpyridine complexes. It is certainly possible to discern aromatic and aza-aromatic entities that form arrays of "face-to-face" and "edge-to-face" types, and these are associated with various C···C and CH···C (CH···π) interactions that exceed dispersion and are thus evident on the Hirshfeld surfaces. In the totality of the interactions exceeding dispersion (Figure 4), there is no evidence for nucleophile addition to C-atoms adjacent to N, so it can be said at this point that M(II) cations seem to be distinguishable from M(V) and M(VI) cations in this respect. Interactions of the cations with the counteranions and lattice solvents occur along with aromatic–aromatic interactions, and even in only the eight cases considered, there are variations in both the counteranions and lattice solvents as well as in the space groups, so it is unsurprising that there are differences in the exact patterns of interactions across the eight, but in every one, there are C$\cdots$C and CH$\cdots$C interactions beyond dispersion with both the tpy head groups and at least one of the phenyl groups of the tails. A comparison of the mixed-ligand complexes **5** and **6** shows that aromatic$\cdots$aromatic interactions beyond dispersion are little affected by the difference in anion, indicating that aromatic$\cdots$aromatic interactions in general, including dispersion, may be the dominant factor influencing the array of cations in the crystal. Analysis (Figure S3) of the structures reported [43] for a family of complexes of the ligand 4′-biphenyl-2,2′:6′,2″-terpyridine (**bptpy**) confirms the observation that for a similar range of M(II) species (M = Co, Ni, Cu, Zn, Cd and Ru), none provides evidence of significant weak interaction with C-atoms bound to N, though perhaps because of the differences in temperature for the present and these earlier determinations, the C$\cdots$C and CH$\cdots$C interactions beyond dispersion between aromatic/aza-aromatic units are generally less numerous for the bptpy derivatives. The known structures for the Co(II) [37] and Zn(II) [35] complexes of the very long-tailed 4′-quaterphenyl-2,2′:6′,2″-terpyridine (**qptpy**) ligand provide no exceptions to the observation of a lack of any enhanced electrophilicity of C-atoms bound to coordinated N, just like the very numerous structures of complexes of 2,2′:6′,2″-terpyridine itself and its 4′-phenyl and 4-methylphenyl (tolyl) derivatives, where, however, there are examples of M(III) complexes where the generalisation appears to be valid (M = Co, Rh) [67,68] and at least one (M = Cr) [69] where it is not. It is, of course, well-known [70] that the synthesis of functionalised aza-aromatic ligands can be based on nucleophilic attack on precisely the C-atoms adjacent to N.

(a) **[Fe(tptpy)₂](ClO₄)₂·CH₃OH** (Fe = orange), **2**.

(b) **[Co(tptpy)₂](BF₄)₂·CH₃OH** (Co = pink; B = brown) [37].

Figure 4. *Cont.*

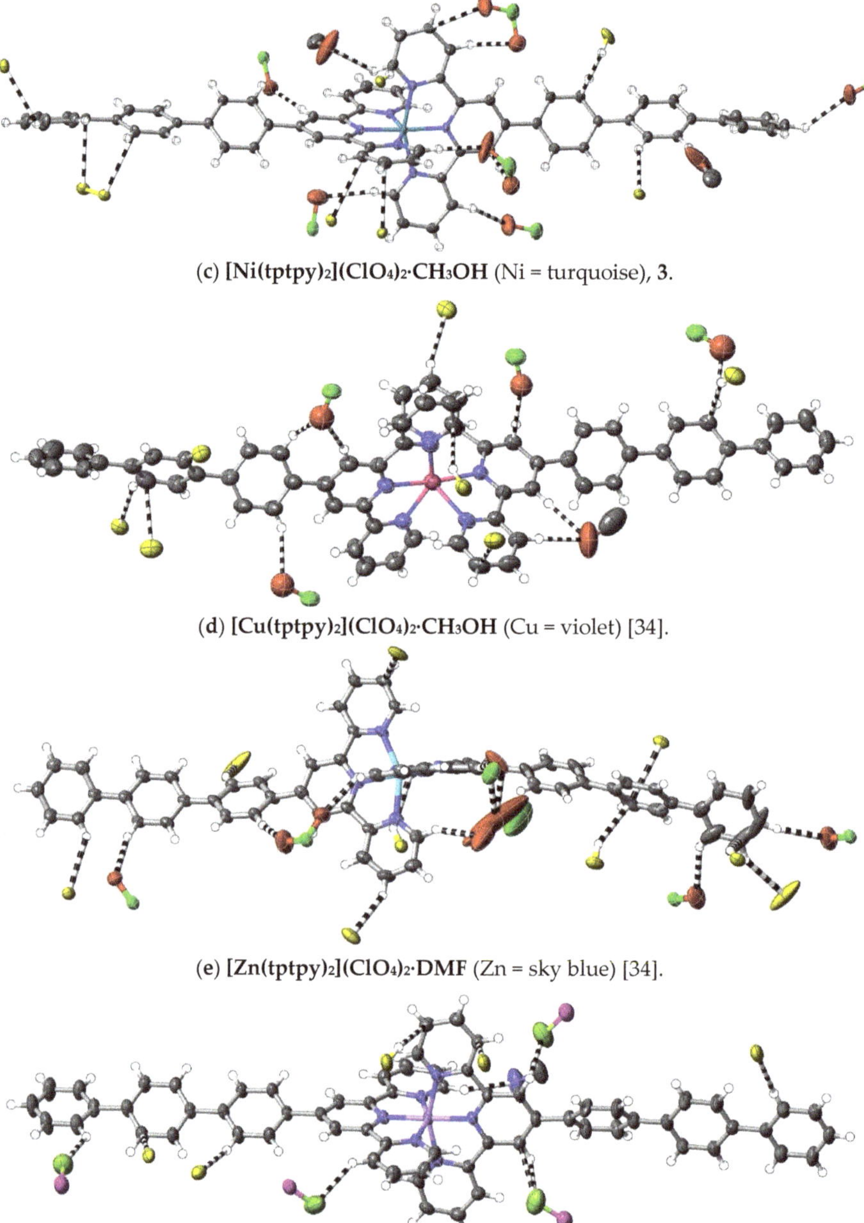

(c) **[Ni(tptpy)$_2$](ClO$_4$)$_2$·CH$_3$OH** (Ni = turquoise), **3**.

(d) **[Cu(tptpy)$_2$](ClO$_4$)$_2$·CH$_3$OH** (Cu = violet) [34].

(e) **[Zn(tptpy)$_2$](ClO$_4$)$_2$·DMF** (Zn = sky blue) [34].

(f) **[Ru(tptpy)$_2$](PF$_6$)$_2$** (Ru = mauve), **4**.

Figure 4. *Cont.*

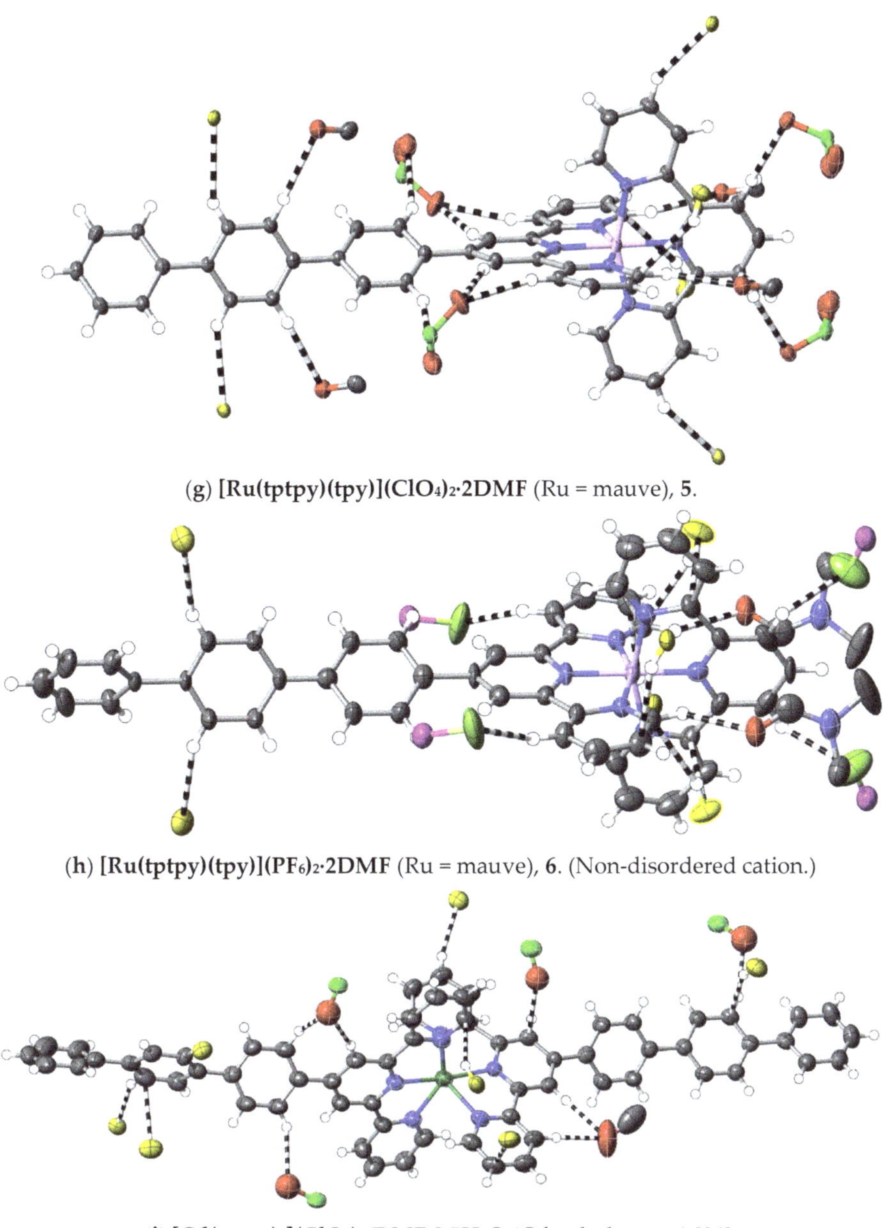

(**g**) **[Ru(tptpy)(tpy)](ClO₄)₂·2DMF** (Ru = mauve), **5**.

(**h**) **[Ru(tptpy)(tpy)](PF₆)₂·2DMF** (Ru = mauve), **6**. (Non-disordered cation.)

(**i**) **[Cd(tptpy)₂](ClO₄)₂·DMF·0.5H₂O** (Cd = dark green) [34].

Figure 4. Interactions exceeding dispersion (dashed lines) of cations complexed by 4′-terphenyl-2,2′:6′,2″-terpyridine. As in Figure 1, C-atoms of ligands not directly bound to the one metal ion involved are shown in yellow. Different O-atoms involved are identified by their attached atoms (e.g., Cl for perchlorate-O).

A different means of extending an apolar 4′ tail on 2,2′:6′,2″-terpyridine is seen in complexes **7–16**, where ethynyl units serve to link phenyl rings. Although initially envisaged as sites of polarisable electron density that might give rise to significant interactions beyond dispersion, the ethynyl units have, at most, a limited role of this type in any of the 10 complexes characterised (as well as in the complexes **32–34** of the further-functionalised phenylethynyl ligands—see below). Thus, as shown in Figure 5, in the Co(III) complex **9** of **papatpy**, one outer ethynyl group is involved in C···C interactions, and the other, in one CH···C interaction (with both C-atoms), while in the Ni(II) complex **10**, one inner unit is involved in a C···HC interaction (again with both C-atoms to one H), but in the Cu(II) complex **11** and the Zn(II) complex **12**, there are no interactions beyond dispersion with any ethynyl unit, as is the case for the Zn(II) complex **7** of **patpy**. In the Ru(II) complex **13**, there are C···C interactions with one outer ethynyl unit, but in the mixed-ligand **papatpy-tpy** Ru(II) complex **14**, there are no interactions, as is the case in the 1:2 and 1:1 complexes **15** and **16**, respectively, of **papatpy** with Cd(II) and also in the complex **17** of **pfpatpy** with Fe(II). All the structures **7–16** were determined at 100(2) K, so temperature, for which there is some evidence of influence in the determination of the visibility of weak interactions (see ahead), cannot be considered as a factor determining these observed variations in ethynyl group interactions. What, of course, should be noted is that the multiple prominent interactions apparent in all the cases are those with anion atoms (O, F or Cl), and while these tend to be concentrated close to the formal positive site of the metal ion, they also occur with the extremities of the complexes, again consistent with the notion that both dispersive interactions and those that exceed them are involved in a delicate balance that may vary with changes such as that of the anion or uncoordinated solvent but also minor factors such as bond length changes and stereochemical preferences (for example, where the Jahn–Teller effect renders the coordination sphere of Cu(II) somewhat more irregular than that for other metal ions). Thus, the pattern of interactions exceeding dispersion is different in every complex and shows no obvious correlation with electron configurations, though the Ni(II) complex **10** does provide an example where there are (perchlorate)O···C interactions involving C-atoms bound to N-donors. This is an unusual example where a single O-atom is in interaction with four C-atoms, two of which are directly bound to N and two of which are not, so each individual interaction may be reinforced by the other three. A clearer situation of nucleophile interaction with C bound to an N-donor, however, is found in complex **17** (Figure 6), where an aryl-F atom bridges C-atoms adjacent to N on separate ligand units. In fact, the F-atom substituents of the **pfpatpy** ligand are also involved in multiple interactions beyond dispersion that link different cations together, providing the first example in the present study of a marked effect of ligand substituents on secondary interactions.

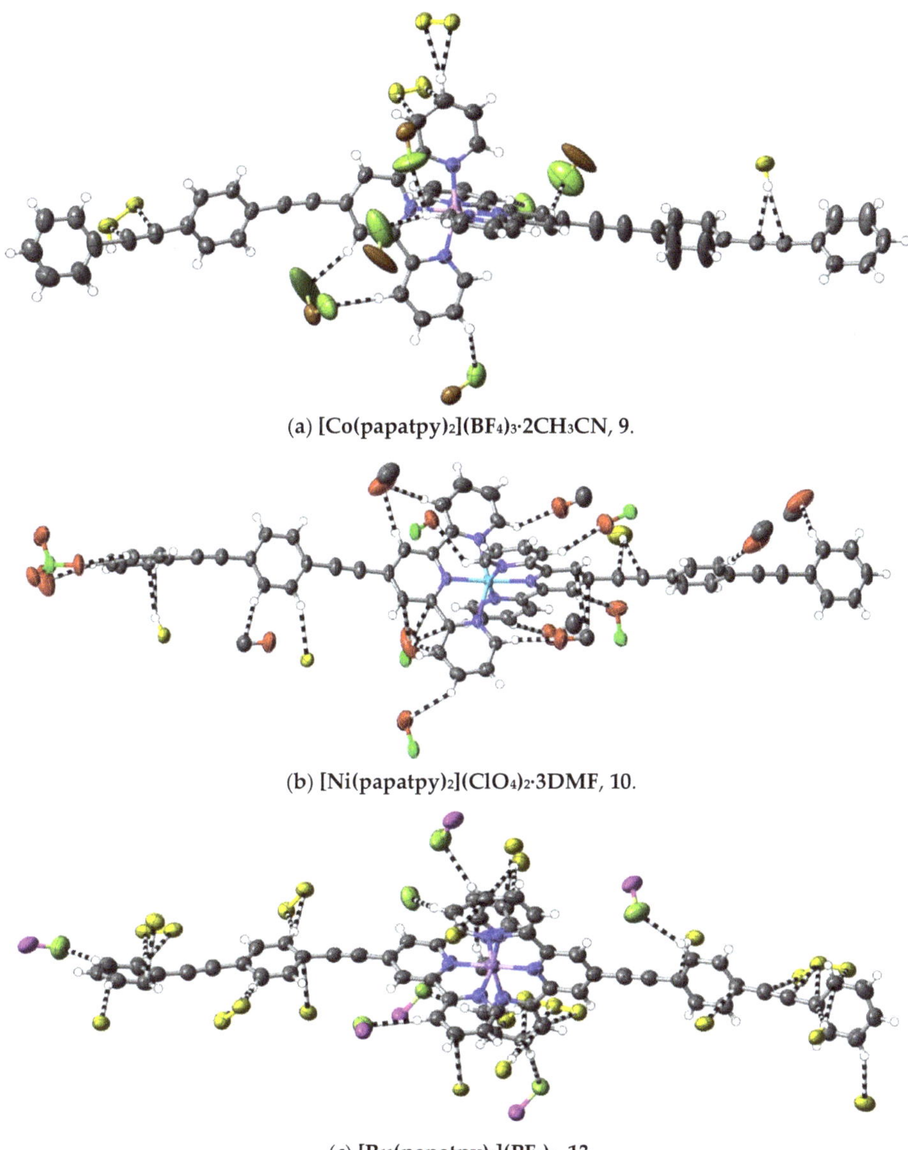

(a) [Co(papatpy)₂](BF₄)₃·2CH₃CN, **9**.

(b) [Ni(papatpy)₂](ClO₄)₂·3DMF, **10**.

(c) [Ru(papatpy)₂](PF₆)₂, **13**.

Figure 5. Interactions beyond dispersion involving the ethynyl subunits in complexes of phenylethynyl derivatives of 2,2′:6′,2″-terpyridine. Colour coding as in Figure 1.

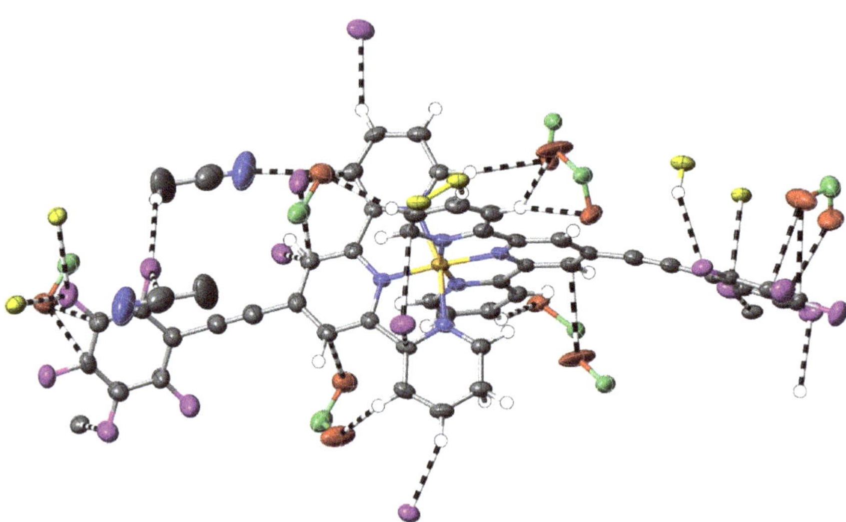

Figure 6. Organofluorine interactions exceeding dispersion in [Fe(pfpatpy)$_2$](ClO$_4$)$_2$·CH$_3$CN·H$_2$O, **17**.

More subtle consequences of the introduction of a polar substituent in the ligand are seen in the interactions observed in the complexes **18–20** of **pybptpy**. These structures augment those of our earlier work [14] on related Zn(II)/**pyptpy** and Cd(II)/**pybptpy** complexes, although that of the Cu(II) complex **19** is significantly disordered, and its weak interactions cannot be fully analysed. All add to the family of complexes of (4-pyridyl)-substituted 2,2′:6′,2″-terpyridine complexes [71,72], for which a thorough study [72] of complexes of the "parent" ligand 4′-(4-pyridyl)-2,2′:6′,2″-terpyridine, **pytpy**, has shown not only that the bound ligand may both twist and bow (as observed here) but that the H-bonding capacity of the pyridyl unit has a major influence upon the crystal structure, to the point that it may exclude stacking interactions. In the structures of the extended-ligand complexes [Zn(**pyptpy**)$_2$](ClO$_4$)$_2$·2CH$_3$OH·CHCl$_3$ and [Cd(**pybptpy**)$_2$](ClO$_4$)$_2$·DMF·2.8H$_2$O [14], various units with aromatic stacking are apparent, as they are in complexes **18–20**. As is usual with such arrays, any interactions appear to be largely dispersive in nature, but examination of the interactions exceeding dispersion reveals some interesting changes in the pyridine-N interactions as a function of the substituent chain length as it passes from 4-pyridyl through 4-pyridyl-phenyl to 4-pyridyl-biphenyl. Thus, in the bis(ligand) complexes of Fe(II), Ni(II) and Ru(II) with **pytpy** [72], one pyridine-N forms a "classical" H-bond with methanol, and the other, a "non-classical" bond with HC from a separate complex unit, while in [Zn(**pyptpy**)$_2$]·2CH$_3$OH·CHCl$_3$ [14], both pyridine-N atoms form H-bonds with methanol, but one also forms a bond with HC, and in [Cd(**pybptpy**)$_2$](ClO$_4$)$_2$·DMF·2.8H$_2$O [14] and complexes **18–20,** both pyridine-N atoms form N···HC bonds only. In the polymeric chains involving the stacking of the 4′ substituents in the **pybptpy** complexes, these double interactions result in the 4-pyridylbiphenyl units being much closer to planar and lying closer to parallel with their pair than is the case in the otherwise rather similar structures of complexes of the **tptpy** and **papatpy** ligands (Figure 7).

An alternative means of introducing polarity in the same sense as that induced by the pyridine-N of pybptpy complexes is seen in the complexes **21–24** of **Brbptpy** with Co(II), Ni(II), Cu(II) and Ru(II). Their structures add to those of Zn(II) and Cd(II) previously described [35], creating another small family of M(II) complexes of the one 2,2′:6′,2″-terpyridine derivative. The presence of a bromo-substituent raises the prospect of "halogen bonding" [73], but only in the structures of the Ni(II) and Ru(II) complexes is there any evidence for such interactions, and regardless of their nature, interactions of the bromine atoms do not result in any obvious restriction of the twisting of the biphenyl tails. There does appear to be a difference in the interactions beyond dispersion when the first-row transition metal complexes are compared with those of the second-row, with C⋯C and CH⋯C interactions being much more prominent for the former and CH⋯O interactions (involving the counteranions) being almost completely dominant in [Cd(**Brbptpy**)$_2$](ClO$_4$)$_2$·CH$_3$CN, though of course, the number of examples available for comparison is rather small (Figure 8). In no case here is there evidence of nucleophile interaction with C-atoms bound to N.

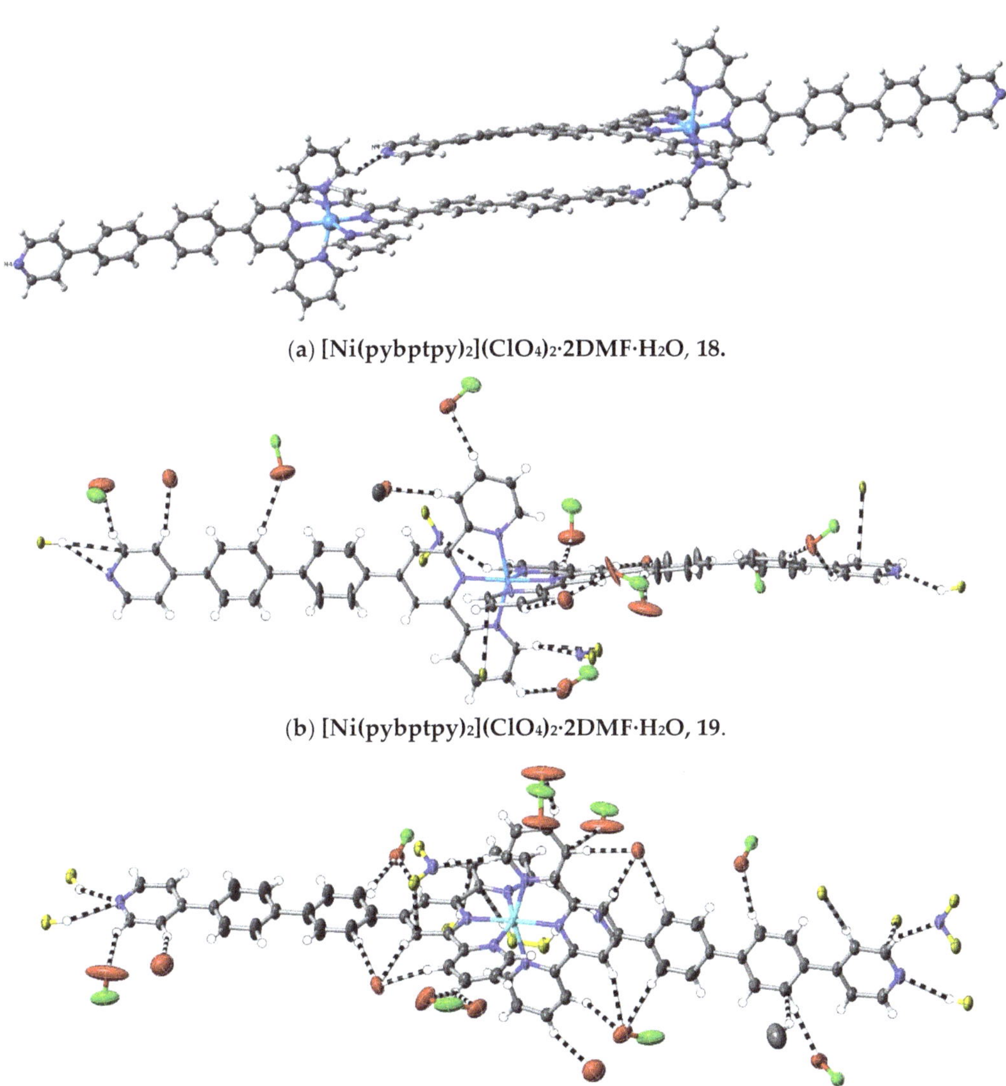

(a) [Ni(pybptpy)$_2$](ClO$_4$)$_2$·2DMF·H$_2$O, 18.

(b) [Ni(pybptpy)$_2$](ClO$_4$)$_2$·2DMF·H$_2$O, 19.

(c) [Zn(pybptpy)$_2$](ClO$_4$)$_2$·2DMF·H$_2$O, 20.

Figure 7. Interactions exceeding dispersion in complexes of **pybptpy,** showing, (a) in the Ni(II) complex as an example, a partial view of the stacking, which appears to be reinforced by N⋯HC interactions, and (b–c) the full range of cation interactions in the Ni(II) and Zn(II) complexes.

Adding an ethoxycarbonyl group as the terminus of a 4′-polyphenyl tail introduces an entity with more than one interaction site and possibly capable of directional interactions. Interactions beyond dispersion of the ethoxycarbonyl groups in the complexes [Ni(**ebptpy**)$_2$](ClO$_4$)$_2$·3DMF (**25**), [Cu(**ebptpy**)$_2$](ClO$_4$)$_2$·CH$_3$CN (**26**) and [Zn(**etptpy**)$_2$] (ClO$_4$)$_2$·2DMF·H$_2$O (**27**) are not, however, unusually prominent compared to others such as those involving the counteranions, and the more significant aspect of the structures is the evidence they provide for the importance of dispersive interactions in the arrays adopted by the terpyridine substituents. All three structures are layered, and within the layers, the full interactions are complicated, but in each, it is possible to identify pairs of cations where the substituent tails are in close proximity. In the **ebptpy** complexes, the tails are twisted such that the phenyl rings of adjacent cations are in arrays closer to edge to face rather than face to face, but in the **etptpy** complex, the greater extent of the aromatic tail is reflected in arrays where there is relatively slight twisting, and the three rings of each cation tail are in close-to-parallel (face-to-face) alignment with three of an adjacent cation (Figure 9). There is, nonetheless, only one C$\cdots$C interaction (C28$\cdots$C54^i; i = 1.5 + x, 1.5 − y, 1 + z; 3.313(3) Å) exceeding dispersion occurring between the three pairs of rings, meaning that dispersion alone is clearly the dominant factor.

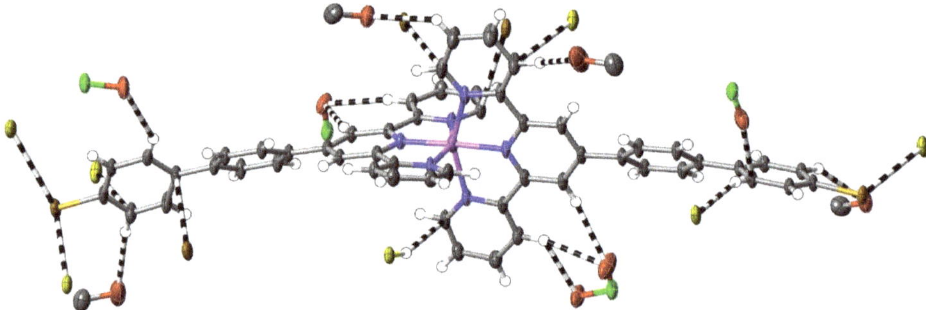

(a) [Cu(**Brbptpy**)$_2$](ClO$_4$)$_2$·2CH$_3$OH, **23**.

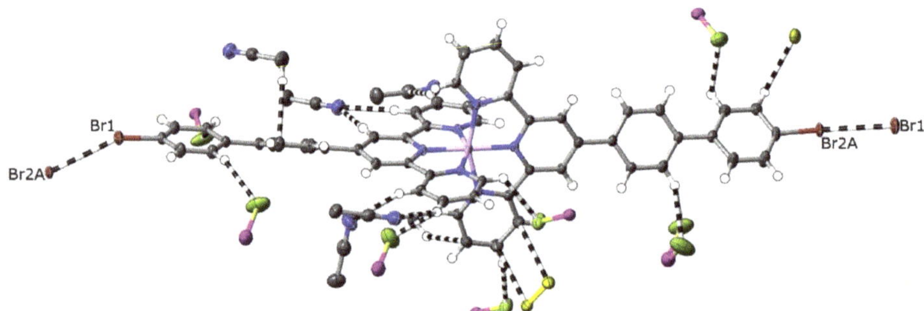

(b) [Ru(**Brbptpy**)$_2$](PF$_6$)$_2$·3CH$_3$CN, **24**.

Figure 8. *Cont.*

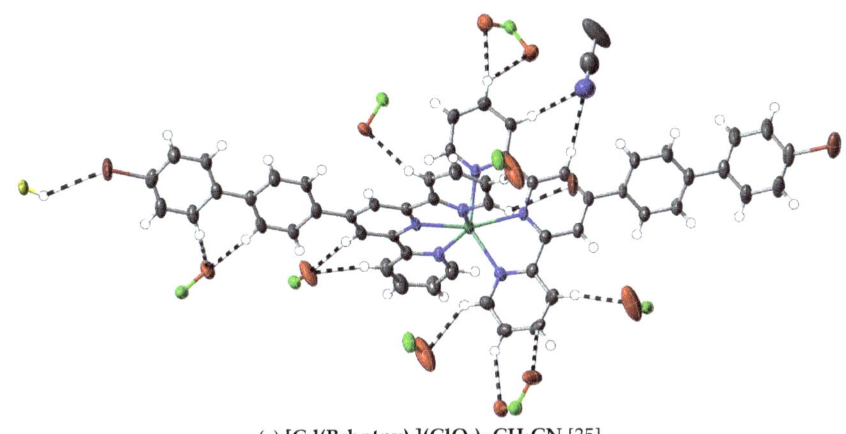

(c) **[Cd(Brbptpy)₂](ClO₄)₂·CH₃CN** [35].

Figure 8. A comparison of interactions beyond dispersion in complexes of first- and second-row transition metal(II) ion complexes of **Brbptpy**.

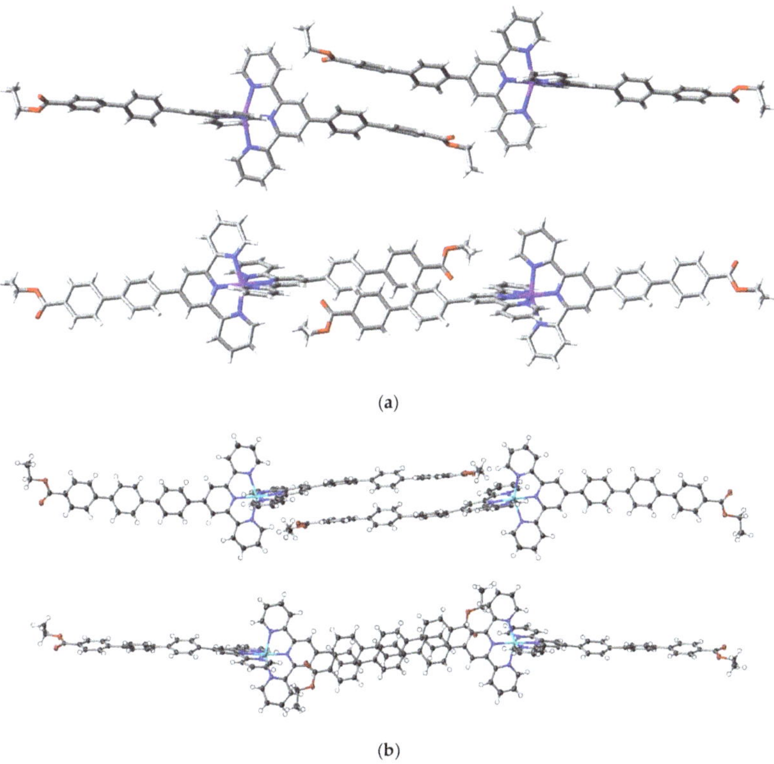

(a)

(b)

Figure 9. Effect of polyphenyl chain length on stacking arrays in complexes of ethoxycarbonylphenyl-substituted terpyridine complexes. Orthogonal views of proximal arms in (**a**) [Cu(**ebptpy**)₂](ClO₄)₂·CH₃CN, **26**, and (**b**) [Zn(**etptpy**)₂](ClO₄)₂·2DMF·H₂O, **27**.

The complexes **28–34** all contain alkoxy-functionalised terpyridine ligands, [Fe(**bzOtpy**)₂](ClO₄)₂ (**28**) and [Ni(**bzOtpy**)₂](ClO₄)₂ (**29**) being additions to the fairly numerous known family [36,64,74], where the alkoxo group is attached directly to the 4′ position of 2,2′:6′,2″-terpyridine. The Hirshfeld surfaces (Figure S4) obtained for the structures of **28** and **29** show marked differences in the number of interactions beyond dispersion, with, in particular, a complete absence of apparent C··· C interactions in **29**. Such loss can be explained as arising from the nearly 200 K difference in the temperatures (**28**, 100 K; **29**, 296 K) at which the structures were determined, with the enhanced vibrational motion at the higher temperature leading to some washing out of the difference between dispersion and other weak interactions. Similar differences have been observed for Co(II) complexes studied in their low-spin and high-spin forms, necessarily at quite different temperatures [37], and the Hirshfeld surface for the cation present in the structure of the Zn(II) complex of the homologous ligand 4′-(4-benzoyloxyphenyl)-2,2′:6′,2″-terpyridine determined at 291 K [75] also shows little evidence of interactions exceeding dispersion. In **28**, interactions beyond the dispersion of the cation are multiple and various, and CH··· C or C··· C interactions are not limited to only one close pair. Even a cursory inspection of the Hirshfeld surfaces (Figure S5; all for low-temperature determinations) of **28** and the Co(II) and Cu(II) complexes of the triply methoxylated ligand 4′-(1,2,3-trimethoxy-5-phenyl)-2,2′:6′,2″-terpyridine, **tmbzOtpy** [36], shows that the introduction of the methoxyl substituents quite dramatically modifies all interactions beyond the dispersion of the cations, further supporting the contention that these interactions are generally a reflection of equilibria involving multivariate influences.

The structures of the complexes **31–34** provide additions to those related examples also concerning terpyridines with trimethoxyphenyl (gallate) substituents previously considered [36] in relation to their possible use as precursors to mesogenic species. While in the protonated **tmtpy** ligand (structure **30**), all the interactions beyond dispersion with the perchlorate anions are of the CH··· O type, in the Zn(II) complex **31** (Figure 10a), there are two methoxyl-O··· C contacts (C12··· O2^i 3.20(6) Å; C13··· O1^i 3.19(6); i = x + 0.5, 2 − y, 3 − z), perhaps indicative of the greater nucleophilicity of methoxyl-O compared to perchlorate-O, although neither C12 nor C13 is bound directly to coordinated N. In all the three complexes **32–34** (Figure 10b–d), however, there are methoxyl-O contacts with C bound to N, and an explanation for this unexpectedly great activity of methoxyl-O may be that it is induced by a multiplicity of other interactions exceeding the dispersion of the 1,2,3-trimethoxyphenyl units. As noted for the earlier examples, these structures display clear stacking arrays of the trimethoxyphenyl rings with terpyridine units, but the interactions beyond dispersion within the stacked assemblies are of CH··· O or C··· O types only, not C··· C.

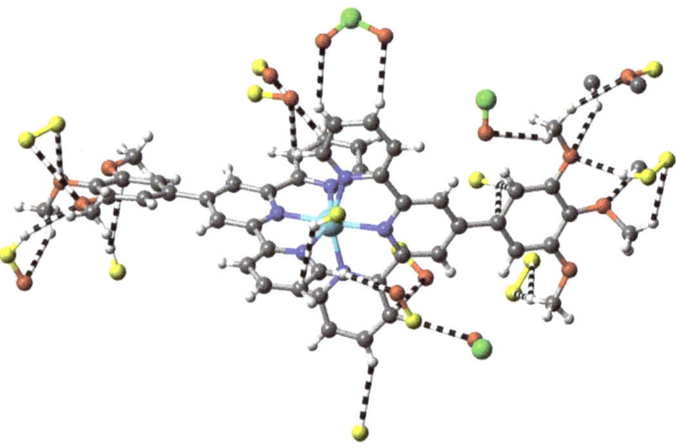

(a) [Zn(**tmptpy**)₂](ClO₄)₂·CH₃CN, **31**.

Figure 10. *Cont.*

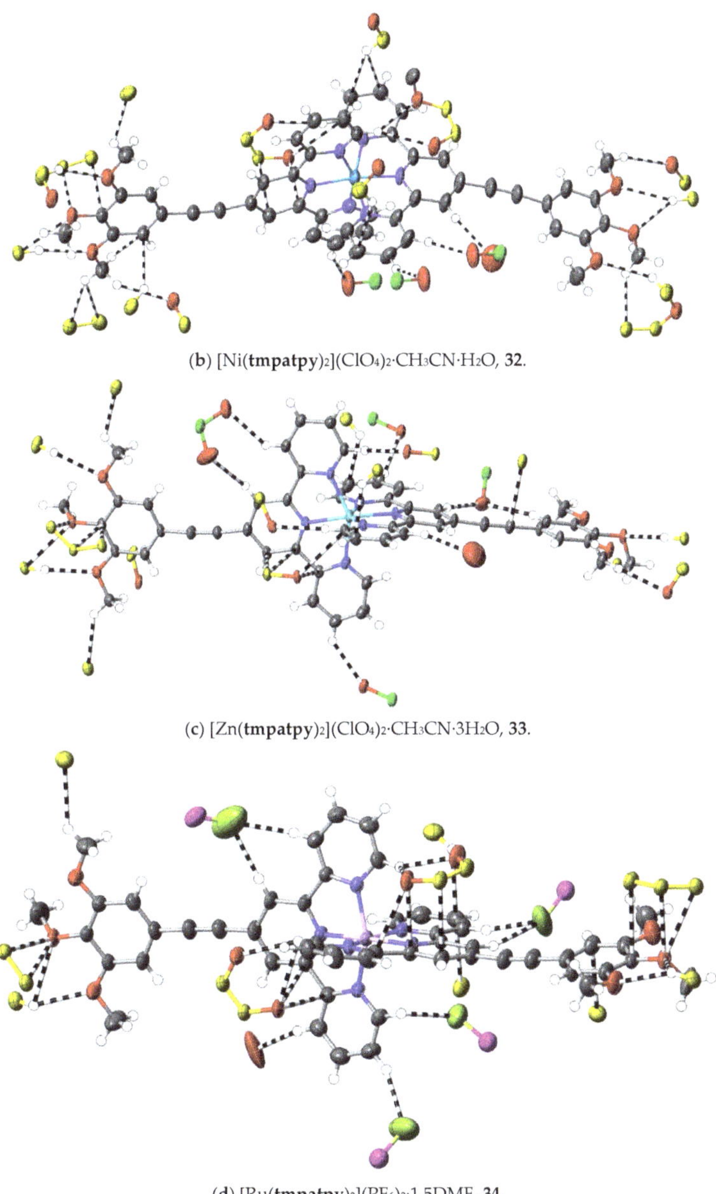

(**b**) [Ni(**tmpatpy**)₂](ClO₄)₂·CH₃CN·H₂O, **32**.

(**c**) [Zn(**tmpatpy**)₂](ClO₄)₂·CH₃CN·3H₂O, **33**.

(**d**) [Ru(**tmpatpy**)₂](PF₆)₂·1.5DMF, **34**.

Figure 10. Interactions beyond dispersion of complex cations incorporating trimethoxyphenyl-substituted 2,2′:6′,2″-terpyridine ligands.

5. Conclusions

Given that the structures reported herein, even with the addition of various known relatives in the literature, represent only a small fraction of the total number known (5200 for terpyridines generally and 1798 for complexes of 4′-substituted terpyridines in the Cambridge Structural Database [76]), it would be unwise to draw general conclusions from what has been seen in the Hirshfeld surface analysis conducted. Ideally, comparisons might be better made of isostructural species with common counteranions for structures determined at the same low temperature, but given the vagaries of growing crystals suitable for structure determination, this is probably not a feasible task, at least as the basis of an extensive investigation. What can be said from the present observations is that a consideration of the interactions exceeding dispersion alone in the crystals of terpyridine complexes offers a rather complicated description of any structure that is not readily related, in an experimental sense, to the dispersive interactions, which may be dominant in stacked arrays of aromatic units. It is, of course, unremarkable that cation$\cdots$anion interactions appear as major influences exceeding dispersion, but they are clearly in competition with other interactions, resulting in balance points that vary considerably, even for closely related species. The possibility that weak interactions exceeding dispersion may reflect ligand polarisation due to coordination—specifically, that with coordinated poly(aza-aromatic) ligands, carbon bound to nitrogen becomes significantly more electrophilic than in the free ligand—has some support in what is seen with highly charged (V(V) and U(VI)) cations but needs further investigation, perhaps through efforts to crystallise the complexes as simple hydroxide or fluoride salts. In relation, generally, to how the properties of any given complex might be influenced by its supramolecular (weak) interactions, it must be noted that the structures studied reveal the importance of, for example, non-classical CH interactions but show little involvement of ethynyl units, perhaps only because of the limited natures of other species (anions and solvents) present within the structures, making this an issue also needing further study.

Supplementary Materials: The following are available online at https://www.mdpi.com/2624-8549/3/1/16/s1. Figure S1: One cation of the structure of $[VO_2(\textbf{bptpy})]ClO_4$, **1**, showing its environment of other cations involved in $C\cdots C$ and $CH\cdots C$ interactions beyond dispersion with that cation. For clarity, only the ligand units of the surrounding cations involved in the interactions are shown (in yellow) and no H-atoms other than those involved in interactions are shown.; Figure S2: $O\cdots C$ contacts indicative of interactions beyond dispersion in stacked uranyl ion complexes. For clarity, stick representations are used and H-atoms are not shown.; Figure S3: Interactions beyond dispersion for cations in complexes of 4'-biphenyl-2,2':6',2''-terpyridine (**bptpy**). Structures reported in reference [63]. Colour coding as for Figure 1; Figure S4: Hirshfeld surfaces, d_{norm} representations in non-transparent mode to render more obvious the red regions indicative of interactions beyond dispersion, for (a) $[Fe(\textbf{bzOtpy})_2](ClO_4)_2$, **28**, and (b) $[Ni(\textbf{bzOtpy})_2](ClO_4)_2$, **29**, along with representations of the full interactions of the cations; Figure S5: Hirshfeld surface diagrams, d_{norm} representations, transparent mode, as obtained with CrystalExplorer for complexes of benzyloxy-substituted terpyridine ligands with or without methoxyl functionalisation; Table S1: Volume of voids per unit cell, number of electrons per void, and possible solvent molecules occupying the voids, as deduced from SQUEEZE results.

Author Contributions: Conceptualisation, Y.H.L., J.H. and Y.K.; methodology, Y.H.L., Y.K. and J.H.; supervision, J.H., Y.K. and S.H.; funding acquisition, Y.K. and S.H.; synthesis, Y.H.L., J.Y.K., S.K., H.O. and M.H.; crystallography and data analysis, P.T., Y.H.L. and J.H.; writing, J.H. All authors have read and agreed to the published version of the manuscript.

Funding: Partial funding for this work was provided by a KAKENHI Grant-in-Aid for Scientific Research (A) JP17H01200.

Data Availability Statement: CCDC depositions 2055133-2055135, 2055137-2055142, 2055148-2055161 and 2055219-2055228 contain the supplementary crystallographic data for this paper. These data can be obtained free of charge via www.ccdc.cam.ac.uk/data_request/cif, or by e-mailing data_request@ccdc.cam.ac.uk, or by contacting The Cambridge Crystallographic Data Centre, 12 Union Road, CambridgeCB2 1EZ, UK; Fax: +44-12-2333-6033.

Acknowledgments: Shinya Hayami is grateful for the support provided by the KAKENHI Grant-in-Aid for Scientific Research. Jack Harrowfield thanks Jean-Marie Lehn for the privilege of continuing, after retirement, to work within his group.

Conflicts of Interest: The authors declare no conflict of interest.

References

1. Hofmeier, H.; Schubert, U.S. Recent developments in the supramolecular chemistry of terpyridine–metal complexes. *Chem. Soc. Rev.* **2004**, *33*, 373–399. [CrossRef] [PubMed]
2. Schubert, U.S.; Hofmeier, H.; Newkome, G.R. *Modern Terpyridine Chemistry*; Wiley-VCH: Weinheim, Germany, 2006.
3. Constable, E.C. 2,2′:6′,2″-Terpyridines: From chemical obscurity to common supramolecular motifs. *Chem. Soc. Rev.* **2007**, *36*, 246–253. [CrossRef] [PubMed]
4. Constable, E.C.; Housecroft, C.E. 'Simple' Oligopyridine Complexes – Sources of Unexpected Structural Diversity. *Aust. J. Chem.* **2020**, *73*, 390–398. [CrossRef]
5. Wild, A.; Winter, A.; Schlütter, F.; Schubert, U.S. Advances in the field of π-conjugated 2,2′:6′,2″-terpyridines. *Chem. Soc. Rev.* **2011**, *40*, 1459–1511. [CrossRef]
6. Winter, A.; Newkome, G.R.; Schubert, U.S. Catalytic Applications of Terpyridines and their Transition Metal Complexes. *ChemCatChem* **2011**, *3*, 1384–1406. [CrossRef]
7. Hancock, R.D. The pyridyl group in ligand design for selective metal ion complexation and sensing. *Chem. Soc. Rev.* **2013**, *42*, 1500–1524. [CrossRef]
8. Mutai, T.; Satou, H.; Araki, K. Reproducible on–off switching of solid-state luminescence by controlling molecular packing through heat-mode interconversion. *Nat. Mater.* **2005**, *4*, 685–687. [CrossRef]
9. Scudder, M.L.; Goodwin, H.A.; Dance, I.G. Crystal supramolecular motifs: Two-dimensional grids of terpy embraces in $[ML_2]^z$ complexes (L = terpy or aromatic N_3-tridentate ligand). *New J. Chem.* **1999**, *23*, 695–705. [CrossRef]
10. McMurtrie, J.; Dance, I. Engineering grids of metal complexes: Development of the 2D M(terpy)$_2$ embrace motif in crystals. *CrystEngComm* **2005**, *7*, 216–219. [CrossRef]
11. Dance, I.G.; Scudder, M.L. Molecules embracing in crystals. *CrystEngComm* **2009**, *11*, 2233–2247. [CrossRef]
12. Figgis, B.N.; Kucharski, E.S. Crystal structure of Bis(2,2′:6′,2″-terpyridyl)cobalt(II) iodide dihydrate at 295 K and at 120 K. *Aust. J. Chem.* **1983**, *36*, 1527–1535. [CrossRef]
13. McMurtrie, J.; Dance, I. Alternative metal grid structures formed by $[M(terpy)_2]^{2+}$ and $[M(terpyOH)_2]^{2+}$ complexes with small and large tetrahedral dianions, and by $[Ru(terpy)_2]^0$. *CrystEngComm* **2010**, *12*, 2700–2710. [CrossRef]
14. Kubota, E.; Lee, Y.H.; Fuyuhiro, A.; Kawata, S.; Harrowfield, J.M.; Kim, Y.; Hayami, S. Synthesis, structure, and luminescence properties of arylpyridine-substituted terpyridine Zn(II) and Cd(II) complexes. *Polyhedron* **2013**, *52*, 435–441. [CrossRef]
15. Janiak, C. A critical account on π–π stacking in metal complexes with aromatic nitrogen-containing ligands. *J. Chem. Soc. Dalton Trans.* **2000**, 3885–3896. [CrossRef]
16. Jennings, W.B.; Farrell, B.M.; Malone, J.F. Attractive Intramolecular Edge-to-Face Aromatic Interactions in Flexible Organic Molecules. *Acc. Chem. Res.* **2001**, *34*, 885–894. [CrossRef] [PubMed]
17. Waters, M.L. Aromatic interactions in model systems. *Curr. Opin. Chem. Biol.* **2002**, *6*, 736–741. [CrossRef]
18. Grimme, S. Do special noncovalent π-π stacking interactions really exist? *Angew. Chem. Int. Ed.* **2008**, *47*, 3430–3434. [CrossRef]
19. Ehrlich, S.; Moellmann, J.; Grimme, S. Dispersion-Corrected Density Functional Theory for Aromatic Interactions in Complex Systems. *Acc. Chem. Res.* **2013**, *46*, 916–926. [CrossRef] [PubMed]
20. Bloom, J.W.G.; Wheeler, S.E. Taking the Aromaticity out of Aromatic Interactions. *Angew. Chem. Int. Ed.* **2011**, *50*, 7847–7849. [CrossRef]
21. Martinez, C.R.; Iverson, B.L. Rethinking the term "pi-stacking". *Chem. Sci.* **2012**, *3*, 2191–2201. [CrossRef]
22. Waters, M.L. Aromatic Interactions. *Acc. Chem. Res.* **2013**, *46*, 873. [CrossRef]
23. Gavezzotti, A. The "Sceptical Chymist": Intermolecular Doubts and Paradoxes. *CrystEngComm* **2013**, *15*, 4027–4035. [CrossRef]
24. Malenov, D.P.; Zaric, S.D. Stacking Interactions between Indenyl Ligands of Transition Metal Complexes: Crystallographic and Density Functional Study. *Cryst. Growth Des.* **2020**, *20*, 4491–4502. [CrossRef]
25. Gavezzotti, A. Calculation of Intermolecular Interaction Energies by Direct Numerical Integration over Electron Densities. I. Electrostatic and Polarization Energies in Molecular Crystals. *J. Phys. Chem. B* **2002**, *106*, 4145–4154. [CrossRef]
26. Dunitz, J.D.; Gavezzotti, A. Molecular Recognition in Organic Crystals: Directed Intermolecular Bonds or Nonlocalized Bonding? *Angew. Chem. Int. Ed.* **2005**, *44*, 1766–1787. [CrossRef] [PubMed]
27. Dunitz, J.D.; Gavezzotti, A. Supramolecular Synthons: Validation and Ranking of Intermolecular Interaction Energies. *Cryst. Growth Des.* **2012**, *12*, 5873–5877. [CrossRef]
28. Henry, M. Nonempirical Quantification of Molecular Interactions in Supramolecular Assemblies. *ChemPhysChem* **2002**, *3*, 561–569. [CrossRef]
29. Spackman, M.A.; Jayatilaka, D. Hirshfeld Surface Analysis. *CrystEngComm* **2009**, *11*, 19–32. [CrossRef]
30. Wolff, S.K.; Grimwood, D.J.; McKinnon, J.J.; Turner, M.J.; Jayatilaka, D.; Spackman, M.A. (Eds.) *CrystalExplorer 3.1*; University of Western Australia: Perth, Australia, 2012.
31. Spackman, M.A. Molecules in Crystals. *Phys. Scr.* **2013**, *87*, 048103. [CrossRef]
32. McKenzie, C.F.; Spackman, P.R.; Jayatilaka, D.; Spackman, M.A. *CrystalExplorer* model energies and energy frameworks: Extension to metal coordination compounds, organic salts, solvates and open-shell systems. *IUCrJ* **2017**, *4*, 575–587. [CrossRef]
33. Lee, Y.H.; Kim, J.Y.; Kim, Y.; Hayami, S.; Shin, J.W.; Harrowfield, J.M.; Stefankiewicz, A.R. Lattice Interactions of Terpyridines and Their Derivatives – Free Terpyridines and Their Protonated Forms. *CrystEngComm* **2016**, *18*, 8059–8071. [CrossRef]

34. Lee, Y.H.; Kubota, E.; Fuyuhiro, A.; Kawata, S.; Harrowfield, J.M.; Kim, Y.; Hayami, S. Synthesis, structure and luminescence properties of Cu(II), Zn(II) and Cd(II) complexes with 4′-terphenylterpyridine. *Dalton Trans.* **2012**, *41*, 10825–10831. [CrossRef]

35. Lee, Y.H.; Fuyuhiro, A.; Harrowfield, J.M.; Kim, Y.; Sobolev, A.N.; Hayami, S. Quaterphenylterpyridine: Synthesis, structure and metal-ion-enhanced charge transfer. *Eur. J. Inorg. Chem.* **2013**, 5862–5870. [CrossRef]

36. Lee, Y.H.; Harrowfield, J.M.; Shin, J.W.; Won, M.S.; Rukmini, E.; Hayami, S.; Min, K.S.; Kim, Y. Supramolecular interactions of terpyridine-derived cores of metallomesogen precursors. *Int. J. Mol. Sci.* **2013**, *14*, 20729–20743. [CrossRef]

37. Lee, Y.H.; Won, M.S.; Harrowfield, J.M.; Kawata, S.; Hayami, S.; Kim, Y. Spin crossover in Co(II) metallorods – replacing aliphatic tails by aromatic. *Dalton Trans.* **2013**, *42*, 11507–11521. [CrossRef]

38. Janiak, C. Engineering coordination polymers towards applications. *Dalton Trans.* **2003**, 2781–2804. [CrossRef]

39. Wu, M.; Mao, J.; Guo, J.; Ji, S. The Use of a Bifunctional Copper Catalyst in the Cross-Coupling Reactions of Aryl and Heteroaryl Halides with Terminal Alkynes. *Eur. J. Org. Chem.* **2008**, 4050–4054. [CrossRef]

40. Constable, E.C.; Ward, M.D. Synthesis and co-ordination behaviour of 6′,6″-bis(2-pyridyl)-2,2′:4,4″:2″,2‴-quaterpyridine; 'back-to-back' 2,2′:6′,2″-terpyridine. *J. Chem. Soc. Dalton Trans.* **1990**, 1405–1409. [CrossRef]

41. Potts, K.T.; Konwar, D. Synthesis of 4′-vinyl-2,2′:6′,2″-terpyridine. *J. Org. Chem.* **1991**, *56*, 4815–4816. [CrossRef]

42. Grosshenny, V.; Romero, F.; Ziessel, R. Construction of Preorganized Polytopic Ligands via Palladium-Promoted Cross-Coupling Reactions. *J. Org. Chem.* **1997**, *62*, 1491–1500. [CrossRef]

43. Alcock, N.W.; Barker, P.R.; Haider, J.M.; Hannon, M.J.; Painting, C.L.; Pikramenou, Z.; Plummer, E.A.; Rissanen, K.; Saarenketo, P. Red and blue luminescent metallo-supramolecular coordination polymers assembled through π–π interactions. *J. Chem. Soc. Dalton Trans.* **2000**, 1447–1461. [CrossRef]

44. Collin, J.-P.; Guillerez, S.; Sauvage, J.-P.; Barigelletti, F.; De Cola, L.; Flamigni, L.; Balzani, V. Photoinduced processes in dyads and triads containing a ruthenium(II)-bis(terpyridine) photosensitizer covalently linked to electron donor and acceptor groups. *Inorg. Chem.* **1991**, *30*, 4230–4238. [CrossRef]

45. Tu, S.; Jia, R.; Jiang, B.; Zhang, J.; Zhang, Y.; Yao, C.; Ji, S. Kröhnke reaction in aqueous media: One-pot clean synthesis of 4′-aryl-2,2′:6′,2″-terpyridines. *Tetrahedron* **2007**, *63*, 381–388. [CrossRef]

46. Benniston, A.C.; Chapman, G.; Harriman, A.; Mehrabi, M.; Sams, C.A. Electron Delocalization in a Ruthenium(II) Bis(2,2′:6′,2″-terpyridyl) Complex. *Inorg. Chem.* **2004**, *43*, 4227–4233. [CrossRef] [PubMed]

47. Muro, M.L.; Castellano, F.N. Room temperature photoluminescence from [Pt(4′-CCR-tpy)Cl]⁺ complexes. *Dalton Trans.* **2007**, 4659–4665. [CrossRef]

48. Du, P.; Schneider, J.; Brennessel, W.W.; Eisenberg, R. Synthesis and Structural Characterization of a New Vapochromic Pt(II) Complex Based on the 1-Terpyridyl-2,3,4,5,6-pentaphenylbenzene (TPPPB) Ligand. *Inorg. Chem.* **2008**, *47*, 69–77. [CrossRef]

49. Popović, Z.; Busby, M.; Huber, S.; Calzaferri, G.; De Cola, L. Assembling Micro Crystals through Cooperative Coordinative Interactions. *Angew. Chem. Int. Ed.* **2007**, *46*, 8898–8902. [CrossRef]

50. Constable, E.C.; Cargill-Thompson, A.M.W.; Tocher, D.A.; Daniels, M.A.M. Synthesis, characterisation and spectroscopic properties of ruthenium(II)-2,2′:6′,2″terpyridine cordination triads. X-Ray structures of 4′-(N,N-dimethylamino)-2,2′:6′,2″-terpyridine and bis(4′-(N,N-dimethylamino)-2,2′:6′,2″-terpyridine)ruthenium(II) hexafluorophosphate acetonitrile solvate. *New J. Chem.* **1992**, *16*, 855–867.

51. Arvai, A.J.; Nielsen, C. *ADSC Quantum-210 ADX Program*; Area Detector System Corporation: Poway, CA, USA, 1983.

52. Otwinowski, Z.; Minor, W. Processing of X-ray diffraction data collected in oscillation mode. *Methods Enzymol.* **1997**, *276*, 307–326. [PubMed]

53. Sheldrick, G.M. Crystal Structure Refinement with SHELXL. *Acta Cryst.* **2015**, *C71*, 3–8.

54. Spek, A.L. *PLATON* SQUEEZE: A tool for the calculation of the disordered solvent contribution to the calculated structure factors. *Acta Cryst.* **2015**, *C71*, 9–18.

55. *CrystalMaker 8.7*; CrystalMaker Software Ltd.: Woodstock, UK, 2014.

56. Florio, P.; Coghlan, C.J.; Lin, C.-P.; Saito, K.; Campi, E.M.; Jackson, W.R.; Hearn, M.T.W. Isolation and Structure of a Hydrogen-bonded 2,2′:6′,2″-Terpyridin-4′-one Acetic Acid Adduct. *Aust. J. Chem.* **2014**, *67*, 651–656. [CrossRef]

57. Murguly, E.; Norsten, T.B.; Branda, N. Tautomerism of 4-hydroxyterpyridine in the solid, solution and gas phases: An X-ray, FT-IR and NMR study. *J. Chem. Soc. Perkin Trans. II* **1999**, 2789–2794. [CrossRef]

58. Fallahpour, R.-A.; Constable, E.C. Novel synthesis of substituted 4′-hydroxy-2,2′:6′,2″-terpyridines. *J. Chem. Soc. Perkin Trans. I* **1997**, 2263–2264. [CrossRef]

59. Liu, Y.; Hong, X.; Lu, W.-G.; Lai, S. A vanadium(V) terpyridine complex: Synthesis, characterization, cytotoxicity in vitro and induction of apoptosis in cancer cells. *Transition Met. Chem.* **2017**, *42*, 459–467.

60. Thangavelu, S.G.; Andrews, M.B.; Pope, S.J.A.; Cahill, C.L. Synthesis, Structures, and Luminescent Properties of Uranyl Terpyridine Aromatic Carboxylate Coordination Polymers. *Inorg. Chem.* **2013**, *52*, 2060–2069. [CrossRef] [PubMed]

61. Carter, K.P.; Kalaj, M.; Cahill, C.L. Probing the Influence of N-Donor Capping Ligands on Supramolecular Assembly in Molecular Uranyl Materials. *Eur. J. Inorg. Chem.* **2016**, 126–137. [CrossRef]

62. Lyczko, K.; Steczek, L. Crystal structure of a hydroxo-bridged dimeric uranyl complex with a 2,2′:6′,2″-terpyridine ligand. *J. Struct. Chem.* **2017**, *58*, 102–106. [CrossRef]

63. Gomez, G.E.; Ridenour, J.A.; Byrne, N.M.; Shevchenko, A.P.; Cahill, C.L. Novel Heterometallic Uranyl-Transition Metal Materials: Structure, Topology, and Solid State Photoluminescence Properties. *Inorg. Chem.* **2019**, *58*, 7243–7254. [CrossRef] [PubMed]

64. Junk, P.C.; Kepert, C.J.; Semenova, L.I.; Skelton, B.W.; White, A.H. The Structural Systematics of Protonation of Some Important Nitrogen-base Ligands. I. Some Univalent Anion Salts of Doubly Protonated 2, 2′:6′, 2″-Terpyridyl. *Zeit. Anorg. Allg. Chem.* **2006**, *632*, 1293–1302. [CrossRef]

65. Maslen, E.N.; Raston, C.L.; White, A.H. Crystal structure of bis(2,2′:6′,2″-terpyridyl)cobalt(II) bromide trihydrate. *J. Chem. Soc. Dalton Trans.* **1974**, 1803–1807. [CrossRef]

66. Nielsen, P.; Toftlund, H.; Bond, A.D.; Boas, J.F.; Pillbrow, J.R.; Hansen, G.R.; Noble, C.; Riley, M.J.; Neville, S.M.; Moubaraki, B.; et al. Systematic study of spin crossover and structure in [Co(terpyRX)$_2$](Y)$_2$ systems (terypyRX = 4′-alkoxy-2,2′:6′,2″-terpyridine, X = 4, 8,12, Y = BF$_4^-$, ClO$_4^-$, PF$_6^-$, BPh$_4^-$). *Inorg. Chem.* **2009**, *48*, 7033–7047. [CrossRef]

67. Figgis, B.N.; Kucharski, E.S.; White, A.H. Crystal structure of Bis(2,2′:6′,2″-terpyridyl)cobalt(III) chloride. *Aust. J. Chem.* **1983**, *36*, 1563–1571. [CrossRef]

68. Paul, J.; Spey, S.; Adams, H.; Thomas, J.A. Synthesis and structure of rhodium complexes containing extended terpyridyl ligands. *Inorg. Chim. Acta* **2004**, *357*, 2827–2832. [CrossRef]

69. Wickramasinghe, W.A.; Bird, P.H.; Serpone, N. Interligand pockets in polypyridyl complexes. Crystal and molecular structure of the bis(terpyridyl)chromium(III) cation. *Inorg. Chem.* **1982**, *21*, 2694–2698. [CrossRef]

70. Sugihara, H.; Hiratani, K. 1,10-phenanthroline derivatives as ionophores for alkali metal ions. *Coord. Chem. Rev.* **1996**, *148*, 285–299. [CrossRef]

71. Granifo, J.; Garland, M.T.; Baggio, R. Hydrogen bonding and π-stacking interactions in the zipper-like supramolecular structure of the monomeric cadmium(II) complex [Cd(pyterpy)(H$_2$O)(NO$_3$)$_2$] (pyterpy = 4′-(4-pyridyl)-2,2′:6′,2″-terpyridine). *Inorg. Chem. Commun.* **2004**, *7*, 77–81. [CrossRef]

72. Beves, J.E.; Bray, D.J.; Clegg, J.K.; Constable, E.C.; Housecroft, C.E.; Jolliffe, K.A.; Kepert, C.J.; Lindoy, L.F.; Neuburger, M.; Price, D.J.; et al. Expanding the 4,4′-bipyridine ligand: Structural variation in {M(pytpy)$_2$}$^{2+}$ complexes (pytpy = 4′-(4-pyridyl)-2,2′:6′,2″-terpyridine, M = Fe, Ni, Ru) and assembly of the hydrogen-bonded, one-dimensional polymer {[Ru(pytpy)(Hpytpy)]}$_n^{3n+}$. *Inorg. Chim. Acta* **2008**, *361*, 2582–2590. [CrossRef]

73. Metrangolo, P.; Resnati, G. Halogen Bonding: A Paradigm in Supramolecular Chemistry. *Chem. Eur. J.* **2001**, *7*, 2511–2519. [CrossRef]

74. Constable, E.C.; Harris, K.; Housecroft, C.E.; Neuburger, M.; Zampese, J.A. Turning {M(tpy)$_2$}$^{n+}$ embraces and CH$\cdots\pi$ interactions on and off in homoleptic cobalt(II) and cobalt(III) bis(2,2′:6′,2″-terpyridine) complexes. *CrystEngComm* **2010**, *12*, 2949–2961. [CrossRef]

75. Li, W.; Lu, Z.G. Diacetato{4′-[4-(benzyloxy)phenyl]-2,2′:6′,2″-terpyridine}zinc(II). *Acta Cryst.* **2009**, *E65*, m1672. [CrossRef]

76. Groom, C.R.; Bruno, I.J.; Lightfoot, M.P.; Ward, S.C. The Cambridge Structural Database. *Acta Cryst.* **2016**, *B72*, 171–179. [CrossRef] [PubMed]

Article

Manipulating the Conformation of 3,2′:6′,3″-Terpyridine in $[Cu_2(\mu\text{-}OAc)_4(3,2':6',3''\text{-tpy})]_n$ 1D-Polymers

Dalila Rocco, Samantha Novak, Alessandro Prescimone, Edwin C. Constable and Catherine E. Housecroft *

Department of Chemistry, University of Basel, BPR 1096, Mattenstrasse 24a, CH-4058 Basel, Switzerland;
dalila.rocco@unibas.ch (D.R.); samantha.novak@stud.unibas.ch (S.N.); alessandro.prescimone@unibas.ch (A.P.);
edwin.constable@unibas.ch (E.C.C.)
* Correspondence: catherine.housecroft@unibas.ch; Tel.: +41-612-071-008

Abstract: We report the preparation and characterization of 4′-([1,1′-biphenyl]-4-yl)-3,2′:6′,3″-terpyridine (**1**), 4′-(4′-fluoro-[1,1′-biphenyl]-4-yl)-3,2′:6′,3″-terpyridine (**2**), 4′-(4′-chloro-[1,1′-biphenyl]-4-yl)-3,2′:6′,3″-terpyridine (**3**), 4′-(4′-bromo-[1,1′-biphenyl]-4-yl)-3,2′:6′,3″-terpyridine (**4**), and 4′-(4′-methyl-[1,1′-biphenyl]-4-yl)-3,2′:6′,3″-terpyridine (**5**), and their reactions with copper(II) acetate. Single-crystal structures of the $[Cu_2(\mu\text{-}OAc)_4L]_n$ 1D-coordination polymers with L = **1–5** have been determined, and powder X-ray diffraction confirms that the single crystal structures are representative of the bulk samples. $[Cu_2(\mu\text{-}OAc)_4(\mathbf{1})]_n$ and $[Cu_2(\mu\text{-}OAc)_4(\mathbf{2})]_n$ are isostructural, and zigzag polymer chains are present which engage in π-stacking interactions between [1,1′-biphenyl]pyridine units. 1D-chains nest into one another to give 2D-sheets; replacing the peripheral H in **1** by an F substituent in **2** has no effect on the solid-state structure, indicating that bifurcated contacts (H...H for **1** or H...F for **2**) are only secondary packing interactions. Upon going from $[Cu_2(\mu\text{-}OAc)_4(\mathbf{1})]_n$ and $[Cu_2(\mu\text{-}OAc)_4(\mathbf{2})]_n$ to $[Cu_2(\mu\text{-}OAc)_4(\mathbf{3})]_n$, $[Cu_2(\mu\text{-}OAc)_4(\mathbf{4})]_n$, and $[Cu_2(\mu\text{-}OAc)_4(\mathbf{5})]_n \cdot n\text{MeOH}$, the increased steric demands of the Cl, Br, or Me substituent induces a switch in the conformation of the 3,2′:6′,3″-tpy metal-binding domain, and a concomitant change in dominant packing interactions to py–py and py–biphenyl face-to-face π-stacking. The study underlines how the 3,2′:6′,3″-tpy domain can adapt to different steric demands of substituents through its conformational flexibility.

Keywords: 3,2′:6′,3″-terpyridine; coordination polymer; copper(II) acetate; paddle-wheel building block; X-ray diffraction

Citation: Rocco, D.; Novak, S.; Prescimone, A.; Constable, E.C.; Housecroft, C.E. Manipulating the Conformation of 3,2′:6′,3″-Terpyridine in $[Cu_2(\mu\text{-}OAc)_4(3,2':6',3''\text{-tpy})]_n$ 1D-Polymers. *Chemistry* **2021**, *3*, 182–198. https://doi.org/10.3390/chemistry3010015

Received: 11 December 2020
Accepted: 26 January 2021
Published: 2 February 2021

Publisher's Note: MDPI stays neutral with regard to jurisdictional claims in published maps and institutional affiliations.

1. Introduction

For many chemists, the word "terpyridine" is synonymous with 2,2′:6′,2″-terpyridine (tpy), the coordination chemistry and applications of which are exceptionally well developed [1–5]. However, from the perspective of assembling coordination polymers and networks, the bis-chelating nature of tpy tends to restrict its use to {M(tpy)_2} units bearing peripheral functionalities that can act as metal-binding domains or polytopic ligands containing multiple tpy metal-binding domains. Such "expanded ligands" with the {M(tpy)_2} unit on the "inside" of the metalloligand have gained significant attention [6]. Ditopic bis(tpy) ligands designed to assemble metallomacrocycles with predetermined internal angles feature in the innovative work of Newkome and coworkers [7]. However, while retaining a terpyridine building block, the most efficient way to access 1D-coordination polymers and 2D- and 3D-coordination networks is to turn to other isomers of terpyridine [8]. The 4,2′:6′,4″- and 3,2′:6′,3″-isomers are synthetically accessible using either Kröhnke methodology [9] or the one-pot approach of Wang and Hanan [10]. Over the last decade, the coordination chemistry of 4,2′:6′,4″-terpyridines (4,2′:6′,4″-tpy (**I**), Scheme 1) has gained in popularity [8]. Functionalization in the 4′-position with coordinatively non-innocent substituents increases the connectivity of the building block, taking it from a V-shaped linker to a 3- (or higher) connecting node. It is noteworthy that the introduction of a 4′-(pyridin-4-yl) unit (**II**, Scheme 1) produces a 3-connecting building block analogous

115

to 2,4,6-tris(pyridyl)-1,3,5-triazine (**III**, Scheme 1) employed by Fujita and coworkers for the assembly of molecular capsules [11]. Like the nitrogen atoms of the triazine ring, the N-atom of the central pyridine in 4,2′:6′,4″-tpy does not bind a metal ion. The same is true in 3,2′:6′,3″-terpyridine (3,2′:6′,3″-tpy).

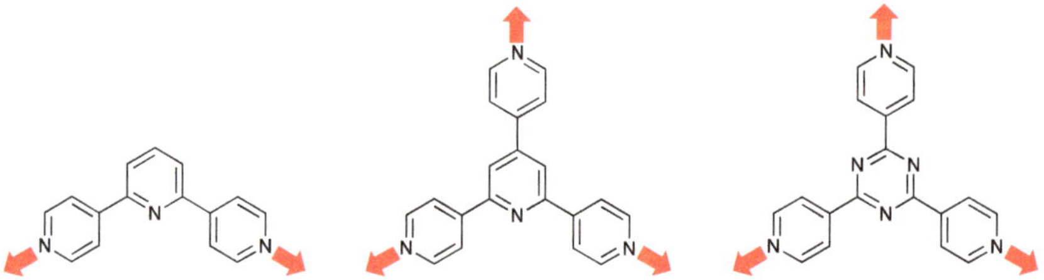

Scheme 1. 4,2′:6′,4″-Terpyridine as a ditopic ligand (**I**), and 4′-(pyridin-4-yl)-4,2′:6′,4″-terpyridine (**II**) as a 3-connecting building block compared to 2,4,6-tris(pyridyl)-1,3,5-triazine (**III**).

The coordination chemistry of 3,2′:6′,3″-tpy has received less attention than that of the 4,2′:6′,4″-isomer [8]. One aspect of 3,2′:6′,3″-tpy that makes predictive crystal engineering difficult is that rotation about the inter-ring C–C bonds (Scheme 2) leads to different ligand conformations and, therefore, to variable vectorial properties of the ditopic ligand. Scheme 2 illustrates the three possible planar conformations of 3,2′:6′,3″-tpy. We have recently demonstrated conformational switching in free 4′-(4-*n*-alkyloxyphenyl)-3,2′:6′,3″-terpyridines [12], and in [Cu$_2$(μ-OAc)$_4$L]$_n$ 1D-chains [13] and [Co(NCS)$_2$L$_2$]$_n$ 2D-networks where L is a 4′-(4-*n*-alkyloxyphenyl)-3,2′:6′,3″-tpy [14,15] as a function of the alkyloxy chain length. In these structures, the 3,2′:6′,3″-tpy domain adopts either conformation **I** or **II** (Scheme 2). Although conformation **III** is suited to the formation of discrete molecular architectures [16–20], it also appears in several infinite assemblies [21–24]. In one 3D-assembly, the 3,2′:6′,3″-tpy domain is locked into conformation **III** by virtue of a bridging cyano ligand between the coordinated metal centres of a single 3,2′:6′,3″-tpy unit, and propagation of the coordination network into 3-dimensions relies on the presence of a 4′-pyridin-4-yl substituent [25].

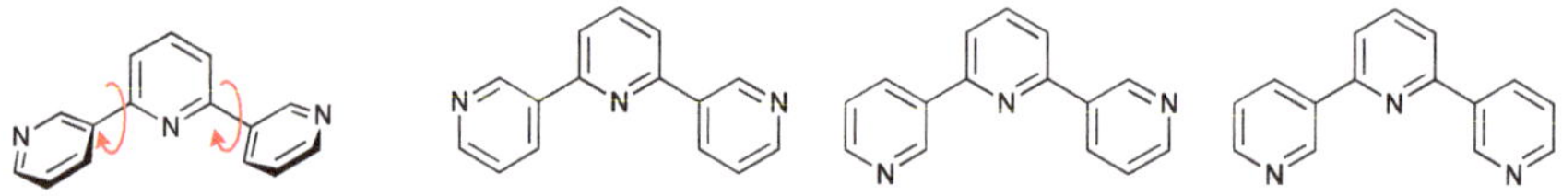

Scheme 2. Rotation about the inter-ring C–C bonds in 3,2′:6′,3″-tpy leads to three planar conformations **I–III**.

In the [Cu$_2$(μ-OAc)$_4${4′-(4-*n*-alkyloxyphenyl)-3,2′:6′,3″-tpy}]$_n$ 1D-polymers with alkyloxy groups being methoxy, butyloxy, pentyloxy, hexyloxy, or heptyloxy [13,26], the conformational variation is more complex than a switch from **I** to **II** [13]. With the ligand in conformation **II**, the arrangement in the two axial sites of the {Cu$_2$(μ-OAc)$_4$} paddle-wheel [27] can follow one of three assembly algorithms as shown in Scheme 3. The labels *in* and *out* refer to the orientation of the lone pair of each coordinating N atom with respect to the central N atom of the 3,2′:6′,3″-tpy unit. Both *in/out/in/out...* and *out/out/in/in...* sequences are observed in [Cu$_2$(μ-OAc)$_4${4′-(4-*n*-alkyloxyphenyl)-3,2′:6′,3″-tpy}]$_n$ chains, and we have proposed that this is related to the growing importance of inter-chain van der Waals forces as the length of the alkyloxy chain increases. These interactions complement π-stacking interactions between phenyl/pyridine and pyridine/pyridine rings [13]. Following from these results, we were interested in exploring the effects of replacing the alkyloxy tails by substituents in which π-stacking interactions would be dominant. We chose to focus on the 3,2′:6′,3″-tpy ligands featuring 1,1′-biphenyl units, and selected ligands **1**–**5** (Scheme 4). We have previously observed that π-stacking between pairs of 1,1′-biphenyl units in the solid state structures of 1D-coordination polymers involving 4′-([1,1′-biphenyl]-4-yl)-4,2′:6′,4″-tpy ligands is a dominant packing interaction, even when the peripheral phenyl ring is perfluorinated [28,29].

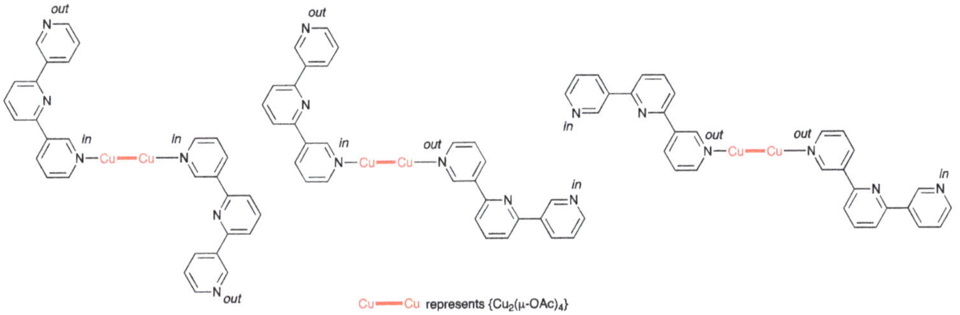

Scheme 3. In conformation **II**, there are three arrangements of the 3,2′:6′,3″-tpy units in the axial sites of the {Cu$_2$(μ-OAc)$_4$} unit if the two ligands are coplanar.

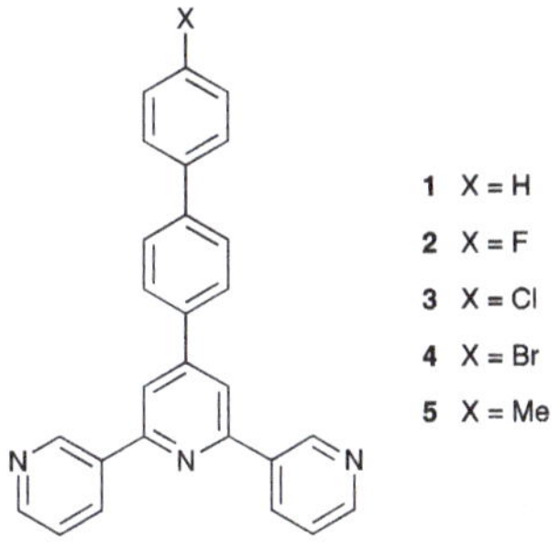

Scheme 4. Structures of ligands **1**–**5**.

2. Materials and Methods

2.1. General

^{1}H, ^{13}C{^{1}H}, ^{19}F{^{1}H}, and 2D NMR spectra were recorded on a Bruker Avance III-500 spectrometer equipped with a BBFO probehead (Bruker BioSpin AG, Fällanden, Switzerland) at 298 K. The ^{1}H and ^{13}C NMR chemical shifts were referenced with respect to residual solvent peaks (δ TMS = 0). A Shimadzu LCMS-2020 instrument (Shimadzu Schweiz GmbH, Roemerstr., Switzerland) was used to record electrospray ionization (ESI) mass spectra. A PerkinElmer UATR Two instrument (Perkin Elmer, 8603 Schwerzenbach, Switzerland) was used to record FT-infrared (IR) spectra, and a Shimadzu UV2600 (Shimadzu Schweiz GmbH, 4153 Reinach, Switzerland) spectrophotometer was used to record absorption spectra. 3-Acetylpyridine and [1,1′-biphenyl]-4-carbaldehyde were purchased from Acros Organics (Fisher Scientific AG, 4153 Reinach, Switzerland), 4′-chloro-[1,1′-biphenyl]-4-carbaldehyde and 4′-methyl-[1,1′-biphenyl]-4-carbaldehyde from Fluorochem (Chemie Brunschwig AG, 4052 Basel, Switzerland), 4′-fluoro-[1,1′-biphenyl]-4-carbaldehyde from Combi-Blocks (Chemie Brunschwig AG, 4052 Basel, Switzerland), and 4′-bromo-[1,1′-biphenyl]-4-carbaldehyde from Apollo (Chemie Brunschwig AG, 4052 Basel, Switzerland). Copper(II) acetate monohydrate was bought from Fluka (Fluka Chemie GmbH, 9471 Buchs, Switzerland). All chemicals were used as received.

All single-crystal growth experiments were carried out under ambient conditions using identical crystallization tubes (i.d. = 13.6 mm, 24 mL).

2.2. Compound 1

[1,1′-Biphenyl]-4-carbaldehyde (1.84 g, 10.0 mmol) was dissolved in EtOH (50 mL), and then 3-acetylpyridine (2.42 g, 2.20 mL, 20.0 mmol) and crushed KOH (1.12 g, 20.0 mmol) were added to the solution. Aqueous NH_3 (32%, 38.5 mL) was slowly added to the reaction mixture. This was stirred at room temperature (ca. 22 °C) overnight. The solid that formed was collected by filtration, washed with H_2O (3 × 10 mL) and EtOH (3 × 10 mL), recrystallized from MeOH, and dried in vacuo. Compound **1** was isolated as a colorless solid (1.54 g, 4.00 mmol, 40.0%). M.p. = 218 °C. ^1H NMR (500 MHz, $CDCl_3$): δ/ppm 9.40 (d, J = 2.2 Hz, 2H, H^{A2}), 8.72 (dd, J = 4.7, 1.7 Hz, 2H, H^{A6}), 8.53 (dt, J = 7.9, 2.2 Hz, 2H, H^{A4}), 8.01 (s, 2H, H^{B3}), 7.85 (d, J = 8.4 Hz, 2H, H^{C2}), 7.79 (d, J = 8.4 Hz, 2H, H^{C3}), 7.68 (m, 2H, H^{D2}), 7.50 (m, 2H, H^{D3}), 7.47 (m, 2H, H^{A5}), 7.42 (m, 1H, H^{D4}). $^{13}C\{^1H\}$ NMR (126 MHz, $CDCl_3$): δ/ppm 155.6 (C^{A3}), 150.6 (C^{B4}), 150.4 (C^{A6}), 148.6 (C^{A2}), 142.6 (C^{C4}), 140.3 (C^{D1}), 137.1 (C^{C1}), 134.8 (C^{B2}), 134.7 (C^{A4}), 129.1 (C^{D3}), 128.2 (C^{C3}), 128.0 (C^{D4}), 127.7 (C^{C2}), 127.3 (C^{D2}), 123.8 (C^{A5}), 117.7 (C^{B3}). UV-VIS (MeCN, 2.0 × 10^{-5} mol dm^{-3}) λ/nm 228 (ε/dm^3 mol^{-1} cm^{-1} 28,400), 265 sh (34,600), 284 (41,100). ESI-MS m/z 386.13 $[M + H]^+$ (calc. 386.17). Found C 83.91, H 4.81, N 11.04; required for $C_{27}H_{19}N_3$ C 84.13, H 4.97, N 10.90.

2.3. Compound 2

4′-Fluoro-[1,1′-biphenyl]-4-carbaldehyde (2.00 g, 10.0 mmol) was dissolved in EtOH (50 mL). 3-Acetylpyridine (2.42 g, 2.20 mL, 20.0 mmol) and crushed KOH (1.12 g, 20.0 mmol) were added, followed by the slow addition of aqueous NH_3 (32%, 38.5 mL). The reaction mixture was stirred at room temperature overnight, and the solid that formed was collected by filtration, washed with H_2O (3 × 10 mL) and EtOH (3 × 10 mL). The product was recrystallized from MeOH, and dried in vacuo. Compound **2** was isolated as a colorless solid (1.62 g, 4.03 mmol, 40.3%). M.p. = 215 °C. ^1H NMR (500 MHz, $CDCl_3$): δ/ppm 9.39 (d, J = 2.2 Hz, 2H, H^{A2}), 8.71 (dd, J = 4.7, 1.5 Hz, 2H, H^{A6}), 8.53 (dt, J = 8.0, 2.2 Hz, 2H, H^{A4}), 8.00 (s, 2H, H^{B3}), 7.84 (m, 2H, H^{C2}), 7.74 (m, 2H, H^{C3}), 7.63 (m, 2H, H^{D2}), 7.48 (ddd, J = 8.0, 4.7, 0.9 Hz, 2H, H^{A5}), 7.18 (m, 2H, H^{D3}). $^{13}C\{^1H\}$ NMR (126 MHz, $CDCl_3$): δ/ppm 162.9 (d, J_{CF} = 239 Hz, C^{D4}), 155.6 (C^{A3}), 150.5 (C^{B4}), 150.3 (C^{A6}), 148.5 (C^{A2}), 141.6 (C^{C4}), 137.1 (C^{C1}), 136.4 (d, J_{CF} = 2.5 Hz, C^{D1}), 134.85 (C^{B2}), 134.8 (C^{A4}), 128.9 (d, J_{CF} = 7.6 Hz, C^{D2}), 128.0 (C^{C3}), 127.8 (C^{C2}), 123.8 (C^{A5}), 117.7 (C^{B3}), 116.1 (d, J_{CF} = 22.5 Hz, C^{D3}). $^{19}F\{^1H\}$ NMR (470 MHz, $CDCl_3$): δ/ppm −114.6. UV-VIS (MeCN, 2.0 × 10^{-5} mol dm^{-3}) λ/nm 227 (ε/dm^3 mol^{-1} cm^{-1} 28,200), 264 sh (33,100), 284 (38,800). ESI-MS m/z 404.12 $[M + H]^+$ (calc. 404.16). Found C 79.08, H 4.59, N 10.42; required for $C_{27}H_{18}FN_3$ C 80.38, H 4.50, N 10.42.

2.4. Compound 3

4′-Chloro-[1,1′-biphenyl]-4-carbaldehyde (2.17 g, 10.0 mmol) was dissolved in EtOH (50 mL), and then 3-acetylpyridine (2.42 g, 2.20 mL, 20.0 mmol) and crushed KOH (1.12 g, 20.0 mmol) were added to the solution. Aqueous NH_3 (32%, 38.5 mL) was slowly added to the reaction mixture, which was then stirred at room temperature overnight. The solid that formed was collected by filtration, washed with H_2O (3 × 10 mL) and EtOH (3 × 10 mL), recrystallized from EtOH, and dried in vacuo. Compound **3** was isolated as a colorless solid (1.43 g, 3.40 mmol, 34.0%). M.p. = 240 °C. ^1H NMR (500 MHz, $CDCl_3$): δ/ppm 9.40 (dd, J = 2.2, 0.9 Hz, 2H, H^{A2}), 8.72 (dd, J = 4.8, 1.6 Hz, 2H, H^{A6}), 8.54 (dt, J = 8.1, 2.2 Hz, 2H, H^{A4}), 8.00 (s, 2H, H^{B3}), 7.84 (m, 2H, H^{C2}), 7.75 (m, 2H, H^{C3}), 7.60 (m, 2H, H^{D2}), 7.50–7.44 (overlapping m, 4H, H^{A5+D3}). $^{13}C\{^1H\}$ NMR (126 MHz, $CDCl_3$): δ/ppm 155.6 (C^{A3}), 150.4 (C^{B4}), 150.3 (C^{A6}), 148.5 (C^{A2}), 141.3 (C^{C4}), 138.7 (C^{D4}), 137.5 (C^{C1}), 134.84 (C^{B2}), 134.8 (C^{A4}), 134.2 (C^{D1}), 129.3 (C^{D3}), 128.5 (C^{D2}), 128.0 (C^{C3}), 127.8 (C^{C2}), 123.8 (C^{A5}), 117.7 (C^{B3}). UV-VIS (MeCN, 2.0 × 10^{-5} mol dm^{-3}) λ/nm 228 (ε/dm^3 mol^{-1} cm^{-1} 29,000), 264 sh (35,800), 288 (45,900). ESI-MS m/z 420.09 $[M + H]^+$ (calc. 420.13). Found C 76.58, H 4.12, N 9.87; required for $C_{27}H_{18}ClN_3$ C 77.23, H 4.32, N 10.01.

2.5. Compound 4

$4'$-Bromo-[1,1$'$-biphenyl]-4-carboxaldehyde (0.809 g, 3.1 mmol) was dissolved in EtOH (50 mL), then 3-acetylpyridine (0.751 g, 0.683 mL, 6.2 mmol) and crushed KOH (0.348 g, 6.2 mmol) were added. Aqueous NH_3 (32%, 11.9 mL) was slowly added to the reaction mixture, and this was stirred at room temperature overnight. The solid product was collected by filtration, washed with H_2O (3 × 10 mL) and EtOH (3 × 10 mL), and dried in vacuo. Compound **4** was isolated as a colorless solid (0.645 g, 1.39 mmol, 44.8%). M.p. = 248 °C. ^{1}H NMR (500 MHz, CDCl$_3$): δ/ppm 9.40 (dd, J = 2.3, 0.9 Hz, 2H, H^{A2}), 8.72 (dd, J = 4.8, 1.7 Hz, 2H, H^{A6}), 8.54 (dt, J = 8.0, 2.3 Hz, 2H, H^{A4}), 8.00 (s, 2H, H^{B3}), 7.84 (m, 2H, H^{C2}), 7.75 (m, 2H, H^{C3}), 7.62 (m, 2H, H^{D3}), 7.54 (m, 2H, H^{D2}), 7.44 (m, 2H, H^{A5}). ^{13}C{^{1}H} NMR (126 MHz, CDCl$_3$): δ/ppm 155.6 (C^{A3}), 150.4 (C^{B4}), 150.3 (C^{A6}), 148.5 (C^{A2}), 141.3 (C^{C4}), 139.2 (C^{D1}), 137.5 (C^{C1}), 134.83 (C^{B2}), 134.81 (C^{A4}), 132.3 (C^{D3}), 128.9 (C^{D2}), 128.0 (C^{C3}), 127.9 (C^{C2}), 123.9 (C^{A5}), 122.4 (C^{D4}), 117.7 (C^{B3}). UV-VIS (MeCN, 2.0×10^{-5} mol dm^{-3}) λ/nm 228 (ε/dm^3 mol^{-1} cm^{-1} 29,000), 264 sh (35,800), 292 (47,000). ESI-MS m/z 464.05 [M + H]$^+$ (calc. 464.08). Found C 69.55, H 3.76, N 8.89; required for C$_{27}$H$_{18}$ BrN$_3$ C 69.84, H 3.91, N 9.05.

2.6. Compound 5

$4'$-Methyl-[1,1$'$-biphenyl]-4-carbaldehyde (1.00 g, 5.09 mmol) was dissolved in EtOH (50 mL), and 3-acetylpyridine (1.23 g, 1.12 mL, 10.2 mmol) was added, followed by crushed KOH (0.571 g, 10.2 mmol). Aqueous NH_3 (32%, 19.6 mL) was added slowly to the reaction mixture, which was then stirred at room temperature overnight. The solid product was collected by filtration, washed with H_2O (3 × 10 mL) and EtOH (3 × 10 mL), recrystallized from EtOH, and dried in vacuo. Compound **5** was isolated as a colorless solid (0.670 g, 1.68 mmol, 32.9%). M.p. = 191 °C. ^{1}H NMR (500 MHz, CDCl$_3$): δ/ppm 9.40 (d, J = 0.9 Hz, 2H, H^{A2}), 8.72 (dd, J = 4.8, 1.7 Hz, 2H, H^{A6}), 8.53 (dt, J = 8.0, 2.0 Hz, 2H, H^{A4}), 8.01 (s, 2H, H^{B3}), 7.83 (d, J = 8.2 Hz, 2H, H^{C2}), 7.78 (d, J = 8.2 Hz, 2H, ^{HC3}), 7.58 (d, J = 8.0 Hz, 2H, H^{D2}), 7.48 (m, 2H, H^{A5}), 7.31 (d, J = 8.0 Hz, 2H, H^{D3}), 2.43 (s, 3H, H^{Me}). ^{13}C{^{1}H} NMR (126 MHz, CDCl$_3$): δ/ppm 155.6 (C^{A3}), 150.7 (C^{B4}), 150.3 (C^{A6}), 148.5 (C^{A2}), 142.5 (C^{C4}), 138.0 (C^{D4}), 137.3 (C^{D1}), 136.8 (C^{C1}), 134.9 (C^{B2}), 134.8 (C^{A4}), 129.8 (C^{D3}), 127.9 (C^{C3}), 127.7 (C^{C2}), 127.1 (C^{D2}), 123.8 (C^{A5}), 117.7 (C^{B3}), 21.3 (C^{Me}). UV-VIS (MeCN, 2.0×10^{-5} mol dm^{-3}) λ/nm 230 (ε/dm^3 mol^{-1} cm^{-1} 30,500), 261 sh (31,100), 295 (38,000). ESI-MS m/z 400.16 [M + H]$^+$ (calc. 400.18). Found C 83.89, H 5.30, N 10.39; required for C$_{28}$H$_{21}$N$_3$ C 84.18, H 5.30, N 10.52.

2.7. Crystal Growth of [Cu$_2$(μ-OAc)$_4$(**1**)]$_n$ and Preparative Scale Reaction

A solution of Cu$_2$(OAc)$_4$·2H$_2$O (12.0 mg, 0.030 mmol) in MeOH (5 mL) was layered over a CHCl$_3$ solution (4 mL) of **1** (11.6 mg, 0.030 mmol). Blue block-like crystals grew after 11 days. A single crystal was selected for X-ray diffraction, and the remaining crystals were washed with MeOH and CHCl$_3$, dried under vacuum, and analyzed by powder X-ray diffraction (PXRD) and FT-IR spectroscopy.

A blue solution of Cu$_2$(OAc)$_4$·2H$_2$O (79.9 mg, 0.200 mmol) in MeOH (30 mL) was added to a colorless CHCl$_3$ solution (10 mL) of **1** (77.1 mg, 0.200 mmol) in a round-bottomed flask. The blue solution was stirred at room temperature and after 1 h, a fine light-green suspension had formed. After 2 h, the suspension was centrifuged, and the solid was collected and dried in vacuo until it was a constant weight (6 h). [Cu$_2$(μ-OAc)$_4$(**1**)]$_n$ (24 mg, 0.032 mmol, 16%) was isolated as a light green powder. Found C 56.11, H 4.17, N 5.28; required for C$_{35}$H$_{31}$Cu$_2$N$_3$O$_8$: C 56.15, H 4.17, N 5.61. PXRD analysis was performed (see text).

2.8. Crystal Growth of [Cu$_2$(μ-OAc)$_4$(**2**)]$_n$ and Preparative Scale Reaction

A solution of Cu$_2$(OAc)$_4$·2H$_2$O (12.0 mg, 0.030 mmol) in MeOH (5 mL) was layered over a CHCl$_3$ solution (4 mL) of **2** (12.1 mg, 0.030 mmol), and after 20 days, blue blocks of X-ray quality had grown. A single crystal was selected for X-ray diffraction, and the rest of the crystals were washed with MeOH and CHCl$_3$, dried in vacuo, and were analyzed by PXRD and FT-IR spectroscopy.

A blue solution of Cu$_2$(OAc)$_4$·2H$_2$O (79.9 mg, 0.200 mmol) in MeOH (30 mL) was added to a colorless CHCl$_3$ solution (15 mL) of **2** (80.7 mg, 0.200 mmol) in a round-bottomed flask. The blue solution was stirred at room temperature, and after 1 h, a fine light green suspension had formed. After 2 h, the solid was collected using a centrifuge and was dried in vacuo until a constant weight was achieved (6 h). [Cu$_2$(μ-OAc)$_4$(**2**)]$_n$ (101 mg, 0.132 mmol, 66.0%) was isolated as a light-green solid. Found C 54.43, H 3.86, N 5.43; required for C$_{35}$H$_{30}$Cu$_2$FN$_3$O$_8$: C 54.83, H 3.94, N 5.48. See text for PXRD.

2.9. Crystal Growth of [Cu$_2$(μ-OAc)$_4$(3)]$_n$ and Preparative Scale Reaction

A MeOH (4 mL) solution of Cu$_2$(OAc)$_4$·2H$_2$O (12.0 mg, 0.030 mmol) was layered over a CHCl$_3$ solution (4 mL) of **3** (12.6 mg, 0.030 mmol). After eight days, X-ray quality green plate-like crystals had grown. One crystal was selected for single-crystal X-ray diffraction, and the remaining crystals were washed with MeOH and CHCl$_3$, dried under vacuum, and analyzed by PXRD and FT-IR spectroscopy.

A solution of Cu$_2$(OAc)$_4$·2H$_2$O (79.9 mg, 0.200 mmol) in MeOH (30 mL) was added to a solution of **3** (84.0 mg, 0.200 mmol) in CHCl$_3$ (15 mL) in a round-bottomed flask. The blue solution was stirred at room temperature, and after about 5 min, a fine light green suspension had formed. After 2 h, the suspension was centrifuged, and the solid was dried in vacuo until it was a constant weight (6 h). [Cu$_2$(μ-OAc)$_4$(**3**)]$_n$ (126 mg, 0.161 mmol, 80.5%) was isolated as a light green powder. Found C 53.13, H 3.74, N 5.30; required for C$_{35}$H$_{30}$Cu$_2$ClN$_3$O$_8$: C 53.68, H 3.86, N 5.37. See text for PXRD.

2.10. Crystal Growth of [Cu$_2$(μ-OAc)$_4$(4)]$_n$ and Preparative Scale Reaction

Cu$_2$(OAc)$_4$·2H$_2$O (12.0 mg, 0.030 mmol) was dissolved in MeOH (4 mL), and the solution was layered over a CHCl$_3$ solution (4 mL) of **4** (13.9 mg, 0.030 mmol). Green plate-like crystals had grown after 20 days, and a single crystal was selected for X-ray diffraction. The remaining crystals were washed with MeOH and CHCl$_3$, dried under vacuum, and analyzed by PXRD and FT-IR spectroscopy.

A solution of Cu$_2$(OAc)$_4$·2H$_2$O (79.9 mg, 0.200 mmol) in MeOH (30 mL) was added to a CHCl$_3$ solution (15 mL) of **4** (92.9 mg, 0.200 mmol) in a round-bottomed flask. The blue solution was stirred at room temperature, and a light-green suspension was observed after about 5 min. After 2 h, the suspension was centrifuged, and the solid was dried in vacuo to a constant weight (6 h). [Cu$_2$(μ-OAc)$_4$(**4**)]$_n$ (132 mg, 0.0797 mmol, 79.7%) was isolated as a light green powder. Found C 49.87, H 3.51, N 4.94; required for C$_{70}$H$_{60}$Cu$_4$Br$_2$N$_6$O$_{16}$: C 50.79, H 3.65, N 5.08. See text for PXRD.

2.11. Crystal Growth of [Cu$_2$(μ-OAc)$_4$(5)]$_n$·nMeOH and Preparative Scale Reaction

Cu$_2$(OAc)$_4$·2H$_2$O (12.0 mg, 0.030 mmol) was dissolved in MeOH (4 mL), and the solution was layered over a CHCl$_3$ solution (4 mL) of **5** (11.9 mg, 0.030 mmol). After 25 days, X-ray quality green plates had grown, and a single crystal was selected for X-ray diffraction. The rest of the crystals were washed with MeOH and CHCl$_3$, dried in vacuo, and analyzed by PXRD and FT-IR spectroscopy.

A solution of Cu$_2$(OAc)$_4$·2H$_2$O (79.9 mg, 0.200 mmol) in MeOH (30 mL) was added to a CHCl$_3$ solution (10 mL) of **5** (79.9 mg, 0.200 mmol) in a round-bottomed flask. The blue solution was stirred at room temperature, and a light-green suspension was observed after about 10 min. After 2 h, the suspension was centrifuged, and the solid was dried under vacuum until the weight was constant (6 h). [Cu$_2$(μ-OAc)$_4$(**5**)]$_n$ (108 mg, 0.0708 mmol, 70.8%) was isolated as a light green powder. Found C 56.30, H 4.27, N 5.50; required for C$_{72}$H$_{66}$Cu$_4$N$_6$O$_{16}$: C 56.69, H 4.36, N 5.51. See text for PXRD.

2.12. Crystallography

Single crystal data were collected on a Bruker APEX-II diffractometer (CuKα radiation) with data reduction, solution, and refinement using the programs APEX [30], ShelXT [31], Olex2 [32], and ShelXL v. 2014/7 [33], or using a STOE StadiVari diffractometer equipped with a Pilatus300K detector and with a Metaljet D2 source (GaKα radiation) and solving the structure using Superflip [34,35] and Olex2 [32]. See Sections 2.13–2.17 for the radiation type (Cu or Ga). The model was refined with ShelXL v. 2014/7 [33]. Structure analysis including the ORTEP-type representations used CSD Mercury 2020.1 [36]. In [Cu$_2$(μ-OAc)$_4$(**5**)]$_n$·nMeOH, the MeOH solvent molecule was disordered over two orientations and was modelled with 75% and 25% occupancy of the sites.

Powder X-Ray diffraction (PXRD) patterns were collected at room temperature in transmission mode using a Stoe Stadi P diffractometer equipped with a Cu Kα1 radiation (Ge(111) monochromator) and a DECTRIS MYTHEN 1K detector. Whole-pattern decomposition (profile matching) analysis [37–39] of the diffraction patterns was performed with the package FULLPROF SUITE [39,40] (version July-2019) using a previously determined instrument resolution function based on a NIST640d standard. The structural models were taken from the single crystal X-Ray diffraction refinements. Refined parameters in Rietveld were scale factor, zero shift, lattice parameters, Cu and halogen atomic positions, background points, and peaks shapes as a Thompson–Cox–Hastings pseudo-Voigt function. Preferred orientations as a March–Dollase multi-axial phenomenological model were incorporated into the analysis.

2.13. [Cu$_2$(μ-OAc)$_4$(**1**)]$_n$

C$_{35}$H$_{31}$Cu$_2$N$_3$O$_8$, M_r = 748.71, blue block, monoclinic, space group $C2/c$, a = 27.7823(14), b = 15.4445(11), c = 7.9423(4) Å, β = 102.301(4)°, V = 3329.7(3) Å^3, D_c = 1.494 g cm^{-3}, T = 150 K, Z = 4, μ(GaKα) = 7.203 mm^{-1}. Total 20,589 reflections, 3485 unique (R_{int} = 0.0693). Refinement of 2756 reflections (222 parameters) with $I > 2\sigma(I)$ converged at final R_1 = 0.0656 (R_1 all data = 0.0979), wR_2 = 0.1287 ($wR2$ all data = 0.1526), gof = 1.212. CCDC 2042171.

2.14. [Cu$_2$(μ-OAc)$_4$(**2**)]$_n$

C$_{35}$H$_{30}$Cu$_2$FN$_3$O$_8$, M_r = 766.70, blue block, monoclinic, space group $C2/c$, a = 27.6940(16), b = 15.9902(7), c = 7.8753(6) Å, β = 102.343(5)°, V = 3406.8(4) Å^3, D_c = 1.495 g cm^{-3}, T = 150 K, Z = 4, μ(GaKα) = 7.077 mm^{-1}. Total 26,608 reflections, 3531 unique (R_{int} = 0.0845). Refinement of 2854 reflections (227 parameters) with $I > 2\sigma(I)$ converged at final R_1 = 0.0907 (R_1 all data = 0.1116), wR_2 = 0.2310 ($wR2$ all data = 0. 2590), gof = 1.151. CCDC 2042172.

2.15. [Cu$_2$(μ-OAc)$_4$(**3**)]$_n$

C$_{35}$H$_{30}$Cu$_2$ClN$_3$O$_8$, M_r = 783.15, green plate, triclinic, space group P–1, a = 8.0001(3), b = 9.3586(3), c = 23.7240(7) Å, α = 99.3320(10), β = 96.370(2), γ = 98.442(2), V = 1717.11(10) Å^3, D_c = 1.515 g cm^{-3}, T = 150 K, Z = 2, Z' = 1, μ(CuKα) = 2.714 mm^{-1}. Total 20,409 reflections, 6355 unique (R_{int} = 0.0293). Refinement of 5729 reflections (446 parameters) with $I > 2\sigma(I)$ converged at final R_1 = 0.0457 (R_1 all data = 0.0502), wR_2 = 0.1286 ($wR2$ all data = 0.1340), gof = 1.054. CCDC 2042175.

2.16. [Cu$_2$(μ-OAc)$_4$(**4**)]$_n$

C$_{70}$H$_{60}$Cu$_4$Br$_2$N$_6$O$_{16}$, M_r = 1655.22, green plate, triclinic, space group P–1, a = 7.9898(5), b = 9.3656(5), c = 23.6072(13) Å, α = 98.626(2), β = 96.301(2), γ = 97.555(2)°, V = 1716.19(17) Å^3, D_c = 1.602 g cm^{-3}, T = 150 K, Z = 1, μ(CuKα) = 3.363 mm^{-1}. Total 27,701 reflections, 6338 unique (R_{int} = 0.0236). Refinement of 6120 reflections (446 parameters) with $I > 2\sigma(I)$ converged at final R_1 = 0.0482 (R_1 all data = 0.0493), wR_2 = 0.1317 ($wR2$ all data = 0.1328), gof = 1.072. CCDC 2042174.

2.17. [Cu$_2$(μ-OAc)$_4$(5)]$_n$·nMeOH

C$_{74}$H$_{74}$Cu$_4$N$_6$O$_{18}$, M_r = 1589.55, green plate, triclinic, space group *P*–1, *a* = 8.0240(4), *b* = 9.4295(5), *c* = 23.8856(11) Å, *α* = 100.7670(10), *β* = 95.673(2), *γ* = 97.244(2)°, *V* = 1747.30(15) Å^3, D_c = 1.511 g cm^{-3}, *T* = 150 K, *Z* = 1, μ(Cu*K*α) = 2.008 mm^{-1}. Total 21,488 reflections, 6397 unique (R_{int} = 0.0280). Refinement of 6197 reflections (477 parameters) with *I* > 2σ(*I*) converged at final R_1 = 0.0366 (R_1 all data = 0.0374), wR_2 = 0.1030 (*wR2* all data = 0.1038), gof = 1.053. CCDC 2042173.

2.18. Density Functional Theory (DFT) Calculations

DFT calculations on ligands **2–4** were carried out using Spartan'18 [41] with a B3LYP 6-31G* basis set with geometry optimization first carried out at the semi-empirical PM3 level.

3. Results and Discussion

3.1. Ligand Synthesis and Characterization

Compounds **1–5** were prepared using the one-pot method of Hanan [10] as shown in Scheme 5. The products precipitated from the reaction mixtures and were isolated in yields varying from 32.9% (for **5**) to 44.8% (for **4**). No attempts were made to optimize the reaction conditions. In the electrospray mass spectrum of each compound, the base peak arose from the [M + H]$^+$ ion (Figures S1–S5 in the Supporting Material) with characteristic isotope patterns observed for compounds **3** (chloro derivative) and **4** (bromo substituent).

The ^{1}H and ^{13}C{^{1}H} NMR spectra of compounds **1–5** were assigned with the aid of COSY, NOESY, HMQC, and HMBC techniques, and ^{1}H NMR, NOESY, HMQC, and HMBC spectra are shown in Figures S6–S25 in the Supporting Material. Figure 1 displays a comparison of the ^{1}H NMR spectra of **1–5**. The signals arising from the protons in rings A, B, and C (see Scheme 5 for ring labels) are unaffected by the change in the substituent in ring D. Assignments of the signals for H^{D2} and H^{D3} (Figure 1) were confirmed from the NOESY cross peaks between H^{C3} and H^{D2}, protons H^{C2} and H^{C3} being first distinguished using the NOESY H^{B3}/H^{C2} crosspeaks (compare Figures S7, S11, S15, S19 and S23). The change from X = H in **1** (Scheme 5) to the halogen substituents in **2**, **3**, and **4** and Me group in **5** has the most significant effect on H^{D3}, consistent with expectations [42]. The IR spectra of the new ligands are shown in Figures S26–S30.

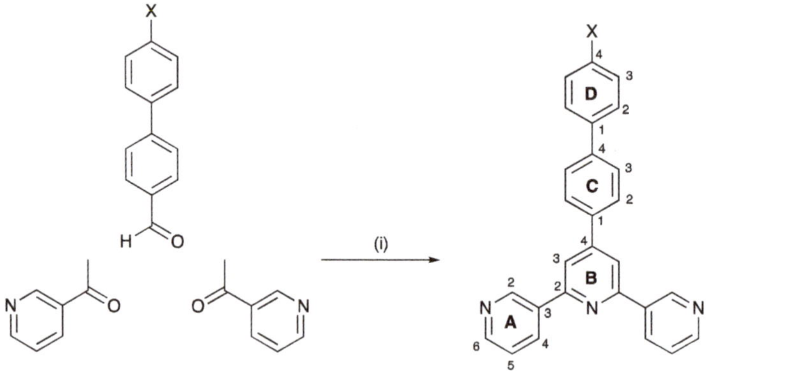

1 X = H; 2 X = F; 3 X = Cl; 4 X = Br; 5 X = Me

Scheme 5. Synthetic route to compounds **1–5**. Conditions: (i) KOH, EtOH; NH$_3$ (aqueous), room temperature, ca. 15 h. Atom numbering for the NMR spectroscopic assignments is given.

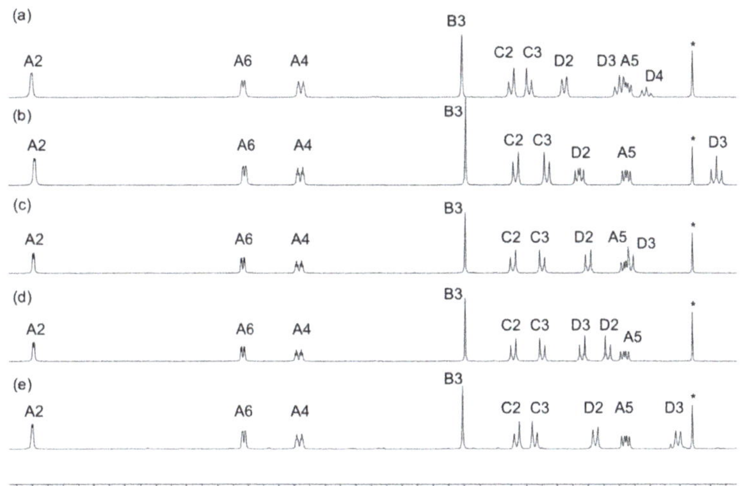

Figure 1. ^{1}H NMR spectra (aromatic region) for compounds (**a**) **1**, (**b**) **2**, (**c**) **3**, (**d**) **4** and (**e**) **5** (500 MHz, CDCl$_3$, 298 K). * = residual CHCl$_3$.

The absorption spectra of acetonitrile solutions of compounds **1–5** are shown in Figure 2a. Each spectrum is dominated by a broad and intense band arising principally from $\pi^*\leftarrow\pi$ transitions. For the three halogen-substituted compounds, the value of λ_{max} shifts from 284 nm (F) to 288 nm (Cl) to 292 nm (Br), consistent with a stabilization of the highest-occupied molecular orbital(s) for the more electron-withdrawing substituent. DFT calculations on ligands **2**, **3**, and **4** revealed that the highest occupied molecular orbital (HOMO) of each complex is localized on the 4′-halo-[1,1′-biphenyl] domain, while the lowest unoccupied molecular orbital (LUMO) manifold is largely localized on the 3,2′:6′,3″-tpy unit (Figure 2b). The HOMO–1 is, in each case, localized on the 3,2′:6′,3″-tpy.

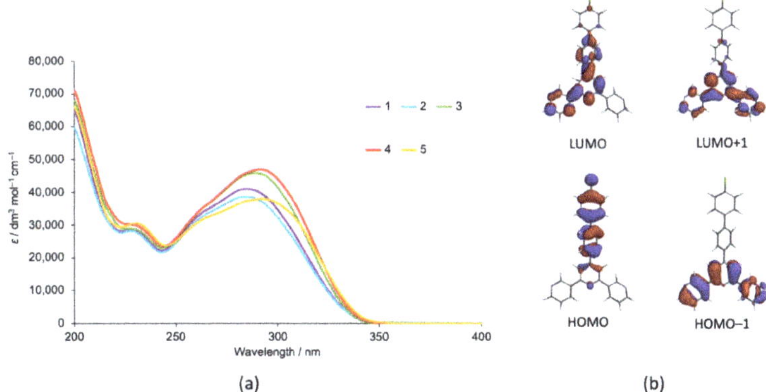

Figure 2. (**a**) Solution absorption spectra (MeCN, 2.0 × 10^{-5} mol dm^{-3}) of compounds **1–5**. (**b**) Highest occupied and lowest unoccupied molecular orbital compositions in compound **3**; similar compositions are seen in **2** and **4**.

3.2. Reactions of Copper(II) Acetate and Ligands **1–5**

Ligands **1–5** were allowed to react with copper(II) acetate under ambient conditions by layering a methanol solution of Cu$_2$(OAc)$_4$·2H$_2$O over a chloroform solution of the appropriate ligand. Single crystals grew within days or several weeks, and after selection of crystals for single crystal X-ray analysis, the remaining crystals were washed with MeOH and CHCl$_3$, dried, and analyzed by PXRD to confirm that the single crystals were representative of the bulk sample (see Section 3.4). The solid-state IR spectra of the bulk materials are all similar and are presented in Figures S31–S35 in the Supporting Material. Yields of the products from the single-crystal growth experiments were not optimized, and were in the range 20–30% if crystal growth was allowed to continue for a month.

The reactions were also carried out on a preparative scale by combining a methanol solution of Cu$_2$(OAc)$_4$·2H$_2$O with a chloroform solution of the respective ligand. The precipitate that formed was separated by centrifugation, dried, and analyzed by elemental analysis and PXRD. Elemental analytical data were in accord with the compositions [Cu$_2$(μ-OAc)$_4$(**L**)]$_n$ with **L** = **1**–**5**. The PXRD data are discussed in Section 3.4.

3.3. Single Crystal Structures

The single-crystal structures of the five copper(II) complexes confirmed the assembly of one-dimensional coordination polymers containing the ubiquitous {Cu$_2$(OAc)$_4$} paddle-wheel motif. The polymers [Cu$_2$(μ-OAc)$_4$(**1**)]$_n$ and [Cu$_2$(μ-OAc)$_4$(**2**)]$_n$ crystallize in the monoclinic space group *C*2/*c* with similar cell dimensions (see Sections 2.13 and 2.14). In contrast, the [Cu$_2$(μ-OAc)$_4$(**3**)]$_n$, [Cu$_2$(μ-OAc)$_4$(**4**)]$_n$, and [Cu$_2$(μ-OAc)$_4$(**5**)]$_n$·nMeOH crystallize in the triclinic space group *P*–1, again with similar cell dimensions for the series of compounds. Only the coordination compound containing ligand **5** contains lattice solvent. ORTEP-type diagrams of the repeating units in each coordination polymer are displayed in Figures 3 and 4, and selected bond lengths are given in Table 1. The bond parameters for the {Cu$_2$(μ-OAc)$_4$} units are unexceptional, and the Cu–N bond distances are typical with the exception of Cu1–N1 in [Cu$_2$(μ-OAc)$_4$(**5**)]$_n$·nMeOH. This bond is somewhat elongated (2.1813(19) Å), and this appears to be associated with the presence of a MeOH molecule, which is hydrogen-bonded to one acetato bridge (Figure 5) and resides in a pocket close to one Cu–N bond. The five coordination polymers fall into two structural classes, which differ in the conformation of the 3,2′:6′,3″-tpy unit. In [Cu$_2$(μ-OAc)$_4$(**1**)]$_n$ and [Cu$_2$(μ-OAc)$_4$(**2**)]$_n$, the asymmetric unit contains half of a ligand molecule **1** or **2**, and the 3,2′:6′,3″-tpy unit adopts conformation **I** (Scheme 2). In the compounds containing ligands **3**, **4**, and **5**, the 3,2′:6′,3″-tpy domain is in conformation **II**. The angles between the planes of pairs of adjacent aromatic rings are compiled in Table 2, and the data reveal that the 3,2′:6′,3″-tpy unit is closer to being planar in [Cu$_2$(μ-OAc)$_4$(**3**)]$_n$, [Cu$_2$(μ-OAc)$_4$(**4**)]$_n$, and [Cu$_2$(μ-OAc)$_4$(**5**)]$_n$·nMeOH than in [Cu$_2$(μ-OAc)$_4$(**1**)]$_n$ and [Cu$_2$(μ-OAc)$_4$(**2**)]$_n$. In addition, the angles between the planes of adjacent rings in the central three-arene ring unit are greater in coordinated ligands **1** and **2** than in **3**–**5**. An inspection of the dominant packing interactions provides an insight into these differences.

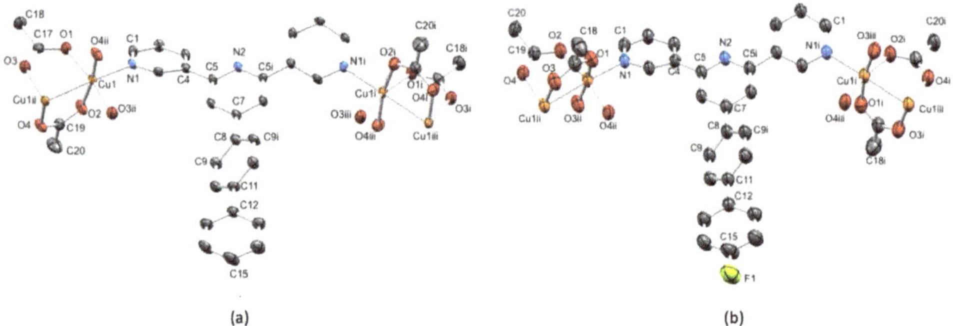

(a) (b)

Figure 3. The repeat units (with symmetry generated atoms) in (**a**) [Cu$_2$(μ-OAc)$_4$(**2**)]$_n$ (ellipsoids plotted at 50% probability level; symmetry codes: i = 1 − *x*, *y*, $^3/_2$ − *z*; ii = $^3/_2$ − *x*, $^1/_2$ − *y*, 1 − *z*; iii = − $^1/_2$ + *x*, $^1/_2$ − *y*, $^1/_2$ + *z*) and (**b**) [Cu$_2$(μ-OAc)$_4$(**2**)]$_n$ (ellipsoids plotted at 40% probability level; symmetry codes: i = 1 − *x*, *y*, $^1/_2$ − *z*; ii = $^1/_2$ − *x*, $^3/_2$ − *y*, 1 − *z*; iii = $^1/_2$ + *x*, $^3/_2$ − *y*, − $^1/_2$ + *z*). H atoms are omitted for clarity.

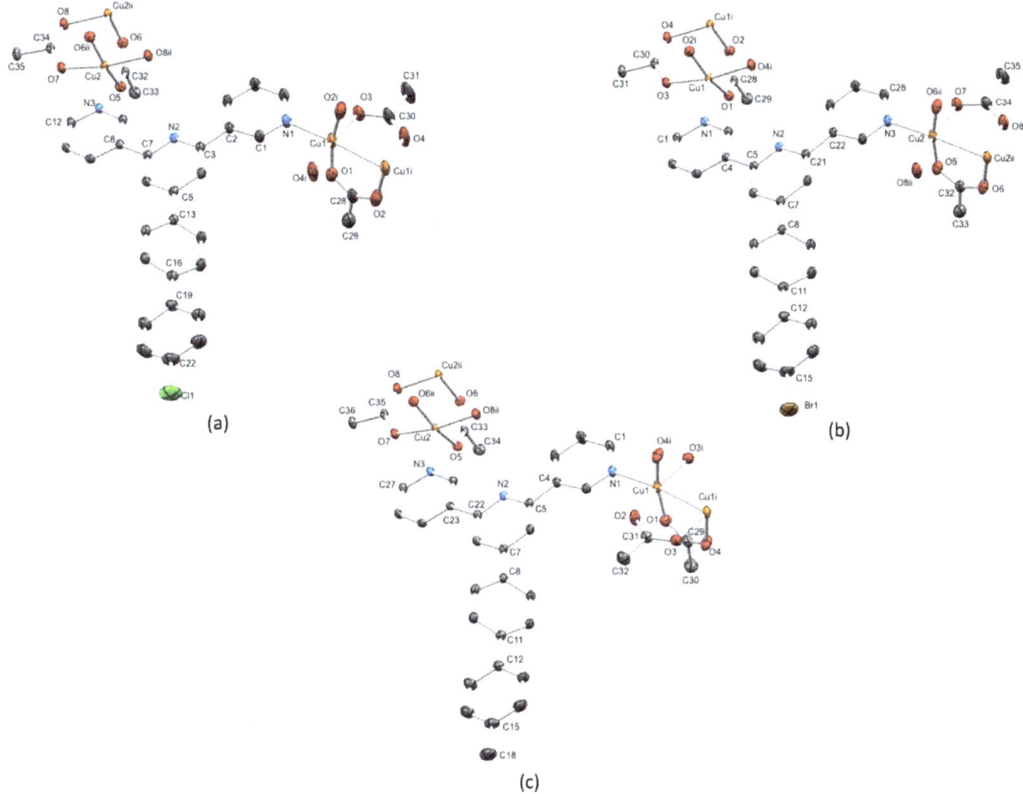

Figure 4. The repeat units (with symmetry generated atoms) in (**a**) [Cu$_2$(μ-OAc)$_4$(**3**)]$_n$ (symmetry codes: i = 1 − x, 2 − y, 2 − z; ii = 2 − x, 2 − y, 1 − z), (**b**) [Cu$_2$(μ-OAc)$_4$(**4**)]$_n$ (symmetry codes: i = 2 − x, 2 − y, 1 − z; ii = 1 − x, 2 − y, 2 − z), and (**c**) [Cu$_2$(μ-OAc)$_4$(**5**)]$_n$·nMeOH (solvent molecule omitted; symmetry codes: i = 1 − x, 2 − y, 2 − z; ii = 2 − x, 2 − y, 1 − z), All ellipsoids are plotted at 40% probability level, and H atoms are omitted for clarity.

Table 1. Selected bond lengths in the copper(II) coordination polymers.

	Cu–O/Å	Cu–N/Å	Cu … Cu/Å
[Cu$_2$(μ-OAc)$_4$(**1**)]$_n$	1.954(4), 1.975(4), 1.990(4), 1.961(4)	2.157(4)	2.6051(13)
[Cu$_2$(μ-OAc)$_4$(**2**)]$_n$	1.953(5), 1.994(5), 1.979(5), 1.959(5)	2.168(5)	2.6149(17)
[Cu$_2$(μ-OAc)$_4$(**3**)]$_n$	1.9760(18), 1.9894(18), 1.9759(18), 1.9789(18), 1.979(3), 1.970(2), 1.966(3), 1.971(3)	2.167(2), 2.151(2)	2.6292(8), 2.6319(7)
[Cu$_2$(μ-OAc)$_4$(**4**)]$_n$	1.973(2), 1.983(2), 1.976(2), 1.979(2), 1.981(3), 1.971(2), 1.972(3), 1.974(3)	2.153(2), 2.158(3)	2.6352(8), 2.6235(9)
[Cu$_2$(μ-OAc)$_4$(**5**)]$_n$·nMeOH	1.9885(15), 1.9787(15), 1.9773(15), 1.9756(15), 1.9887(18), 1.9700(18), 1.9809(19), 1.9644(19)	2.1510(18), 2.1813(19)	2.6551(6), 2.6331(6)

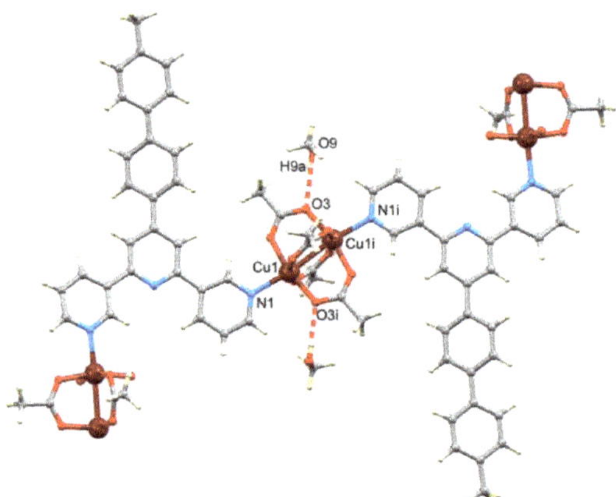

Figure 5. Hydrogen-bonded MeOH molecule in $[Cu_2(\mu\text{-OAc})_4(\mathbf{5})]_n \cdot n\text{MeOH}$, which resides in a pocket close to the Cu1–N1 bond. Symmetry code: $i = 1 - x, 2 - y, 2 - z$.

Table 2. Angles between ring-planes in the copper(II) coordination polymers.

	py–py/°	py$_{N2}$–Phenylene/°	Phenylene–phenyl/°
$[Cu_2(\mu\text{-OAc})_4(\mathbf{1})]_n$	25.1	38.3	41.7
$[Cu_2(\mu\text{-OAc})_4(\mathbf{2})]_n$	21.1	39.8	38.5
$[Cu_2(\mu\text{-OAc})_4(\mathbf{3})]_n$	8.0, 4.5	24.0	27.2
$[Cu_2(\mu\text{-OAc})_4(\mathbf{4})]_n$	8.1, 4.0	22.9	26.5
$[Cu_2(\mu\text{-OAc})_4(\mathbf{5})]_n \cdot n\text{MeOH}$	6.5, 4.0	27.8	28.1

In $[Cu_2(\mu\text{-OAc})_4(\mathbf{1})]_n$ and $[Cu_2(\mu\text{-OAc})_4(\mathbf{2})]_n$, ligand conformation **I** leads to a zigzag profile for each 1D-polymer chain, and adjacent chains are arranged with the biphenyl unit directed into the V-shaped cavity of a neighboring 3,2′:6′,2″-tpy unit (Figure 6). This leads to the assembly of 2D-sheets. Interestingly, the arrangement shown in Figure 6 results in short repulsive H...H contacts in $[Cu_2(\mu\text{-OAc})_4(\mathbf{1})]_n$, while these are replaced by attractive H...F contacts in $[Cu_2(\mu\text{-OAc})_4(\mathbf{2})]_n$. This observation suggests that these contacts are not important in supporting the assembly, and this is reminiscent of the isostructural nature of $[Cu_2(\mu\text{-OAc})_4(\mathbf{6})]_n$ and $[Cu_2(\mu\text{-OAc})_4(\mathbf{7})]_n$, in which **6** is 4′-([1,1′-biphenyl]-4-yl)-4,2′:6′,4″-terpyridine and **7** is 4′-(2′,3′,4′,5′,6′-pentafluoro [1,1′-biphenyl]-4-yl)-4,2′:6′,4″-terpyridine [29]. The X...C separations for the X...H–C contacts in the bifurcated interactions in Figure 6 are 3.11 Å for X = H and 3.16 Å for X = F. The dominant packing forces in $[Cu_2(\mu\text{-OAc})_4(\mathbf{1})]_n$ and $[Cu_2(\mu\text{-OAc})_4(\mathbf{2})]_n$ are the head-to-tail π-stacking between pairs of [1,1′-biphenyl]pyridine units displayed in Figure 7. Each pair of arene rings adopts an offset arrangement, which is optimal for a π–π interaction [43]. For the pyridine...phenyl interaction in $[Cu_2(\mu\text{-OAc})_4(\mathbf{1})]_n$, the centroid...centroid distance is 3.97 Å and the angle between the ring planes is 3.4°; the corresponding parameters in $[Cu_2(\mu\text{-OAc})_4(\mathbf{2})]_n$ are 3.97 Å and 1.4°. For the centrosymmetric pair of phenylene rings, the distances between the ring planes and between their centroids are 3.75 Å and 3.97 Å, respectively, in $[Cu_2(\mu\text{-OAc})_4(\mathbf{1})]_n$, are 3.71 Å, and 3.96 Å in $[Cu_2(\mu\text{-OAc})_4(\mathbf{2})]_n$. The cavities in each sheet visible in Figure 7 are occupied by $\{Cu_2(\mu\text{-OAc})_4\}$ carboxylate groups from an adjacent layer, which protrude above and below each sheet.

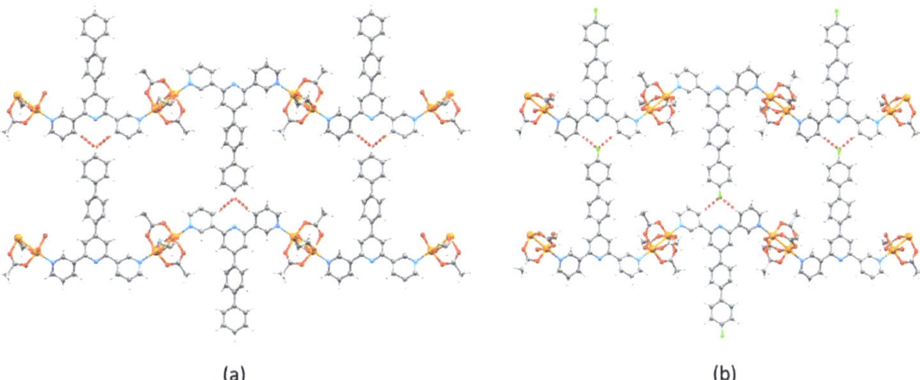

Figure 6. Packing of two adjacent polymer chains in (**a**) $[Cu_2(\mu\text{-OAc})_4(\mathbf{1})]_n$ and (**b**) $[Cu_2(\mu\text{-OAc})_4(\mathbf{2})]$ within one 2D-sheet (see text). The hashed red lines highlight short H...H contacts in $[Cu_2(\mu\text{-OAc})_4(\mathbf{1})]_n$ and complementary short H...F contacts in $[Cu_2(\mu\text{-OAc})_4(\mathbf{2})]_n$.

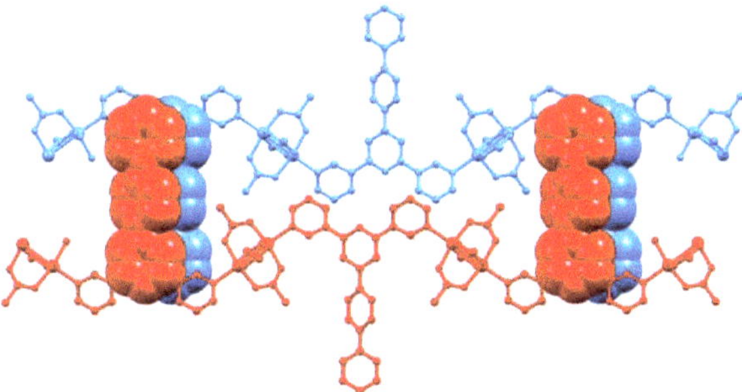

Figure 7. Head-to-tail π-stacking between pairs of [1,1′-biphenyl]pyridine units in adjacent sheets in $[Cu_2(\mu\text{-OAc})_4(\mathbf{1})]_n$. The same motif is present in $[Cu_2(\mu\text{-OAc})_4(\mathbf{2})]_n$. H atoms are omitted.

In each of $[Cu_2(\mu\text{-OAc})_4(\mathbf{3})]_n$, $[Cu_2(\mu\text{-OAc})_4(\mathbf{4})]_n$, and $[Cu_2(\mu\text{-OAc})_4(\mathbf{5})]_n \cdot n\text{MeOH}$, the 3,2′:6′,3″-tpy adopts conformation **II** (Scheme 2), and the coordination arrangement at the paddle-wheel units (defined in Scheme 3) is *in/in/out/out...* Figure 8a illustrates part of one coordination polymer chain in $[Cu_2(\mu\text{-OAc})_4(\mathbf{3})]_n$, and this structure is replicated in $[Cu_2(\mu\text{-OAc})_4(\mathbf{4})]_n$ and $[Cu_2(\mu\text{-OAc})_4(\mathbf{5})]_n \cdot n\text{MeOH}$, as are the packing motifs described below. Figure 8b illustrates the interdigitation of 1D-polymer chains to produce 2D-sheets. The profile of the chain in Figure 8a contrasts with the zigzag nature of the polymers in Figure 7, and packing interactions are necessarily different. The near planarity of the 3,2′:6′,3″-tpy unit (Table 2) reflects the involvement of this domain in crystal packing. Centrosymmetric pairs of pyridine rings containing N3 (N3 and N3^{iii}, symmetry code iii = $1 - x$, $1 - y$, $1 - z$) stack with an interplane distance of 3.34 Å and inter-centroid separation of 3.68 Å. The pyridine ring with N1 engages in a face-to-face contact with the phenyl ring containing C22^{iv} (symmetry code iv = $1 + x$, $1 + y$, z) with a centroid...centroid distance of 3.94 Å and an angle between the ring planes of 11.1°. In addition, the pyridine ring containing N1 also sits over a biphenyl unit, thereby extending the stack of arene rings. The projection shown in Figure 8c illustrates how the layers comprise domains of π-stacked arene rings and columns of $\{Cu_2(\mu\text{-OAc})_4\}$ paddle-wheel units (see also Figure S36 in the Supporting Material).

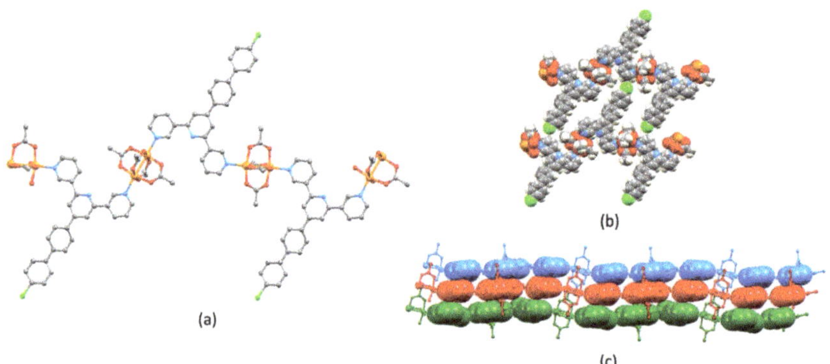

Figure 8. (**a**) Part of one polymer chain in [Cu$_2$(μ-OAc)$_4$(**3**)]$_n$, (**b**) interdigitation of chains generates a 2D-sheet, and (**c**) stacking of three adjacent sheets involves both pyridine and biphenyl rings.

3.4. Characterization by PXRD

To ensure that the single crystal structures were representative of the bulk materials, powder X-ray diffraction patterns were determined for crystals remaining in the crystallization tubes after single crystals had been selected. The refinements (Figures S37–S41) confirmed that the bulk materials of all the compounds were representative of the analyzed single crystals. Each peak in the experimental plots has a corresponding peak in the fitted spectra, and the differences in the intensities can be rationalized in terms of differences in the preferred orientations. Only [Cu$_2$(μ-OAc)$_4$(**2**)]$_n$ (Figure S38) shows minor impurities (ca. 10%). A comparison of Figure S38 with PXRD patterns for the precursors **2** and Cu$_2$(OAc)$_4$·2H$_2$O did not reveal matching peaks.

PXRD was also carried out on the materials obtained from the preparative scale reactions. The powder patterns matched those of the materials obtained from the single-crystal growth experiments. Figure 9 displays the data for [Cu$_2$(μ-OAc)$_4$(**1**)]$_n$ as a representative example, and the superimpositions of the powder patterns for the remaining four coordination polymers are shown in Figures S42–S45 in the Supporting Materials.

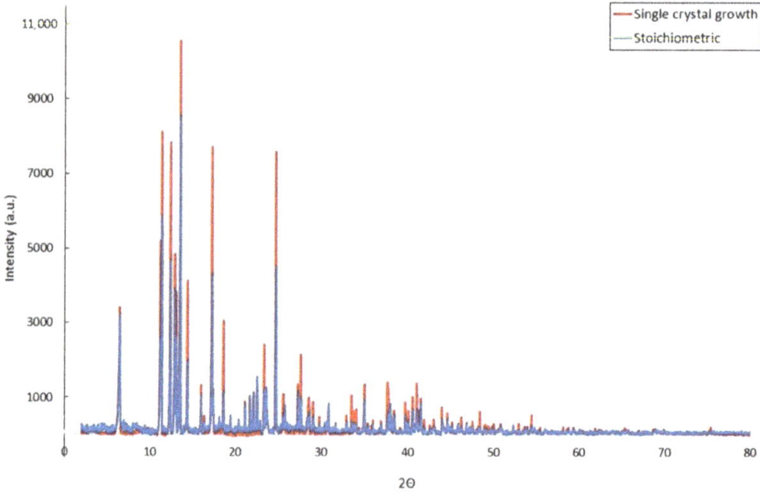

Figure 9. Superimposition of the PXRD pattern for [Cu$_2$(μ-OAc)$_4$(**1**)]$_n$ obtained from a preparative scale reaction (blue) and from the single-crystal growth experiment (red).

4. Conclusions

We have prepared and characterized compounds **1–5**, which feature [1,1′-biphenyl], 4′-fluoro-[1,1′-biphenyl], 4′-chloro-[1,1′-biphenyl], 4′-bromo-[1,1′-biphenyl] and 4′-methyl-[1,1′-biphenyl] attached to the 4′-position of a 3,2′:6′,3″-tpy metal-binding domain. Single-crystal structures of the $[Cu_2(\mu\text{-OAc})_4L]_n$ 1D-coordination polymers with L = **1–5** have been determined. The assembly common to both $[Cu_2(\mu\text{-OAc})_4(\mathbf{1})]_n$ and $[Cu_2(\mu\text{-OAc})_4(\mathbf{2})]_n$ in which ligands **1** and **2** adopt conformation **I** (Scheme 2) is directed by π-stacking interactions between centrosymmetric pairs of [1,1′-biphenyl]pyridine units. Although bifurcated contacts (H...H for **1** or H...F for **2**) are secondary, the two sets of interactions are interdependent, and are also dependent upon the 3,2′:6′,3″-tpy adopting conformation **I**. Increasing the steric demands of the 4′-substituent in the 1,1′-biphenyl group would force the chains (in a sheet) further apart, and if the π-stacking between pairs of [1,1′-biphenyl] pyridine units were to be retained, channels would be introduced into the lattice, reducing the packing efficiency. This hypothesis is consistent with the observed switch in ligand conformation to **II** (Scheme 2) on going from $[Cu_2(\mu\text{-OAc})_4(\mathbf{1})]_n$ and $[Cu_2(\mu\text{-OAc})_4(\mathbf{2})]_n$ to the analogous complexes containing **3**, **4**, and **5**. The switch in ligand conformation leads to the dominant packing interactions involving py–py and py–biphenyl face-to-face π-stacking interactions. We are currently exploring both the coordination behavior of ligands **1–5** with other metal salts and the effects of varying the 4′-arene functionalities.

Supplementary Materials: The following are available online at https://www.mdpi.com/2624-854 9/3/1/15/s1: Figures S1–S5: mass spectra of **1–5**; Figures S6–S25: NMR spectra of **1–5**; Figures S26–S35: IR spectra of ligands and coordination polymers; Figure S36: Packing diagram for $[Cu_2(\mu\text{-OAc})_4(\mathbf{3})]_n$. Figures S37–S45: PXRD figures.

Author Contributions: Project conceptualization, administration, supervision, and funding acquisition: C.E.H. and E.C.C.; investigation and data analysis: D.R. and S.N.; single-crystal X-ray diffraction and PXRD: A.P. and D.R.; manuscript writing: C.E.H. and D.R.; manuscript editing and review: all authors. All authors have read and agreed to the published version of the manuscript.

Funding: This research was partially funded by the Swiss National Science Foundation (grant number 200020_182000).

Institutional Review Board Statement: Not applicable.

Informed Consent Statement: Not applicable.

Data Availability Statement: The data presented in this study are available on request from the corresponding author. The data are not publicly accessible at present.

Acknowledgments: We gratefully acknowledge the support of the University of Basel.

Conflicts of Interest: The authors declare no conflict of interest.

References

1. Constable, E.C. The Coordination Chemistry of 2,2′:6′,2″-Terpyridine and Higher Oligopyridines. *Adv. Inorg. Chem. Radiochem.* **1986**, *30*, 69–121. [CrossRef]
2. Constable, E.C. 2,2′:6′,2″-Terpyridines: From chemical obscurity to common supramolecular motifs. *Chem. Soc. Rev.* **2007**, *36*, 246–253. [CrossRef] [PubMed]
3. Schubert, U.S.; Hofmeier, H.; Newkome, G.R. *Modern Terpyridine Chemistry*; Wiley–VCH: Weinheim, Germany, 2006; ISBN 978-3-527-31475-1.
4. Wei, C.; He, Y.; Shi, X.; Song, Z. Terpyridine-metal complexes: Applications in catalysis and supramolecular chemistry. *Coord. Chem. Rev.* **2019**, *385*, 1–19. [CrossRef]
5. Schubert, U.S.; Winter, A.; Newkome, G.R. *Terpyridine-Based Materials: For Catalytic, Optoelectronic and Life Science Applications*; Wiley–VCH: Weinheim, Germany, 2011; ISBN 9783527330386.
6. Constable, E.C. Expanded ligands: An assembly principle for supramolecular chemistry. *Coord. Chem. Rev.* **2008**, *252*, 842–855. [CrossRef]
7. Chakraborty, S.; Newkome, G.R. Terpyridine-based metallosupramolecular constructs: Tailored monomers to precise 2D-motifs and 3D-metallocages. *Chem. Soc. Rev.* **2018**, *47*, 3991–4016. [CrossRef] [PubMed]

8.   Housecroft, C.E.; Constable, E.C. The Terpyridine Isomer Game: From Chelate to Coordination Network Building Block. *Chem. Commun.* **2020**, *56*, 10786–10794. [CrossRef]
9.   Kröhnke, F. Syntheses Using Pyridinium Salts. *Angew. Chem. Int. Ed.* **1963**, *2*, 225. [CrossRef]
10.   Wang, J.; Hanan, G.S. A facile route to sterically hindered and non-hindered 4′-aryl-2,2′:6′,2″-terpyridines. *Synlett* **2005**, 1251–1254. [CrossRef]
11.   Sato, S.; Murase, T.; Fujita, M. Self-Assembly of Coordination Cages and Spheres. In *Supramolecular Chemistry: From Molecules to Nanomaterials*; Gale, P.A., Steed, J.W., Eds.; Wiley-Blackwell: Oxford, UK, 2012. [CrossRef]
12.   Rocco, D.; Prescimone, A.; Constable, E.C.; Housecroft, C.E. Switching the conformation of 3,2′:6′,3″-tpy domains in 4′-(4-*n*-alkyloxyphenyl)-3,2′:6′,3″-terpyridines. *Molecules* **2020**, *25*, 3162. [CrossRef]
13.   Rocco, D.; Manfroni, G.; Prescimone, A.; Klein, Y.M.; Gawryluk, D.J.; Constable, E.C.; Housecroft, C.E. Single and double-stranded 1D-coordination polymers with 4′-(4-alkyloxyphenyl)-3,2′:6′,3″-terpyridines and {Cu$_2$(μ-OAc)$_4$} or {Cu$_4$(μ$_3$-OH)$_2$(μ-OAc)$_2$(μ$_3$-OAc)$_2$(AcO-κO)$_2$} motifs. *Polymers* **2020**, *12*, 318. [CrossRef]
14.   Rocco, D.; Prescimone, A.; Constable, E.C.; Housecroft, C.E. Straight versus branched chain substituents in 4′-(butoxyphenyl)-3,2′:6′,3″-terpyridines: Effects on (4,4) coordination network assemblies. *Polymers* **2020**, *12*, 1823. [CrossRef] [PubMed]
15.   Rocco, D.; Prescimone, A.; Constable, E.C.; Housecroft, C.E. Directing 2D-coordination networks: Combined effects of a conformationally flexible 3,2′:6′,3″-terpyridine and chain length variation in 4′-(4-*n*-alkyloxyphenyl) substituents. *Molecules* **2020**, *25*, 1663. [CrossRef] [PubMed]
16.   Granifo, J.; Gavino, R.; Freire, E.; Baggio, R. The new sulphur-containing ligand 4′-(4-methylthiophenyl)-3,2′:6′,3″-terpyridine (**L1**) and the supramolecular structure of the dinuclear complex [Zn$_2$(μ-**L1**)(acac)$_4$] (acac = acetylacetonato): The key role of non-covalent S . . . O contacts and C–H . . . S hydrogen bonds. *J. Mol. Struct.* **2011**, *1006*, 684–691. [CrossRef]
17.   Henling, L.M.; Marsh, R.E. Some more space-group corrections. *Acta Crystallogr.* **2014**, *C70*, 834–836. [CrossRef]
18.   Granifo, J.; Vargas, M.; Garland, M.T.; Ibanez, A.; Gavino, R.; Baggio, R. The novel ligand 4′-phenyl-3,2′:6′,3″-terpyridine (L) and the supramolecular structure of the dinuclear complex [Zn$_2$(μ-L)(acac)$_4$]·H$_2$O (acac = acetylacetonato). *Inorg. Chem. Comm.* **2008**, *11*, 1388–1391. [CrossRef]
19.   Klein, Y.M.; Lanzilotto, A.; Prescimone, A.; Krämer, K.W.; Decurtins, S.; Liu, S.-X.; Constable, E.C.; Housecroft, C.E. Coordination behaviour of 1-(3,2′:6′,3″-terpyridin-4′-yl)ferrocene: Structure and magnetic and electrochemical properties of a tetracopper dimetallomacrocycle. *Polyhedron* **2017**, *129*, 71–76. [CrossRef]
20.   Zhao, M.; Tan, J.; Su, J.; Zhang, J.; Zhang, S.; Wu, J.; Tian, Y. Syntheses, crystal structures and third-order nonlinear optical properties of two series of Zn(II) complexes using the thiophene-based terpyridine ligands. *Dyes Pigments* **2016**, *130*, 216–225. [CrossRef]
21.   Zhang, L.; Li, C.-J.; He, J.-E.; Chen, Y.-Y.; Zheng, S.-R.; Fan, J.; Zhang, W.-G. Construction of New Coordination Polymers from 4′-(2,4-disulfophenyl)-3,2′:6′,3″-terpyridine: Polymorphism, pH-dependent syntheses, structures, and properties. *J. Solid State Chem.* **2016**, *233*, 244–254. [CrossRef]
22.   Cheng, Y.; Yang, M.-L.; Hu, H.-M.; Xu, B.; Wang, X.; Xue, G. Syntheses, structures and luminescence for zinc coordination polymers based on a multifunctional 4′-(3-carboxyphenyl)-3,2′:6′,3″-terpyridine ligand. *J. Solid State Chem.* **2016**, *239*, 121–130. [CrossRef]
23.   Wang, T.-T.; Zhang, J.-L.; Hu, H.-M.; Cheng, Y.; Xue, L.-L.; Wang, X.; Wang, B.-Z. Syntheses, structures and luminescent properties of Zn/Cd coordination polymers based on 4′-(2-carboxyphenyl)-3,2′:6′,3″-terpyridine. *Polyhedron* **2018**, *151*, 43–50. [CrossRef]
24.   Lufei, X.; Yupeng, T. CSD Refcode VUKMOF. In *CSD Communication CCDC 1407410*; The Cambridge Crystallographic Data Centre (CCDC): Cambridge, UK, 2015.
25.   Liu, C.; Ding, Y.-B.; Shi, X.-H.; Zhang, D.; Hu, M.-H.; Yin, Y.-G.; Li, D. Interpenetrating Metal−Organic Frameworks Assembled from Polypyridine Ligands and Cyanocuprate Catenations. *Cryst. Growth Des.* **2009**, *9*, 1275–1277. [CrossRef]
26.   Li, L.; Zhang, Y.Z.; Yang, C.; Liu, E.; Golen, J.A.; Zhang, G. One-dimensional copper(II) coordination polymers built on 4′-substituted 4,2′:6′,4″- and 3,2′:6′,3″-terpyridines: Syntheses, structures and catalytic properties. *Polyhedron* **2016**, *105*, 115–122. [CrossRef]
27.   Köberl, M.; Cokoja, M.; Herrmann, W.A.; Kühn, F.E. From molecules to materials: Molecular paddle-wheel synthons of macromolecules, cage compounds and metal-organic frameworks. *Dalton Trans.* **2011**, *40*, 6834–6859. [CrossRef] [PubMed]
28.   Constable, E.C.; Housecroft, C.E.; Neuburger, M.; Schönle, J.; Vujovic, S.; Zampese, J.A. Molecular recognition between 4′-(4-biphenylyl)-4,2′:6′,4″-terpyridine domains in the assembly of d^9 and d^{10} metal ion-containing one-dimensional coordination polymers. *Polyhedron* **2013**, *60*, 120–129. [CrossRef]
29.   Constable, E.C.; Housecroft, C.E.; Vujovic, S.; Zampese, J.A.; Crochet, A.; Batten, S.R. Do perfluoroarene . . . arene and C–H...F interactions make a difference to 4,2′:6′,4″-terpyridine-based coordination polymers? *CrystEngComm* **2013**, *15*, 10068–10078. [CrossRef]
30.   Software for the Integration of CCD Detector System Bruker Analytical X-ray Systems, Bruker axs, Madison, WI (after 2013).
31.   Sheldrick, G.M. ShelXT-Integrated space-group and crystal-structure determination. *Acta Cryst.* **2015**, *A71*, 3–8. [CrossRef] [PubMed]
32.   Dolomanov, O.V.; Bourhis, L.J.; Gildea, R.J.; Howard, J.A.K.; Puschmann, H. Olex2: A Complete Structure Solution, Refinement and Analysis Program. *J. Appl. Cryst.* **2009**, *42*, 339–341. [CrossRef]
33.   Sheldrick, G.M. Crystal Structure Refinement with ShelXL. *Acta Cryst.* **2015**, *C27*, 3–8. [CrossRef]

34. Palatinus, L.; Chapuis, G. Superflip-A Computer Program for the Solution of Crystal Structures by Charge Flipping in Arbitrary Dimensions. *J. Appl. Cryst.* **2007**, *40*, 786–790. [CrossRef]
35. Palatinus, L.; Prathapa, S.J.; van Smaalen, S. EDMA: A Computer Program for Topological Analysis of Discrete Electron Densities. *J. Appl. Cryst.* **2012**, *45*, 575–580. [CrossRef]
36. Macrae, C.F.; Sovago, I.; Cottrell, S.J.; Galek, P.T.A.; McCabe, P.; Pidcock, E.; Platings, M.; Shields, G.P.; Stevens, J.S.; Towler, M.; et al. Mercury 4.0: From visualization to analysis, design and prediction. *J. Appl. Cryst.* **2020**, *53*, 226–235. [CrossRef] [PubMed]
37. LeBail, A.; Duroy, H.; Fourquet, J.L. Ab-initio structure determination of $LiSbWO_6$ by X-ray powder diffraction. *Mater. Res. Bull.* **1988**, *23*, 447–452. [CrossRef]
38. Pawley, G.S. Unit-cell refinement from powder diffraction scans. *J. Appl. Cryst.* **1981**, *14*, 357–361. [CrossRef]
39. Rodríguez-Carvajal, J. Recent Advances in Magnetic Structure Determination by Neutron Powder Diffraction. *Physica B* **1993**, *192*, 55–69. [CrossRef]
40. Roisnel, T.; Rodríguez-Carvajal, J. WinPLOTR: A Windows tool for powder diffraction patterns analysis Materials Science Forum. In Proceedings of the Seventh European Powder Diffraction Conference (EPDIC 7), Barcelona, Spain, 20–23 May 2000; pp. 118–123.
41. *Spartan '18 Version 1.4.4*; Wavefunction Inc.: Irvine, CA, USA, 2019.
42. Bienz, S.; Bigler, L.; Fox, T.; Meier, H. *Spektroskopische Methoden in der Organischen Chemie*, 9th ed.; Thieme: Stuttgart, Germany, 2016; ISBN 9783135761091.
43. Janiak, C. A critical account on π–π stacking in metal complexes with aromatic nitrogen-containing ligands. *J. Chem. Soc. Dalton Trans.* **2000**, 3885–3896. [CrossRef]

Article

New Syntheses, Analytic Spin Hamiltonians, Structural and Computational Characterization for a Series of Tri-, Hexa- and Hepta-Nuclear Copper (II) Complexes with Prototypic Patterns

Ana Maria Toader [1], Maria Cristina Buta [1], Fanica Cimpoesu [1,*], Andrei-Iulian Toma [2], Christina Marie Zalaru [3], Ludmila Otilia Cinteza [4] and Marilena Ferbinteanu [2,*]

[1]   Institute of Physical Chemistry, Splaiul Independentei 202, 060021 Bucharest, Romania; ancutatoader@yahoo.fr (A.M.T.); butamariacristina@gmail.com (M.C.B.)
[2]   Department of Inorganic Chemistry, Faculty of Chemistry, University of Bucharest, Dumbrava Rosie 23, 020462 Bucharest, Romania; andreitoma119@yahoo.com
[3]   Department of Organic Chemistry, Biochemistry and Catalysis, Faculty of Chemistry, University of Bucharest, Panduri 90-92, 050663 Bucharest, Romania; christina.zalaru@chimie.unibuc.ro
[4]   Physical Chemistry Department, University of Bucharest, 030118 Bucharest, Romania; ocinteza@gw-chimie.math.unibuc.ro
*   Correspondence: cfanica@yahoo.com (F.C.); marilena.cimpoesu@g.unibuc.ro (M.F.); Tel.: +40-21-312-1147 (F.C.); +40-21-210-3495 (M.F.)

Citation: Toader, A.M.; Buta, M.C.; Cimpoesu, F.; Toma, A.-I.; Zalaru, C.M.; Cinteza, L.O.; Ferbinteanu, M. New Syntheses, Analytic Spin Hamiltonians, Structural and Computational Characterization for a Series of Tri-, Hexa- and Hepta-Nuclear Copper (II) Complexes with Prototypic Patterns. *Chemistry* **2021**, *3*, 411–439. https://doi.org/10.3390/chemistry3010031

Received: 28 January 2021
Accepted: 4 March 2021
Published: 15 March 2021

Publisher's Note: MDPI stays neutral with regard to jurisdictional claims in published maps and institutional affiliations.

Abstract: We present a series of pyrazolato-bridged copper complexes with interesting structures that can be considered prototypic patterns for tri-, hexa- and hepta- nuclear systems. The trinuclear shows an almost regular triangle with a μ_3-OH central group. The hexanuclear has identical monomer units, the Cu_6 system forming a regular hexagon. The heptanuclear can be described as two trinuclear moieties sandwiching a central copper ion via carboxylate bridges. In the heptanuclear system, the pyrazolate bridges are consolidating the triangular faces, which are sketching an elongated trigonal antiprism. The magnetic properties of these systems, dominated by the strong antiferromagnetism along the pyrazolate bridges, were described transparently, outlining the energy levels formulas in terms of Heisenberg exchange parameters *J*, within the specific topologies. We succeeded in finding a simple Kambe-type resolution of the Heisenberg spin Hamiltonian for the rather complex case of the heptanuclear. In a similar manner, the weak intermolecular coupling of two trimer units (aside from the strong exchange inside triangles) was resolved by closed energy formulas. The hexanuclear can be legitimately proposed as a case of coordination-based aromaticity, since the phenomenology of the six-spins problem resembles the bonding in benzene. The Broken-Symmetry Density Functional Theory (BS-DFT) calculations are non-trivial results, being intrinsically difficult at high nuclearities.

Keywords: trinuclear complex; hexamer metallacycle; heptanuclear complex; pyrazolato-bridged ligand systems; spin Hamiltonian; spin models; Kambe-type energy formulas; Broken-Symmetry Density Functional Theory calculations

1. Introduction

This work is built on the frame of our deals in structural chemistry [1], putting on equal footing the interest for experimental data from new syntheses (followed by X-ray characterization) and theoretical approaches (phenomenological or computational). We will present a suite of copper complexes having prototypical structural patterns, rising and solving important aspects of their molecular magnetism. The common feature of all the presented systems is the presence of pyrazolato-bridges in their polynuclear edifices.

The coordination chemistry of the pyrazole-based ligands provides a variety of compounds with a large spectrum of molecular geometries, nuclearities and interesting properties [2]. Pyrazole and its derivatives proved to be versatile ligands, exhibiting several

133

coordination modes [3], the bridging functionality allowing to form polynuclear structures [4]. Polynuclear copper complexes attracted special attention due to their magnetic properties and for their relevance to the biomimetics of the copper proteins (e.g., ascorbate oxidase, laccase, ceruloplasmin, methane monooxygenase) [5]. Among them, the triangular trinuclear copper complexes were highlighted for their physical and chemical properties related with blue multicopper oxidases [6–8].

From the magnetic point of view, the cyclic trinuclear copper systems were regarded as geometrically frustrated antiferromagnetic compounds, offering the possibility to test the exchange coupling models [9,10]. In the last decades, density functional theory (DFT) methods and especially the broken-symmetry approach were raised as instruments of choice to evaluate the magnetic coupling parameters for such systems [11]. There are several electronic structure calculations on a cyclic trinuclear copper system, done on model systems [12,13] or on experimental molecules [14,15]. In this work we will corroborate the magnetic measurements with computational quests, performing and discussing non-trivial numeric experiments.

The patterns of the discussed compounds are shown in Scheme 1, the labeling of frames corresponding to those of the presented structures. While there are several known trinuclear [16–18] and hexa-nuclear [19–23] analogs of the reported compounds, the hepta-nuclear congeners are rare [24,25]. The insight offered by the phenomenological and computational analyses is rather unprecedented, fixing landmarks on account of the discussed systems, considering that each compound is a representative of their given structural pattern.

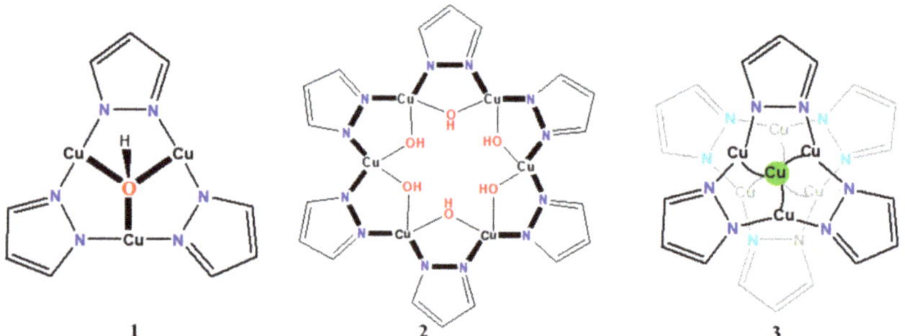

Scheme 1. Structural patterns of compounds **1**, **2** and **3**.

The triangular copper complexes with a hydroxy- or oxo-trinuclear bridge were obtained usually by the pyrazole deprotonation in the presence of a base [16–18], while in the case of using carboxylates copper salts, the pyrazole deprotonation is favored by the presence of carboxylate ions. Our compound **1** parallels literature results in terms of synthetic procedures and outcomes [26,27].

Concretely, we used the reaction of copper (II) acetate with 3,5-dimethyl-4-nitro-pyrazole (dmnpz) in protic solvents to obtain a new triangular trinuclear copper complex, $[Cu_3(\mu_3\text{-}OH)(\mu\text{-}dmnpz)_3(OCOCH_3)_2(H_2O)_3]\cdot H_2O\cdot EtOH$ (**1**). The oxo-trinuclear complex contains two carboxylates coordinated to two Cu(II) ions. It must be pointed out that among the known compounds that have structure **1** from Scheme 1, there are no studies with pyrazole derivatives dressed with different substituents in the 3, 4 and 5 positions, as we approached here. The compounds of type **1** are interesting for the relative simplicity in the interpretation of magnetic properties, the pyrazolate and $\mu_3\text{-}OH$ bridges determining strong antiferromagnetism [28–30].

Performing the reaction with the same starting materials, but changing the solvent with a more basic one, pyridine, the $[Cu_6(\mu\text{-}OH)_6(\mu\text{-}dmnpz)_6(Py)_6]\cdot 6Py\cdot 2EtOH$ (**2**) compound was obtained. The neutral complex belongs to the class of six-membered metallacycles, several known analogs being: $([\{Cu(\mu\text{-}OH)(\mu\text{-}R_1,R_2\text{-}pz)\}_6]$, $R_1 = R_2 = H$ [19,20],

$R_1 = CH_3$, $R_2 = H$ [21], $R_1 = CH_3$, Py, $R_2 = H$ [22], $R_1 = CF_3$, $R_2 = H$ [23,24]. The six-membered rings with unsubstituted pyrazole ligands adopt a crown structure, with a *cis* arrangement, where the six hydroxyl groups of the metallacycle are pointing towards a focal point on an approximate symmetry axis, while the pyrazoles towards the periphery, the ensemble with, consequently, hydrophilic and hydrophobic domains. For substituted pyrazole, or for higher nuclearities of the metallacycles, the crown structures are sketching a *trans* arrangement with half of the OH groups and half of the pyrazolato ligands oriented on the same side, with respect to the mean plane of the metal-ions ring, while the other subsets of these ligands are placed on the opposite hemispace.

The triangular fragment is remarkably stable and can be used as a secondary building unit (SBU) in the construction of several coordination polymers [31,32]. We succeed in connecting two SBU via six carboxylate anions to a central copper ion to obtain a heptanuclear compound $[Cu_7(\mu\text{-}OH)_2(\mu\text{-}dmnpz)_6(\mu\text{-}OCOCH_3)_6(MeOH)_6]$ **(3)**, in the shape of an hourglass. Enlarging, by future syntheses and analyses, the list of the heptanuclears belonging to the structural type **3** (see Scheme 1) can be proposed as a challenging task.

Our attention was focused on the reaction of copper(II) carboxylates with pyrazole derivatives, 3,5-dimethyl-4-nitro-pyrazole (dmnpz), observing that different nuclearities and topologies depend on the employed solvent. The pyrazole (dmnpz) was synthesized according to Scheme 2 [33–35].

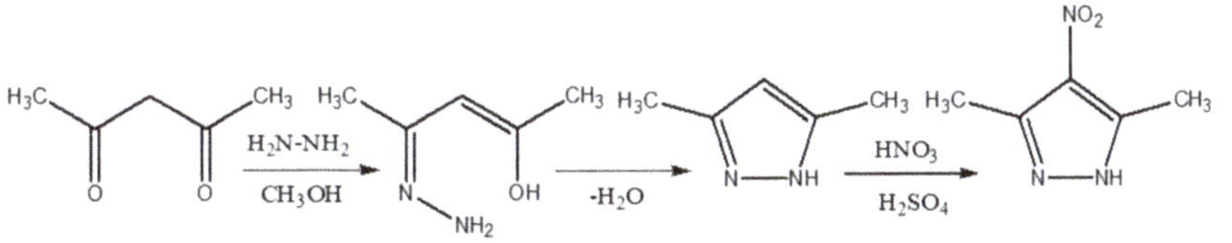

Scheme 2. Synthesis of the used pyrazole ligand.

2. Materials and Methods

2.1. Materials and Physical Measurements

All reagents were of analytical grade, being used without further purification. 3,5-dimethyl-4-nitro-pyrazole (dmnpz) was synthesized according to literature procedures [33–35]. The content in carbon, hydrogen and nitrogen was determined by elemental analysis using a Perkin Elmer PE 2400 analyzer (Perkin Elmer, Boston, MA, USA). The IR spectra were performed by a JASCO 4200 spectrometer with Pike ATR unit in the 400–4000 cm^{-1} range (JASCO, Tokyo, Japan). The magnetic data were collected in the range of 2–300 K at an applied field of 0.1 T on a MPMS-5S SQUID magnetometer (Quantum Design, San Diego, CA, USA). The diamagnetism was corrected by Pascal constants [36].

2.2. Synthesis

2.2.1. $[Cu_3(\mu_3\text{-}OH)(\mu\text{-}dmnpz)_3(OCOCH_3)_2(H_2O)_3]\cdot H_2O\cdot EtOH$ (1)

A solution of copper acetate monohydrate $Cu(CH_3COO)_2\cdot H_2O$ (0.4 g, 2 mmol) in ethanol/water (5 mL/5mL) was added drop wise, under continuous stirring, to a solution of dmnpz (0.282 g, 2 mmol) in ethanol (5 mL). The resulting blue-greenish mixture was stirred at room temperature for 24 h and then filtered. A blue microcrystalline precipitate formed, and was subsequently filtered, washed with EtOH and dried under vacuum. It was recrystallized from EtOH, yielding in few days dark greenish-blue crystals for (**1**), suitable for X-ray crystal structure determination and for all measurements. Yield (based on Cu^{2+}): 73%. Elemental Analysis: calc. for $C_{21}H_{39}Cu_3N_9O_{16}$ (Mr = 864.22): C, 29.19; H, 4.55; N, 14.59; Found: C, 28.78; H, 4.42; N, 14.01%. FT-IR (cm^{-1}): 3517(w), 3446(w), 2916(w), 1673(m), 1538(s), 1474(s), 1404(s), 1354 (vs), 1312 (sh), 1171 (s), 1093 (w), 1008 (sh), 980(m), 952(sh), 832(s), 754(m), 599 (w), 486(w), 457(m), 408(w).

2.2.2. [Cu$_6$(μ-OH)$_6$(μ-dmnpz)$_6$(Py)$_6$]·6Py·2EtOH (2)

A solution of copper acetate monohydrate Cu(CH$_3$COO)$_2$·H$_2$O (0.4 g, 2 mmol) in ethanol (5 mL) was added drop wise, under continuous stirring, to the dmnpz solution (0.282 g, 2 mmol) in ethanol (5 mL). The resulting green mixture was stirred at room temperature for 24 h and then filtered. The filtered clear solution was removed and the separated green solid was recrystallized from pyridine. The blue pyridine solution was slowly evaporated in a cooler. Suitable greenish-blue crystals for (**2**), which appeared in six months, were used for all the measurements. Yield (based on Cu^{2+}): 54%. Elemental Analysis: calc. for C$_{94}$H$_{114}$Cu$_6$N$_{30}$O$_{20}$ (Mr = 2365.38): C, 47.73; H, 4.86; N, 17.76. Found: C, 46.63; H, 4.77; N, 16.02%. FT-IR (cm^{-1}): 3382(w), 3177(m), 1595(w), 1538(m), 1474(m), 1439(sh), 1418(s), 1376 (m), 1348 (vs), 1171 (m), 1100 (w), 1029 (w), 987(m), 902(w), 839(m), 768(w), 690 (m), 606 (w), 535(w), 479(w).

2.2.3. [Cu$_7$(μ-OH)$_2$(μ-dmnpz)$_6$(μ-OCOCH$_3$)$_6$(MeOH)$_6$] (3)

A solution of copper acetate monohydrate Cu(CH$_3$COO)$_2$·H$_2$O (0.4 g, 2 mmol) in methanol (5 mL) was added slowly, under stirring, to a dmnpz solution (0.282 g, 2 mmol) and Et$_3$N (0.202 g, 0.28mL, 2 mmol) in methanol (15 mL). The resulting green solution was stirred at room temperature for 24h and then filtered. The clear dark green solution was slowly evaporated at room temperature. Suitable green crystals for (**3**), appeared in few days, and were used for all the instrumental measurements. Yield (based on Cu^{2+}): 89%. Elemental Analysis: calc. for C$_{48}$H$_{80}$Cu$_7$N$_{18}$O$_{32}$ (Mr = 1866.07): C, 30.89; H, 4.32; N, 13.76. Found: C, 30.47; H, 4.38; N, 13.14%. FT-IR (cm^{-1}): 3586 (m), 3446(m), 2980(w), 1683(w), 1658(m), 1542(vs), 1521(m), 1472(s), 1457(sh), 1416(vs), 1379 (s), 1357 (vs), 1176 (s), 1108 (w), 1029 (w), 990(m), 835(m), 763(w), 690(m), 606(w), 530(w), 484(w), 419(w).

2.3. X-ray Crystallography

A Rigaku Rapid II R-Axis diffractometer (MoKα (λ = 0.71075Å), was used to collect the diffraction data of complexes **1–3**. All calculations were performed using the Crystal-Structure [37] crystallographic software package. The refinement was performed using the Olex1.2 [38] program. The structure was solved by direct methods and refined in a routine manner (SHELXL) [39]. Non-hydrogen atoms were refined anisotropically. The hydrogen atoms were refined using the riding model. Molecular graphics were generated by MERCURY 3.9 [40] and POV-Ray [41] software. The details of the X-ray crystal data and the structure solutions, as well as the refinements, are given in Table S1. A summary of selected bond lengths (Å) and angles (in degrees °) is outlined in Tables S2–S4 (see Supplementary Materials). The Crystallographic data for the structures reported in this paper have been deposited in the Cambridge Crystallographic Data Centre with CCDC reference numbers 2059160 for compound **1**, 2059163 for **2** and 2059182 for **3**.

2.4. Computational Methods

The calculations were realized on the experimental geometries of the complex compounds, obtained from the X-ray diffraction analysis, in the frame of Density Functional Theory (DFT) [42], using the B3LYP [43] functional and 6-31G* basis on copper, nitrogen and oxygen, while 6-31G for the carbon and hydrogen atoms [44]. The calculations were done with the GAMESS [45] suite.

3. Results and Discussions

3.1. Structural Analysis from Crystallographic Data

3.1.1. Structure of Compound **1**

The asymmetric unit of complex **1** contains two molecular complexes with slight mutual differences and disordered solvent molecules, which affect the final refinement parameters. In Figure 1 we present only one molecular species, the couple of closely similar structures being shown in Supplementary Materials (Figure S1). Each molecule (see Figure 1), [Cu$_3$(μ_3-OH)(μ-dmnpz)$_3$(OCOCH$_3$)$_2$(H$_2$O)$_3$], reveals a triangular trinuclear neutral core. The copper ions are connected via a trihapto oxygen from a central hydroxyl group and three bridging pyrazolate ligands. Two acetate ions are coordinated, each to Cu1 and Cu3 sites. Two water molecules are linked to the Cu2 center and another one to Cu1. The structure of compound **1** is special for containing, simultaneously, three distinct stereochemistries of copper ions, which can be quantified by the help of so-called τ_5 [46] and τ_4 [47] indices. The idealized margins are: $\tau_5 = 0$ for a square-pyramidal complex, while $\tau_5 = 1$ for a trigonal bipyramidal frame; $\tau_4 = 0$ for a square planar unit, while $\tau_4 = 1$ for a tetrahedral case. The sites in compound **1** can be characterized as square pyramid for Cu1 (with $\tau_5 = 0.227$), trigonal bipyramid for Cu2 (with $\tau_5 = 0.567$) and square planar for Cu3 ($\tau_4 = 0.098$).

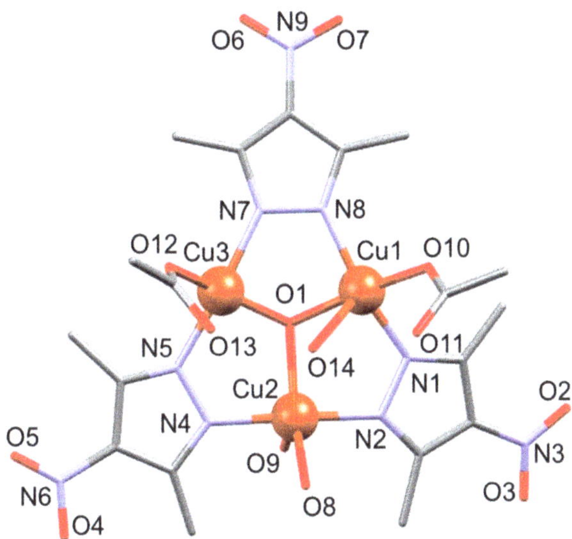

Figure 1. Asymmetric unit of compound **1** with partial atom labeling scheme. Hydrogen atoms are omitted for clarity.

The triangular frame is not symmetric, having different Cu–Cu distances. There are slightly different bond lengths formed by copper and central oxygen (Cu1–O1 1.974(8) Å, Cu2–O1 2.039(8) Å, Cu3–O1 1.985(8) Å), larger than those reported for systems with the same core [48]. The capping μ_3-O(1)H group is out of the plane formed by copper ions, displaced by about 0.752(2) Å. Two of the Cu–O(H)–Cu angles are very close to each other (Cu1–O1–Cu3 and Cu1–O1–Cu2, 105.7(4)° and 105.0(4)°, respectively), quite different in comparison with the third one (Cu2–O1–Cu3 109.3(4)°). Only one pyrazolate ring bearing N4–N5 is coplanar with the copper ions moiety, since the other two pyrazolate rings (N1–N2 and N7–N8) are displaced out of the {Cu$_3$} plane, conferring to the overall structure a flattened bowl-shape. The Cu–N(bridged) distances are all in the range of 1.939(13)–1.984(12) Å. The Cu–N–N(bridged) angles are in the 115.1(9)–119.2(9)° interval.

The triangular bowl-shaped molecules are stacked along the *a* axis (Figure 2) within a columnar arrangement established through hydrogen bonds, the most interesting being the hydrogen bond between the carboxylate group coordinated to Cu1 and the O(1)H group from the next molecule (O1′–O11 = 2.738(4) Å). The distance between the copper ion planes in this packing is slightly alternating, by 7.086 and 7.100 Å. The packing details for **1** are presented in Figures S2–S4 from the Supplementary Materials.

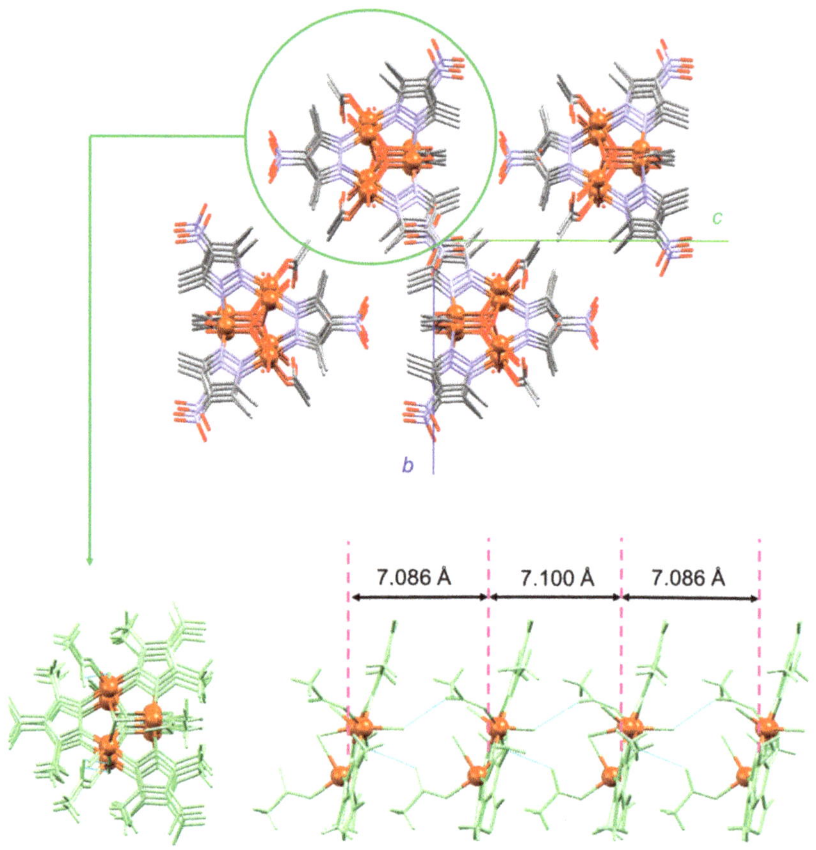

Figure 2. Packing diagram for compound **1** along *a* axis (**top**) and packing detail along *c* axis, showing the supramolecular column arrangement stabilized by hydrogen bonds (**bottom**).

The structural parameters for compound **1** are very poor, due to a sever disorder of two solvent molecules and the poor quality of crystals. Therefore, we consider it as a structural model for the further discussions. However, the reduced resolution, does not imply doubts about the molecular structures themselves. Since compound **1** is important in the red line of this work as preamble to compound **3**, a system with a high degree of novelty, we carry on the discussion, retaining this slight caveat.

3.1.2. Structure of Compound **2**

The compound **2** crystalized in a triclinic system, R-3 space group. As seen in Figure 3, the molecular unit of **2** consists in a six membered metallacycle copper (II) pyrazolate, where the copper atoms are in a hexagonal planar geometry, with a Cu–Cu distance of 3.32 Å. The neutral cyclic complex [Cu$_6$(μ-OH)$_6$(μ-dmnpz)$_6$(Py)$_6$] contain six distorted square-pyramidal Cu ions with τ_5 = 0.44 stereochemical index [46]. The square pyramide basis plane is formed by two nitrogen atoms from two dmnpz ligands arranged in *trans* (Cu1–N1 = 2.018(4) Å and Cu1–N2 = 2.003(4) Å) and two oxygen atoms from the OH (Cu1–O1 = 1.943(4) Å, Cu1–O1 = 1.947(4) Å) bridging groups. The apical position of each site is occupied by pyridine (Cu1–N4 = 2.359(5) Å). The pyridine molecules are coordinated in the plane of the copper ring, while the dmnpz and OH groups are forming two bowl-shaped cavities. Such cavities can accommodate guest molecules, and in our case, two disordered ethanol molecules are involved. The compound **2** shows a hexagonal tubular structure, on the packing along *c* axis, which can be interesting for further debate on the porosity properties. The packings for **2** are presented in Supplementary Materials Figures S6–S8.

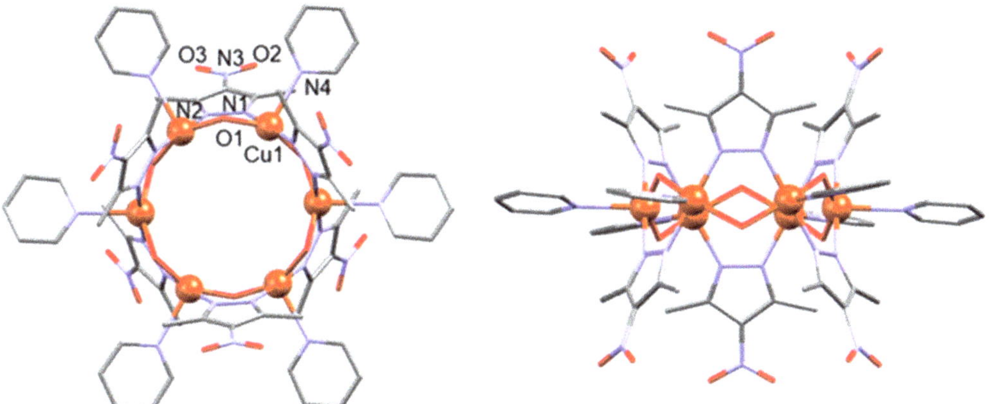

Figure 3. Asymmetric unit of compound **2** with atom numbering scheme (**left** side). Side-view (**right** half). Hydrogen atoms and solvent molecules are omitted for clarity.

3.1.3. Structure of Compound **3**

The compound **3** has a surprising symmetry, crystallizing in the trigonal system, R-3c space group. The structure (Figure 4) shows an "hourglass" pattern with two trinuclear triangular copper complexes at the ends, connected to a central octahedral copper ion via carboxylate anions coordinated in chelating and bridging fashion. The compound can be viewed like consisting of two oxo-trinuclear units, which act here like SBUs.

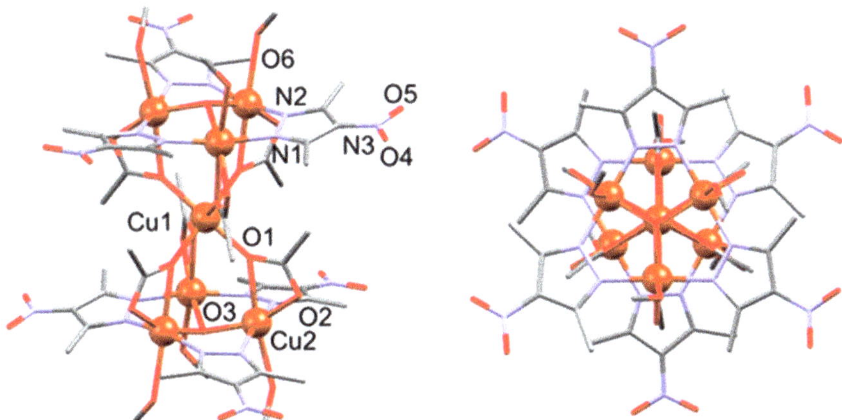

Figure 4. Asymmetric unit of compound **3** with atom numbering scheme (**left**). Side-view (**right**). Hydrogen atoms are omitted for clarity.

The three-fold axis is oriented along O3, O3A and the central Cu1 atom. The copper ions from the trinuclear SBU have an elongated octahedral geometry, along the O1–Cu2–O6 direction with 2.462(4) and 2.506 Å bond lengths, respectively. The current literature knows two similar compounds, one with the same ligand, reported by Lang and Zhu [24], following a single crystal-to-single-crystal transformation, where ligated-EtOH molecules were replaced with MeOH. The mentioned authors did not carry further investigations, while the interesting pattern deserves detailed attention, as we devoted in the following. According to the Cambridge Structural Database, there is a second compound with the same pattern (CCDC 853959, refcod XEMYIZ), including a different pyrazole derivative, respective 3,5-bis-(trifluoromethyl)-pyrazole [25]. This last study includes only the structural analysis.

A synopsis of the discussed structures, highlighting for view and measurements the copper atoms, is given in Figure 5. As advocated previously and proved by the following analyses, the considered structural patterns have prototypic virtues. The conclusions extracted from specific structural analyses show generalized relevance for all the previous and future members of each class. The next section, devoted to the formal spin coupling models, comprises useful methodological advances, extracting new closed formulas, usable for the simulation of the magnetic properties of the considered prototypes.

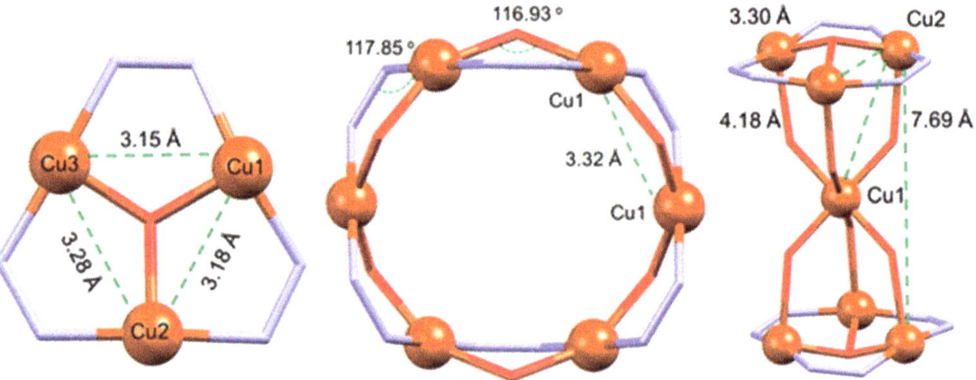

Figure 5. The structural patterns and Cu–Cu distances of the compounds **1** (**left** side), **2** (**center**) and **3** (**right** side).

3.2. Spin Hamiltonians and Magnetic Data

3.2.1. Spin Hamiltonian and Magnetic Susceptibility Equations

In the following, we will rely on the Heisenberg-Dirac-van Vleck (HDvV) Spin Hamiltonian [49,50],

$$\hat{H}_{HDvV} = -2\sum_{j<i} J_{ij}\,\hat{S}_i \cdot \hat{S}_j \tag{1}$$

for accounting for the magnetic properties of the discussed systems. The general treatment of large or non-symmetric systems demands the full numeric approach, namely constructing and resolving Hamiltonian matrices having the dimension of the total number of spin states, taken as the product of local spin multiplicities on the constituent ions. In the case of copper homo-metallic systems with nuclearity N, the total number of states (including the $2S + 1$ spin degeneracies) is 2^N. In the absence of further spin-type interactions (like Zeeman, dipolar, spin-orbit or zero-field-splitting), the whole matrices are factorized in blocks corresponding to total spin projections, S_z, and furthermore by total spin quantum numbers, S, provided that the corresponding basis transformations are made. Solving the Hamiltonian as eigenvectors E_I, associated to total spin quantum numbers S_I, the product between magnetic susceptibility and temperature is given by the van Vleck equation [36]:

$$(\chi T)_{HDVV} = \frac{N_A\beta^2}{3k_B} \cdot \frac{\sum_I g_I^2 \cdot S_I(S_I+1)(2S_I+1)\cdot Exp\left(-\frac{E_I}{k_BT}\right)}{\sum_I (2S_I+1)\cdot Exp\left(-\frac{E_I}{k_BT}\right)} \tag{2}$$

where N_A and k_B are Avogadro and Boltzmann constants, β is the Bohr magneton and g_I are gyromagnetic factors of each state, algebraically formed from g_i parameters of each paramagnetic center. Often, a global unique g^2 factor is taken, which is literally true in the case of equivalent isotropic sites. Sometimes, the practice imposes adding certain empirical correction terms, such as the linear add-on,

$$\chi T = (\chi T)_{HDVV} + PIM + TIP \cdot T \tag{3}$$

with a constant off-set having the meaning of paramagnetic impurity (PIM) and the slope known as temperature independent paramagnetism (TIP) [36]. The paramagnetic impurity can come literally from inherent sample conditions, or can cumulate extended lattice interactions, not accounted by the assumed molecular model. The TIP can result from first order expansion of non-HDvV interactions, such as spin-orbit or dipolar terms, being then a surrogate avoiding higher, possibly unreachable, complexity of the case.

The general procedure for numerical handling of the HDvV Hamiltonian in its matrix representation can be tedious and have a black box nature. Fortunately, the discussed systems can be presented in a transparent fashion, as it follows. Thus, we will draw explicit formulas for the counted E_I states of the systems, as a function of involved HDvV coupling parameters and the defining spin quantum numbers.

3.2.2. The Magnetism of the Trinuclear Complex

The trinuclear compound has three pyrazoles as bridges, having the appearance of an almost isosceles triangle. The slightly different coordination sites make the exchange couplings on the Cu–Cu edges inequivalent, while yet comparable. The case of a general triangle (scalene), with all edges inequivalent, is tractable by closed energy formulas. Putting the lowest spin doublet in the energy origin, we have the following relative spin state energies for the two spin-doublet states and quartet:

$$E\left(S = \frac{1}{2}\right)_1 = 0 \tag{4}$$

$$E\left(S = \frac{1}{2}\right)_2 = 2\sqrt{J_{12}^2 + J_{13}^2 + J_{23}^2 - J_{12}J_{13} - J_{12}J_{23} - J_{13}J_{23}} \tag{5}$$

$$E\left(S = \frac{3}{2}\right) = -J_{12} - J_{13} - J_{23} + \sqrt{J_{12}^2 + J_{13}^2 + J_{23}^2 - J_{12}J_{13} - J_{12}J_{23} - J_{13}J_{23}} \tag{6}$$

However, on the existing data, the general model cannot safely discriminate three different couplings. As the calculations described in the following indicate, the different coupling values are comparable, so that one may go on the simplified hypothesis of equal parameters: $J = J_{12} = J_{13} = J_{23}$. In this case, one may benefit from the simplified formula, as a function of the total spin quantum number:

$$E(S) = -JS(S + 1) \tag{7}$$

One may easily verify that shifting the $E(S = 1/2)$ from Equation (7) to zero, the equilateral triangle is obtained as a particular situation of the above ascribed (4–5) formulas for a scalene topology. The above result is very well-known, belonging to the class of so-called Kambe formulas [51,52]. However, for the sake of completeness, let us recall here the derivation. The spin Hamiltonian of the equilateral triangle is:

$$\hat{H}_{M_3} = -2J\left(\hat{S}_1 \cdot \hat{S}_2 + \hat{S}_1 \cdot \hat{S}_3 + \hat{S}_2 \cdot \hat{S}_3\right) \tag{8}$$

Then, observe that the operator can be rewritten as follows:

$$\begin{aligned}
2\left(\hat{S}_1 \cdot \hat{S}_2 + \hat{S}_1 \cdot \hat{S}_3 + \hat{S}_2 \cdot \hat{S}_3\right) &= \left(\hat{S}_1 + \hat{S}_2 + \hat{S}_3\right)^2 - \hat{S}_1^2 - \hat{S}_2^2 - \hat{S}_3^2 = \\
&= S(S + 1) - S_1(S_1 + 1) - S_2(S_2 + 1) - S_3(S_3 + 1)
\end{aligned} \tag{9}$$

The last equality involves the regularity that the sum of local spin operators yields the total spin operator, $\hat{S}$, and the $\hat{S}^2$ quantum square is $S(S+1)$, in all instances of total or local spins. Since the sum comprising the square of the three local spin operators is a constant shift, it can be eliminated, arriving at Equation (9). In the regular triangle, the two $S = 1/2$ levels are degenerate, in a multiplet with "e" orbital label in the D_{3h} symmetry. Then, this is a case of the Jahn–Teller effect [53,54], namely experiencing the trend to remove the degeneracy by distorting the molecular frame. Probably this is the reason for which the compound forms itself with slightly inequivalent coordination sites. Considering that the rather rigid pyrazole bridges would not easily allow a sensible distortion of the triangle itself, for example, to an isosceles, the formation of different coordination sites is the alternative, instead of distorting a system having identical coordination sites.

A reasoning succeeding to extract analytic energy formulas from the spin Hamiltonian, linear in all the coupling parameters, is possible in a limited number of cases, with high formal symmetries. Fortunately, one may further exploit such a strategy aiming to describe long the long-range lattice interaction between two equilateral trimers, needed to better account for the magnetism of compound **1**. In this view, we are going to enforce the assumption that all the centers of a trimer are coupled by the same parameter, j, with all the nodes of the other triangle. This is not the realistic situation, but it allows, in the Gordian knot-alike strategy, to tackle the problem in a yet acceptable semiquantitative manner. The idea is that, in this way, one may account for a sort of global interaction between molecular units, as monolithic spin carriers. Then, the Hamiltonian for a dimer of trimers is ascribed as follows:

$$\hat{H}_{M_3-M_3} = -2J\left(\hat{S}_1 \cdot \hat{S}_2 + \hat{S}_1 \cdot \hat{S}_3 + \hat{S}_2 \cdot \hat{S}_3\right) - 2J\left(\hat{S}_4 \cdot \hat{S}_5 + \hat{S}_4 \cdot \hat{S}_6 + \hat{S}_5 \cdot \hat{S}_6\right)$$
$$-2j\left(\hat{S}_1 \cdot \hat{S}_4 + \hat{S}_1 \cdot \hat{S}_5 + \hat{S}_1 \cdot \hat{S}_6 + \hat{S}_2 \cdot \hat{S}_4 + \hat{S}_2 \cdot \hat{S}_5 + \hat{S}_2 \cdot \hat{S}_6 + \hat{S}_3 \cdot \hat{S}_4 + \hat{S}_3 \cdot \hat{S}_5 + \hat{S}_3 \cdot \hat{S}_6\right)$$

$$(10)$$

Let us define the spin operators on molecular moieties:

$$\hat{S}_{123} = \hat{S}_1 + \hat{S}_2 + \hat{S}_3 \tag{11}$$

$$\hat{S}_{456} = \hat{S}_4 + \hat{S}_5 + \hat{S}_6 \tag{12}$$

The total spin is, obviously:

$$\hat{S} = \hat{S}_{123} + \hat{S}_{456} \tag{13}$$

Let us observe that the parenthesis factored by the long-range $-2j$ interaction is actually $\hat{S}_{123} \cdot \hat{S}_{345}$, the scalar product of the spins on individual triangles:

$$\hat{H}_{M_3-M_3} = -2J\left(\hat{S}_1 \cdot \hat{S}_2 + \hat{S}_1 \cdot \hat{S}_3 + \hat{S}_2 \cdot \hat{S}_3 + \hat{S}_4 \cdot \hat{S}_5 + \hat{S}_4 \cdot \hat{S}_6 + \hat{S}_5 \cdot \hat{S}_6\right) - 2j\hat{S}_{123} \cdot \hat{S}_{345} \tag{14}$$

Then, using the trick exemplified previously, we can rewrite the Hamiltonian

$$\hat{H}_{M_3-M_3} = -J\left(\hat{S}_{123}^2 + \hat{S}_{456}^2\right) - J\sum_{i=1}^{6} \hat{S}_i^2 - j\left(\hat{S}^2 - \hat{S}_{123}^2 - \hat{S}_{456}^2\right) \tag{15}$$

Dropping the middle constant term (from the summation on local spins and converting the squared operators to their quantum expectation value), one may arrive to a nice simple formula, as a function of the total spin and of spin states on the triangular moieties:

$$E(S, S_{123}, S_{456}) = -(J - j)\left(S_{123}(S_{123} + 1) + S_{456}(S_{456} + 1)\right) - jS(S + 1) \tag{16}$$

Then, to concretely apply these formulas, the triads of (S, S_{123}, S_{456}) quantum numbers must be counted properly. For instance, let us exemplify the arising of five global singlet states expected in a system with six $1/2$ local spins. Thus, we have three $S = 0$ states originating from the interaction of each spin state on a trimer with its mirror image on the other one. In addition, two more singlets must be added coming from the cross-coupling of the first singlet at one triangle with the second singlet on the companion, and vice-versa.

The groundstate of the ensemble is expected to formally originate from antiferromagnetic coupling of the lowest doublets on each triangular moiety. Since the first and second doublets are degenerate in the case of the imposed equilateral triangle, the groundstate has a fourfold degeneracy. This is an artificial situation, determined by the rather strong idealization assumed at the beginning. However, in spite of limitations, the transparency enabled by this treatment is quite precious. To the best of our knowledge, such a rationalization was not presented before. The magnetism of stacked trinuclear copper complexes resembling the discussed system is a problem with a certain degree of generality, given the many similar known complexes [10,28,29,55–61], so that the here proposed model acquires a quite relevant meaning.

The dimer of trimers is needed to account for, in an effective manner, the sudden drop of the χT at low temperature, observed in Figure 6. The slow increase of χT in the high temperature domain, without reaching the paramagnetic plateau, is the signature of strong antiferromagnetic coupling. Indeed, we fitted the value $J = -110\ \mathrm{cm}^{-1}$, for the averaged coupling inside the trimer units, altogether with a $g = 2.1$ Landé factor. The global intermolecular parameter was found slightly antiferromagnetic, $j = -3\ \mathrm{cm}^{-1}$. We avoided introducing a TIP correction, although it may seem necessary, considering that, without this, a small plateau trend may appear at intermediate temperatures, till the on-set of a molecular effective magnetic moment, after reaching the saturation of the intermolecular coupling. However, a paramagnetic impurity correction seemed necessary to describe why the χT does not drop nearby zero at the lowest T point, taking then PIM $= 0.1\ \mathrm{cm}^3\ \mathrm{mol}^{-1}\ \mathrm{K}^{-1}$, as an empiric adjustment. The HDvV part was equated with the energies from Equation (16), introduced in the van Vleck master formula (Equation (2)). The resulted $(\chi T)_{\mathrm{HDvV}}$ for the model hexanuclear was divided into a half, to consider the molecular susceptibility of the trimer itself. The concrete enumeration of states and expressions of energy levels are shown in Table S5 of the Supplementary Materials.

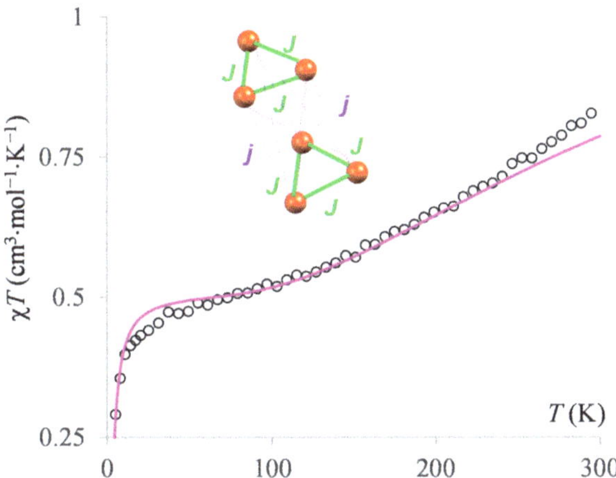

Figure 6. The χT vs. T curve for the trinuclear compound **1**: experimental data (open circles) and fit to model (continuous line). The model is made for the hexanuclear formed as dimer of trimers (shown as sketched inset), in order to account long-range interaction by the unique j coupling parameter, while the χT is taken as a half from the hexamer amount, in order to effectively describe the trimer only.

3.2.3. The Magnetism of the Hexanuclear Complex

The hexamer has a $\{Cu_6O_6\}$ ring resembling the chain conformation of the cyclohexane, with the hydroxo-bridges at vertices and the copper ions in the middle of the almost linear O-Cu-O edges. Then, the $\{Cu_6\}$ subsystem forms a regular planar hexagon. With this note, one may properly say that, from the perspective of exchange coupling effects, the hexanuclear is absolutely similar to the benzene, treated in the frame of a valence bond theory [62], to which the HDvV effective Hamiltonian is explicitly adequate [63]. From this perspective, one may also speculate that the regular shape of the whole system, with identical coordination spheres of the monomeric units and regular $\{Cu_6\}$ frame, is a direct consequence of the aromatic-alike stabilization ensured by the exchange interaction in such a topology.

The associated spin Hamiltonian, considering the idealized D_{6h} symmetry, is obvious:

$$\hat{H}_{M_6(D_{6h})} = -2J(\hat{S}_1 \cdot \hat{S}_2 + \hat{S}_2 \cdot \hat{S}_3 + \hat{S}_3 \cdot \hat{S}_4 + \hat{S}_4 \cdot \hat{S}_5 + \hat{S}_5 \cdot \hat{S}_6 + \hat{S}_1 \cdot \hat{S}_6) \tag{17}$$

The hexagonal hexanuclear does not benefit from a simplified Kambe-like tractability. However, taking as a single significant parameter the coupling J along the proximal copper sites, via hydroxo and pyrazolate bridges, the energy levels can be defined explicitly in a numeric manner, as illustrated in Table S6 from the Supplementary Materials. The neglecting of distant interaction within the ring, between copper ions mutually placed in *meta* and *para* is reasonable, considering the very strong antiferromagnetic coupling along the edges (i.e., between *ortho*-type neighbors).

The fit (see Figure 7) demands empirical adjustment with paramagnetic impurity, since the χT values are not dropping to zero at low temperatures, as expected from the anti-ferromagnetic six-spins system and also a TIP term, to account for a slope in the interval to about 100K, instead of the plateau expected due to large absolute value of the coupling. As will be debated later on, the large coupling is expected from computational hints and also from precedent reported data on comparable systems [64]. We obtained the following fit parameters: $g = 2.1$, $J = -310$ cm^{-1}, PIM = 0.08 cm^3 mol^{-1} K^{-1}, TIP = 0.00011 cm^3 mol^{-1} K^{-2}. The above cited work obtained an even larger antiferromagnetic coupling, $J = 650$ cm^{-1}. The larger absolute value can be correlated with shorter Cu–Cu distance in the previously reported system, about 3.07 Å, in comparison with our spacing, at 3.32 Å. Probably, the positive sign reported by cited authors is a printing error, since according to their data and discussion, the system is recognized as antiferromagnetic, while using the same spin Hamiltonian as in Equation (17). The mentioned work [64] used the Magpack code for fit [65]. Since here we are making explicit the energy levels (see Supplementary Materials, Table S6), we bring further insight to the particularities of hexagonal six-spins systems.

Let us come back to aromaticity-like spin coupling effects [66,67]. Within the HDvV spin Hamiltonian, the resonance energy is $E_{res} = 1.1055 |J|$ [68,69]. Then, with the above fitted coupling value, one obtains $E_{res} = 342.7$ cm^{-1}, i.e., about 26.7 kcal/mol. This is an amount that may count in the balance of deciding the stereochemistry of coordination ensembles. Looking at the larger picture, one may conclude the virtues of the HDvV spin Hamiltonian. It can answer to bonding regime problems (see Reference [1] pp. 325–331 and 411–423), being more valuable than a tool for interpreting the magnetism, as is perceived by its intensive use in magnetochemistry. It can account for other chemical species as well, for instance, the aromatic hydrocarbons with spin [70]. On the other hand, as proven here, it can describe certain bonding aspects in coordination compounds, not only their magnetism.

3.2.4. The Magnetism of the Heptanuclear Complex

The heptanuclear compound can be described as a trigonal antiprism having in center another metal ion. The coordination of the central ion is a trigonally distorted octahedron, while the caps of the embedding antiprism resemble the above-discussed trinuclears, having the same $\{Cu_3(pyrazole)_3\}$ bridged sub-structure.

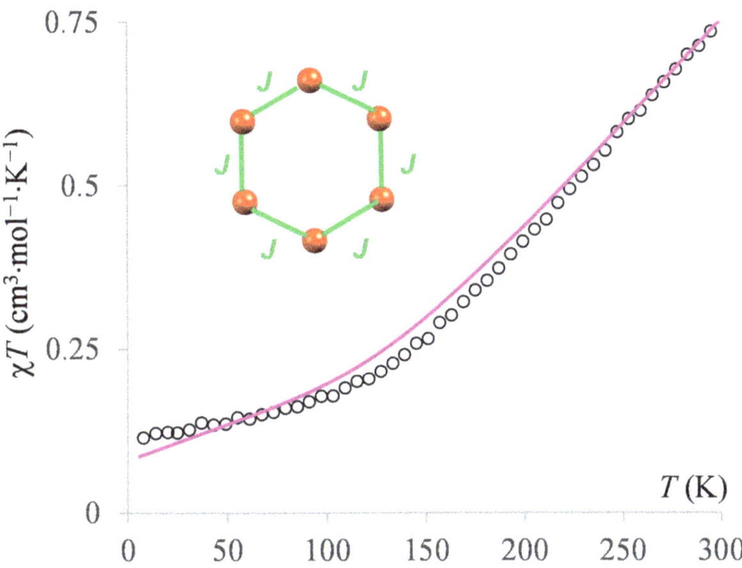

Figure 7. The χT vs. T curve for the hexanuclear compound **2**: experimental data (open circles) and fit to model (continuous line). The model follows the hexagonal pattern, with a J coupling parameter along edges, as illustrated in the inset.

A convenient fact is that the heptamer can be described, in idealized mode, by analytical formulas of the spin of Hamiltonian, provided that the hexamer moiety is taken in a topology as those previously stated and the coupling from the central atom toward all the others is assumed equal, say by the unique parameter J'. Thus, with the M$'$ as the seventh paramagnetic site, having the spin S_7, while S_{1-6} denotes the quantum number of the hexamer unit, the spin Hamiltonian of the heptamer is defined by the following addition to the hexamer formula:

$$\hat{H}_{M_3-M_3-M\prime} = \hat{H}_{M_3-M_3} - 2J'\hat{S}_{1-6} \cdot \hat{S}_7 \tag{18}$$

Then, with the total spin S running from $|S_{1-6} - S_7|$ to $S_{1-6} + S_7$, by transformations resembling the previous ones, one arrives at the following formula of the energy levels:

$$\begin{aligned}E(S, S_{1-6}, S_{123}, S_{456}) = &-(J-j)(S_{123}(S_{123}+1) + S_{456}(S_{456}+1)) \\ &-(j-J')S_{1-6}(S_{1-6}+1) - J'S(S+1),\end{aligned} \tag{19}$$

indexed with the intermediate spins S_{1-6}, S_{123} and S_{456}.

The magnetic data are shown in Figure 8. Although formally kept, inherited from previous derivation, the j parameter will be ignored, to avoid overcharging the fit. Then, the result is: $g = 2.0$, $J = -130$ cm^{-1}, PIM = 0.05 cm^3 mol^{-1} K^{-1}, TIP = 0.0001 cm^3 mol^{-1} K^{-2}. To the best of our knowledge, there are no other literature data on similar compounds to be compared with.

The Supplementary Materials contains in Table S7 details on the spin states in the copper heptanuclear.

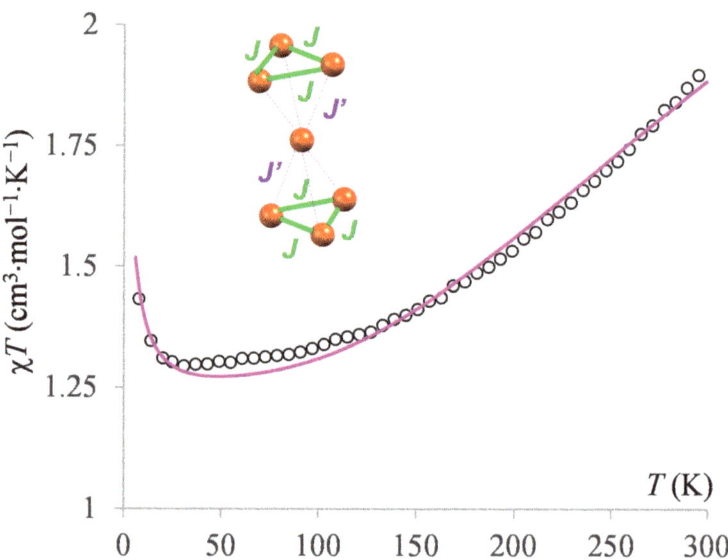

Figure 8. The χT vs. T curve for the heptanuclear compound **3**: experimental data (open circles) and fit to model (continuous line). The model consists of merging the above-defined dimer of trimers, based on J, j couplings with a central node having equivalent parameters, J' towards the vertices of the hexamer. At the end, the j parameter is neglected, imposing $j = 0$ in the actual fit. The inset annotates the considered J and J' interactions.

3.3. Broken Symmetry (BS) Calculations

3.3.1. Realization and Interpretation of BS Calculations

The so-called Broken Symmetry (BS) approach [71–74] represents the strategy usable to obtain estimations of exchange coupling parameters in the frame of Density Functional Theory. The BS-DFT calculations in compounds with many metal centers are laborious, since they are implying the obtaining of single-determinants with unrestricted localized orbitals, having a controlled spin polarization distribution on specific centers. Namely, they must drive the convergence toward defined maps, switching, by proper educated guesses, the local spin population on coordination sites between α or β preponderance. In this view, we proceeded by the following steps: (i) first, do the unrestricted calculation of the system with highest spin multiplicity (HS); (ii) collect the natural orbitals, emulating in this way a restricted-type result; (iii) perform an orbital localization, focusing on functions having singly-occupied nature; (iv) duplicate the sequences of restricted-type localized orbitals, as the initial guess for both α and β unrestricted subsets; (v) initiate the desired number of BS calculations operating the corresponding permutations in α and β orbital lists, in order to achieve the intended spin-flip scheme. The following Schemes 3–5 should be regarded as a roadmap for the desired sequence of HS and BS calculations, customized by the described algorithm, while the subsequent Figures 8–14, with spin density maps, can be seen as a confirmation for achieving the targeted patterns.

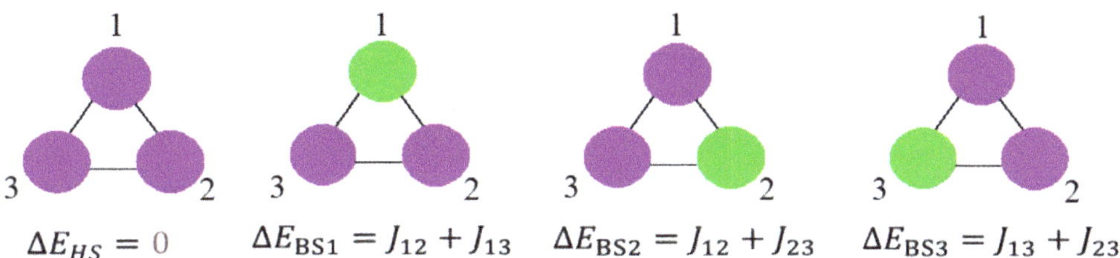

$$\Delta E_{HS} = 0 \qquad \Delta E_{BS1} = J_{12} + J_{13} \qquad \Delta E_{BS2} = J_{12} + J_{23} \qquad \Delta E_{BS3} = J_{13} + J_{23}$$

Scheme 3. The qualitative patterns of spin polarization (α in purple, β in green) and corresponding model energies for Broken Symmetry (BS) configurations in a trinuclear system.

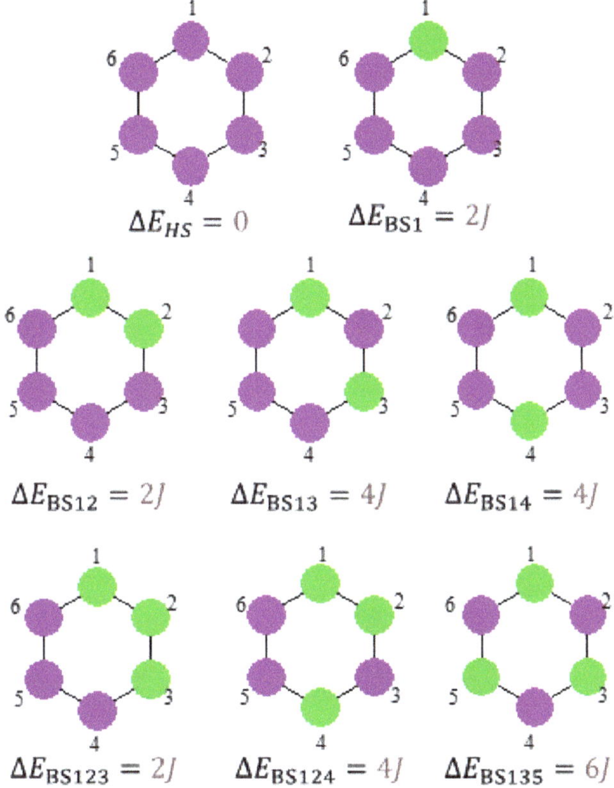

$$\Delta E_{HS} = 0 \qquad \Delta E_{BS1} = 2J$$

$$\Delta E_{BS12} = 2J \qquad \Delta E_{BS13} = 4J \qquad \Delta E_{BS14} = 4J$$

$$\Delta E_{BS123} = 2J \qquad \Delta E_{BS124} = 4J \qquad \Delta E_{BS135} = 6J$$

Scheme 4. The qualitative spin polarization patterns (α in purple, β in green) and model energies for symmetry-independent position isomers of BS configurations in a regular hexagon.

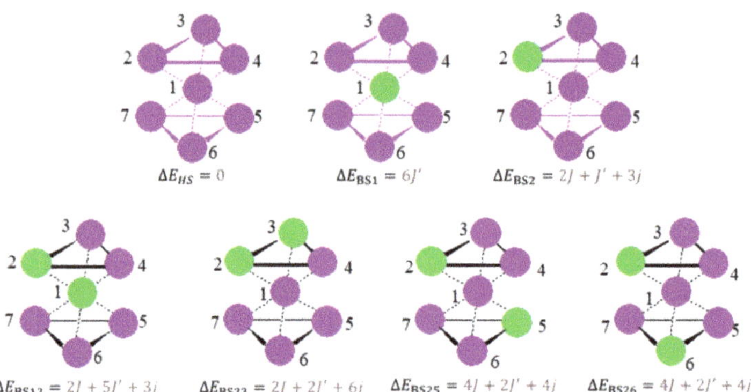

Scheme 5. The spin polarization patterns (α in purple, β in green) and model energies for symmetry-independent position isomers of BS states compound **3**. The $S_z = 1/2$ states are omitted.

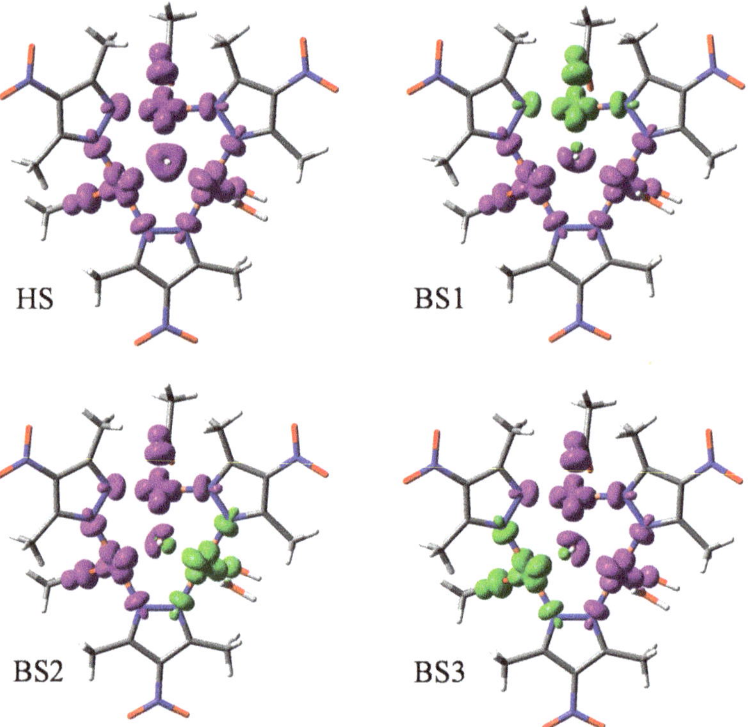

Figure 9. The spin density maps of the unrestricted density functional theory (DFT) calculations for the highest spin (HS) and BS configurations for the trinuclear unit of compound **1**, drawn at 0.05 e/Å^3 isosurfaces. The violet and green colors are representing the α and β spin densities, respectively. Note that the spin swap occurs on the aimed centers, as sketched in green in Scheme 3.

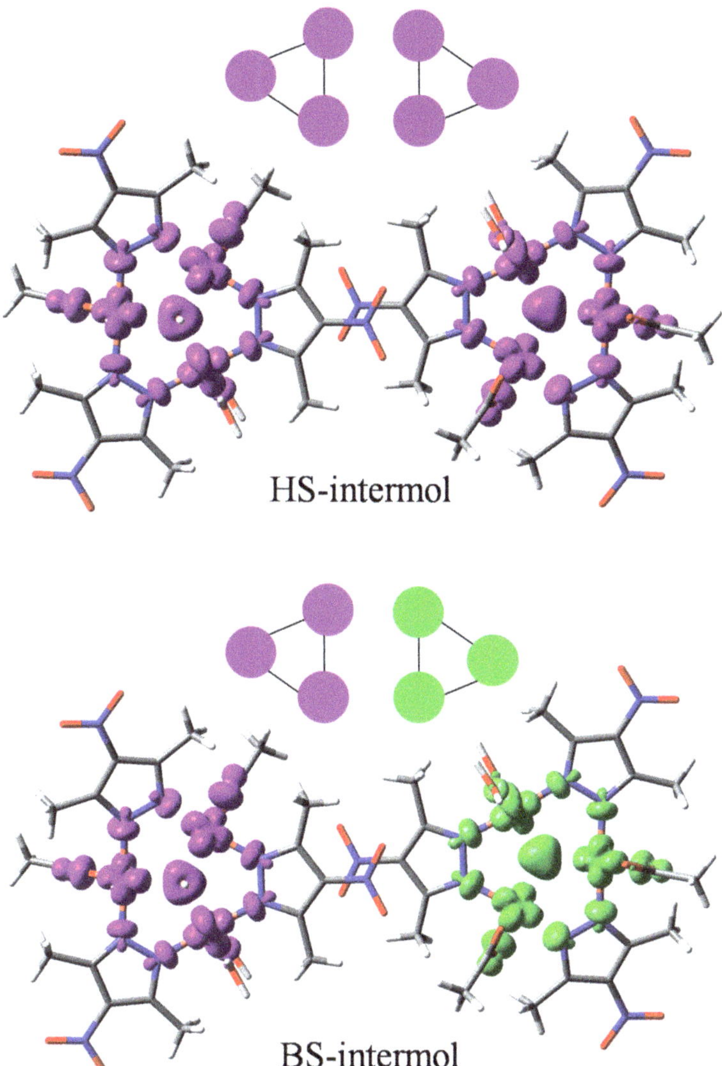

Figure 10. The spin density maps for the BS calculations attempting the description of the intermolecular spin coupling. The couple of dimers is taken as an excerpt from the crystal structure. The qualitative schemes are shown as insets.

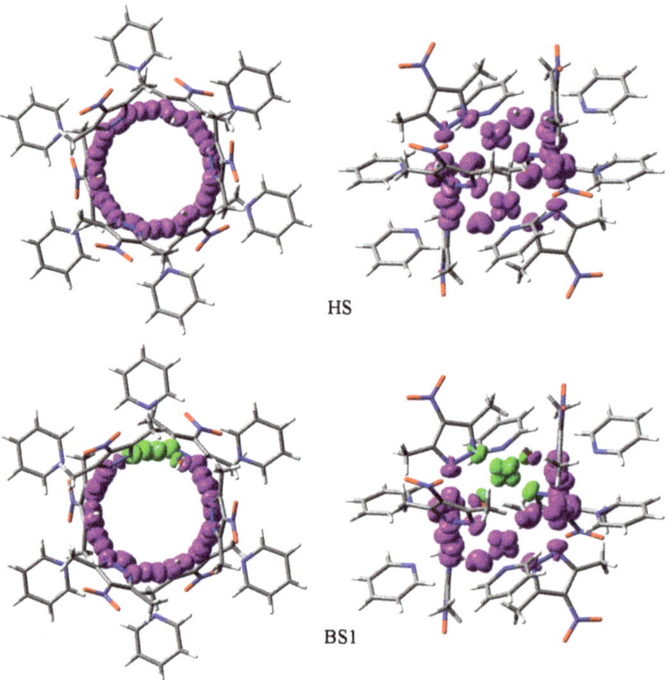

Figure 11. The spin density maps for the HS (**upper** half) and BS1 (**lower** half) for the copper hexamer. The isosurfaces are drawn at the 0.05 e/Å^3 threshold. The violet and green colors represent the α and β spin densities, respectively.

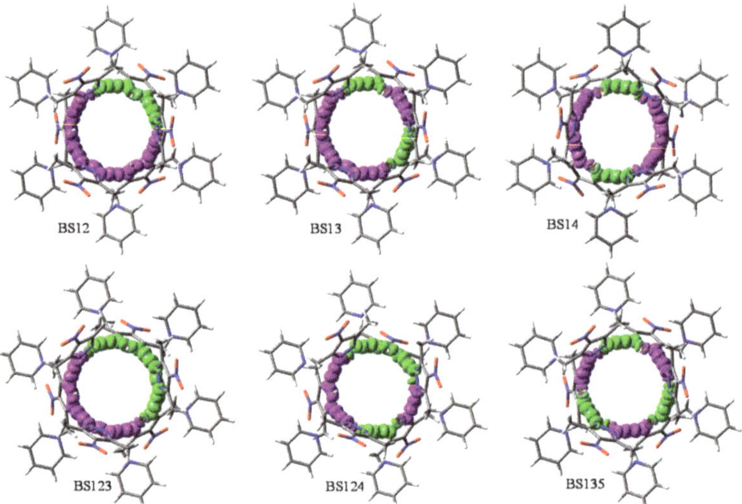

Figure 12. The spin density maps for the BS configurations with $S_z = 1$ (**upper** half) and $S_z = 0$ (**lower** half) for the copper hexamer. The isosurface coloring and drawing threshold is the same as in Figure 9. Note that the green zones are matching the patterns planed in Scheme 4.

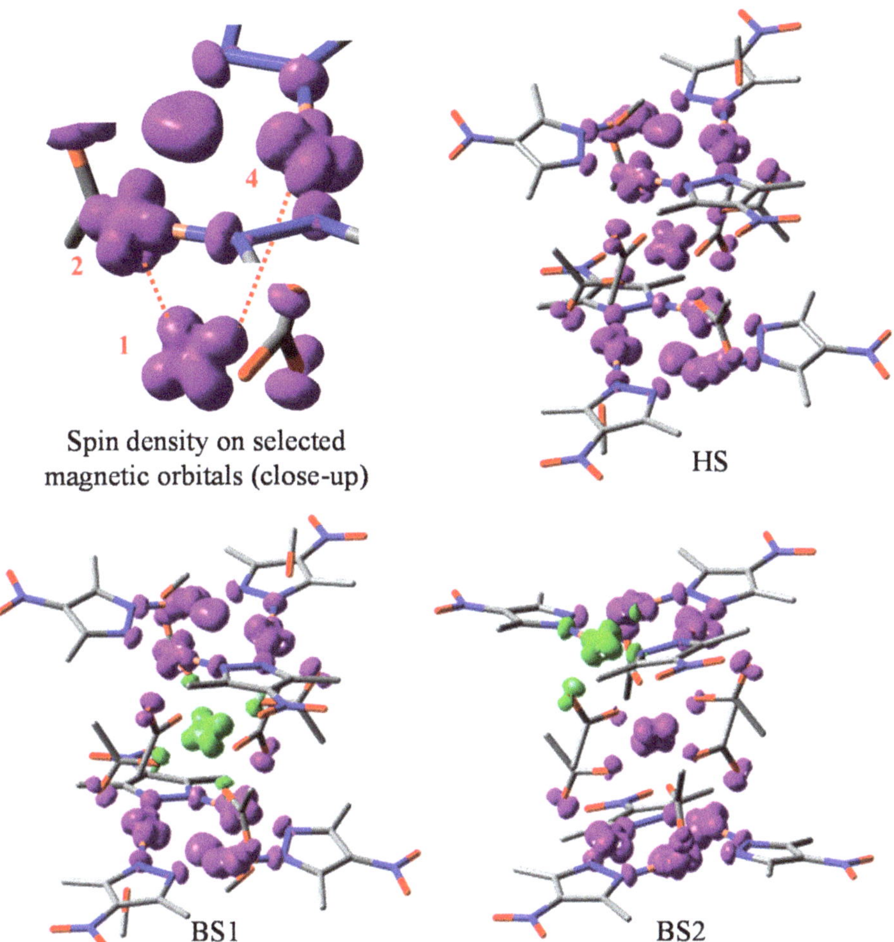

Figure 13. The spin density maps for the $S_z = 7/2$ (HS) and $S_z = 5/2$ (BS1, BS2) configurations of the heptanuclear complex, drawn at 0.01 e/Å^3 isosurfaces. The upper-left corner shows a magnification from the HS spin density map, emphasizing the d-type profiles on the metal ions labeled 1, 2 and 4. A part of the structure components (metal center #3 and certain ligand fragments) were erased to ease the visibility to the portions suggesting the mutual orientation of the magnetic orbitals.

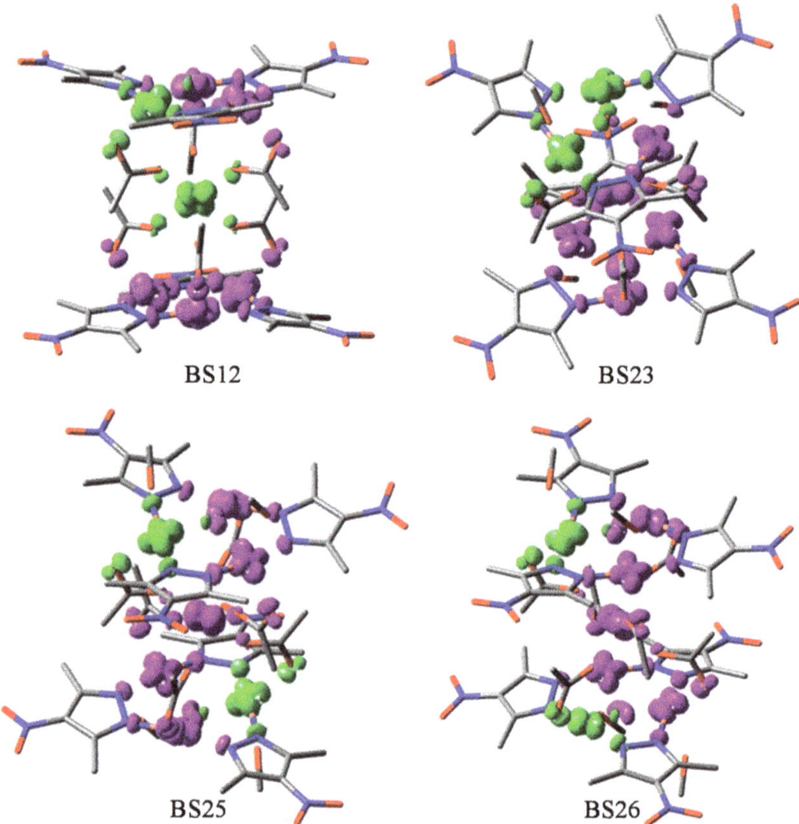

Figure 14. The spin density maps for the $S_z = 5/2$ (BS12, BS23, BS25 and BS26) configurations of the heptanuclear complex, drawn at 0.01 e/Å^3 isosurfaces. Molecular frames are slightly rotated, mutually, in order to ensure better visibility of the sites with β spin density (green surfaces). A part of the half-coordinated methanol molecules was removed to clear the visibility of density surfaces.

The interpretation of BS calculations is straightforward in binuclear systems, the case of higher polynuclears being debated by Shoji et al. [75]. We also outlined a general procedure, exposed in Reference [1] pp. 629–633. Other methodological considerations for equating BS in polynuclears were advanced by Ruiz [76].

The easiest way for interpretation is to directly rely on the so-called Ising spin Hamiltonian, to which the BS calculations are conceptually related. For the sake of clarity, we will try to reformulate now, in a simple way, the handling of BS data in polynuclears, given the chance of the panoply of systems presented here as a quite interesting application. The Ising form takes simple numeric products of spin projections instead of scalar products of true spin operators, met in Equation (1):

$$\hat{H}_{Ising} = -2 \sum_{j<i} J_{ij}\, S_i^z S_j^z \tag{20}$$

By convention, the above formula can be taken without the -2 factor in the front of the summation. The Ising Hamiltonian is invoked in certain dedicated physical problems where pseudospins are adopted as the conventional description of certain site properties [77,78], but actually is of little use in the molecular magnetism itself. The Ising-type expression is nothing else than the form of the diagonal elements in the matrix formulation of the full HDvV Hamiltonian. Therefore, the Ising energies are not directly usable in a physical problem, but it happens that these are parallel to the meaning of energy in BS configurations [76].

The BS treatments imply achieving calculations that reverse the local spin site from projection $S_i^z = +S_i$ to $S_i^z = -S_i$, starting from the reference with all spins up. Note that the whole spin population of a given site should undergo the transmutation from α to β local density, intermediate projections not being admitted as of BS nature. In the case of sites with one unpaired electron, i.e., $S_i = 1/2$, as are our copper polynuclears, the only possibilities are, of course, $S_i^z = \pm 1/2$.

In a practical sense, we shall rely on the BS energies relative to those of the highest spin. The high spin (HS) configuration has the following Ising energy:

$$E_{HS} = -2 \sum_{j<i} J_{ij}\, S_i S_j \tag{21}$$

A BS configuration, labeled ω, can be ascribed as consisting in a sequence of spin-flip factors, $\sigma_{ij} = \pm 1$, as follows:

$$E_{BS(\omega)} = -2 \sum_{j<i} \sigma_{ij}(\omega) J_{ij}\, S_i S_j \tag{22}$$

Obviously, we have $\sigma_{ij} = 1$ for projections with the same sign, $(S_i^z > 0$ and $S_j^z > 0)$ or $(S_i^z < 0$ and $S_j^z < 0)$, and $\sigma_{ij} = -1$ in the case of opposite projections. Let us take the difference between a given BS state, ω, and the HS reference:

$$\Delta E_{BS(\omega)} = E_{BS(\omega)} - E_{HS} = -2 \sum_{j<i} \left(\sigma_{ij}(\omega) - 1 \right) J_{ij}\, S_i S_j \tag{23}$$

In a simplified recipe, the summation on pairs, $j < i$, retains only the terms with $\sigma_{ij} = -1$, i.e., the situations where a relative spin flip occurs at the given i-j contact. The Ising-like interpretation results can be saliently applied (using the DFT computed energies of the different BS(ω) configurations) as long as the corresponding computed expectation values of the squared spin, $\langle S(S+1) \rangle_\omega$, are well retrieving the following regularity:

$$\Delta \langle S(S+1) \rangle_\omega = \langle S(S+1) \rangle_\omega - \langle S(S+1) \rangle_{HS} = -2 \sum_{j<i} \left(\sigma_{ij}(\omega) - 1 \right) S_i S_j \tag{24}$$

All the computations presented in the following are obeying this demand. Note that, while Equation (23) can omit the ij couples for which J_{ij} is assumed negligible, the summation in Equation (24) should include all the possible pairs. For details, see Reference [1] pp. 629–633.

3.3.2. The BS Calculations of the Trinuclear Complex

The BS treatment of a triangle with inequivalent edges implies one HS configuration with $S_z = 3/2$ total spin projection and three configurations with $S_z = 1/2$, labeled BSi, according to the address of the i center lodging the β spin. These are qualitatively represented in Scheme 3, according to the explicit $\Delta E(\text{BS}i)$ general formulas. The computed BS energies of states, relative to the HS, are -287.6, -259.2 and -257.0 cm^{-1}, at the reversal of the spin on the respective coordination sites labeled 1, 2 and 3. With the model equations ascribed in Scheme 3, one may deduce the following distinct coupling parameters, $J_{12} = -144.9$ cm^{-1}, $J_{13} = -142.7$ cm^{-1} and $J_{23} = -114.3$ cm^{-1}.

As initially guessed from qualitative structural considerations, these parameters are mutually comparable, indebting the approximate treatment as equilateral. In the regular triangle, the energy of each BS state is $2J$, the computed average being $J = -134.0$ cm^{-1}. This result is in good relation with the above-reported fit value (see Section 3.2.4). The differences in spin-squared expectation values, $\Delta\langle S(S+1)\rangle_{HS} - \Delta\langle S(S+1)\rangle_{BS}$ are, in all three cases, about 2.03, i.e., close to the ideal quantity, namely 2.

The spin density maps of HS and BS configurations are revealing the shapes of the magnetic orbitals carried by each site (see Figure 9). Obviously, the sign information associated to the wave-function lobes is lost in such representations, having only the orbital contours. Thus, one may observe the "four leaf clover" shapes corresponding to the d-type orbitals and mushroom-like profiles of the lone pairs from pyrazole and acetate ligands. The central oxygen has a donut shape in the HS map, being affected by spin polarization in the BS cases, when it forms asymmetric lobes. One observes that the BS spin reversal is spreading also on the neighboring ligands, which are also colored in green around the β d-type orbital. The four-lobe d-type orbitals are oriented almost in the Cu(pyrazole) local frames, certifying that the N-Cu-N axes are those undergoing the strongest ligand field. Slight tilts out of the {Cu$_3$} plane seem to be determined by the pyramidalization of the coordinated central hydroxo group above the copper triangle fragment. If we define the local coordination axes with x along the N-Cu-N average directions and y along the Cu-O linkages toward the central group, then the magnetic orbitals of each coordination site can be named qualitatively as x^2-y^2. Details on the BS-DFT data of the presented system are in Table S8 from the Supplementary Materials.

Subsequently, we tried to investigate in the BS methodology the intermolecular interaction, roughly described by the j parameter in the model from Section 3.2.2. As discussed, the j factorizes the interaction between the whole trimer moieties. We considered a pair of molecules as a model which, in crystal are related almost by an inversion center (corresponding to Figure S1 in Supplementary Materials). The BS treatment can be simply realized taking one trimer moiety in the HS configuration, while another one has the same appearance, reverted in β spin density, as illustrated in Figure 10. Although the BS intermolecular spin coupling regime is achieved, as certified by the obtained spin density maps (see Figure 10), the energy difference between the HS and BS forms is practically null. In turn, the fit discussed in Section 3.2.2 found a small, yet sizeable, intermolecular antiferromagnetism. There may be multiple reasons for the computational failure. First of all, one may suspect the intrinsic drawbacks of current density functionals in describing long-range effects [79,80]. We refrained from introducing any of the many possible empirical long-range corrections [81–84], since none of these were calibrated for such subtle goals. Another possibility stays in the limitation of the "dimer of trimers" model. In the crystal, such supramolecular contacts are continued as chains and sheets. Then, a very small coupling at a dimeric sequence can be amplified, in the band structure interaction regime, to a sizeable coupling effect. Thus, the fitted j, formally assigned to a dimer, actually describes effects from extended systems. Since the failures are also a part of the fair scientific inquires, we mentioned the case here.

3.3.3. The BS Calculations of the Hexanuclear Complex

The BS calculations on the hexagonal complex are challenging, facing the situation of many position isomers in terms of mutual placement of the α and β sites. Since only one coupling parameter must be found, J, a single BS configuration will suffice. However, treating the whole set implies a challenging technical virtuosity, and besides, retrieving the same J value will certify the validity of the BS approach itself.

Scheme 4 outlines in a qualitative manner the distinct symmetry BS states possible in a regular hexagon. The labeling by BSi, BSij and BSijk denotes the i, j or k sites taking the β spin in the respective situations when one, two or three spin swaps are made with respect to the HS state (with $S_z = 3$ spin projection). There is only one distinct isomer with $S_z = 2$ projection, marked as BS1 (equivalent with one spin reversed on any center). The $S_z = 1$ has the three possibilities of *ortho-*, *meta-* and *para-* BS isomers, labeled BS12, BS13 and BS14, respectively. Finally, the $S_z = 0$ has the BS123, BS124 and BS135 cases. Their energies, relative to the HS are printed in Scheme 4. There is a simple recipe to retrieve the BS formulas: the cofactor of a given type of J is the count of edges carrying circles of different colors, in the drawn insets of Scheme 4. Details on the BS-DFT data of the hexamer are in given Table S9 from the Supplementary Materials.

In order to extract information about the shape of magnetic orbitals, Figure 11 shows the HS and BS1 configurations, each, in two orientations. One view is perpendicular to the mean plane of the metal ions ring, aiming to verify the obtaining, from computation, of the spin polarization pattern charted in Scheme 4. The views presented on the left side of Figure 11 are allowing to view the sites labeled 1 and 4 particularly. Thus, one may observe the four-lobed orbitals oriented perpendicularly to the Cu_6 plane. The lone pairs due to pyrazole ligands have mushroom shapes. The spin delocalized over a hydroxo bridge takes the hearth-like shapes in the HS maps and in the points bridging ions with the same spin. As a bridge between different local spin projections, the OH groups are polarized in two-lobe profiles, with different colors. Interestingly, as a bridge between two spin projections, the pyrazoles get one lone pair colored in purple and one in green, certifying the inner spin polarization. Figure 12 shows the top view of the other computed BS spin density maps.

The relative BS energies for the {BS1, BS12, BS13, BS123, BS124, BS135} sequence are, correspondingly, {−480.6, −482.0, −960.2, −961.0, −481.9, −961.5 and −1439.0} (values in cm^{-1}). Dividing these quantities by coefficients of J ascribed in Scheme 4 as analytic BS energies, the following estimations for the J parameters are, respectively, obtained: {−240.3, −241.0, −240.1, −240.3, −241.0, −240.4 and −239.8} (all values in cm^{-1}). Thus, all the estimations from the different BS calculations are very close to the $J = -240$ cm^{-1}, the whole treatment being validated. The value is reasonably close, in semiquantitative respects, to the J resulted from fit, in Section 3.2.3.

3.3.4. The BS Calculations of the Heptanuclear Complex

The heptamer presents a very large number of BS position isomers, whether the whole set is aimed. We will limit ourselves to the configurations with the following definitions: $S_z = 7/2$ (HS), $S_z = 5/2$ (BS1 and BS2, as symmetry distinct species) and $S_z = 3/2$ (BS12, BS23, BS25 and BS26). Thus, we eluded the case of smallest projection, $S_z = 1/2$, with a large number of positional mutations. The full BS-DFT data are posted in Table S10 from Supplementary Materials. The considered set of six configurations offers sufficient information to extract the three parameters assumed by the model presented in Section 3.2.4, namely J for coupling inside the triangular caps of the structure, J', for all the contacts between these faces and the central atom, and j for the presumably weak effects between triangular faces. The BS configurations and their model energies are presented in Scheme 5. One may observe again the simple recipe for the BS formulas, by summing all the coupling parameters picked from edges encountering centers of opposed spin polarity. This simple rule, based on Ising interpretation is valid if the check assumed by Equation (24) is approximately passed. Thus, the ideal values for the $\langle S(S+1)\rangle_{HS} - \langle S(S+1)\rangle_{BS}$ differences are 6 and 10 for the respective $S_z = 5/2$ and $S_z = 3/2$ subsets. The computational details given in Table S7 from Supplementary Materials are illustrating that this regularity is held with reasonable accuracy, within the second decimal digit.

The relative energies for the (BS1, BS2, BS12, BS23, BS25 and BS26) sequences are {27.7, −313.9, −294.3, −312.0, −631.8 and −643.8}, respectively (all values in cm^{-1}). The least square solutions of the BS equations are: $J = -162.3$ cm^{-1}, $J' = +4.9$ cm^{-1}, $j = +0.75$ cm^{-1}. One may remark that the J coupling corresponds to a strong antiferromagnetism, while the J' is weakly ferromagnetic, in line with the experimental data. The magnitudes of the computed parameters are matching the range of the experimental ones. Somewhat intriguingly, the computed j coupling results as ferromagnetic, too. This parameter was not the object of the fit and is difficult to charge the problem with too many parameters, because of the inherent uncertainties, given the rather large parametric list.

A further detailing of the issue can be reached associated to the BS fit, complicating the model a bit. Thus, relaxing the constraint of a unique coupling parameter inside the hexanuclear subunit, one may propose two sorts of long-range interactions: a j_c, corresponding to the sites placed mutually in *cis*, inside the trigonally distorted octahedron and a j_t, for the *trans* couples. With this detailing, the BS25 and BS26 become inequivalent, the $4j$ term becoming $4\,j_c$ and $2j_c + 2j_t$, respectively. Within the new BS model fit, the following values are found $J = -163.0$ cm^{-1}, $J' = +5.0$ cm^{-1}, $j_c = +2.8$ cm^{-1} and $j_t = -2.7$ cm^{-1}.

Aside from verifying the success of BS attempts, by retrieving in the computed spin density maps the patterns envisaged at the beginning of calculations, according to the chart form Scheme 5, Figures 13 and 14 are offering information about implied magnetic orbitals. Particularly, it is relevant to observe the magnified excerpt from the upper-left corner of Figure 13, showing the d-type profiles for the central atom (#1) and for two atoms from the upper triangle (#2 and #4). One notes that the four-lobe profile of the atom #1 is tilted by about 45° with respect to the planes of the two Cu$_3$ moieties. Inside the trimeric fragments, the four-lobe shapes are approximately perpendicular to the axes formally connecting the given sites to the central atom, being therefore tilted with about 30° from their Cu$_3$ planes. Looking along the dashed line drawn between #1 and #2 copper sites (see the upper-left corner of Figure 13) one may say that, if this is taken as the z axis in a local coordinate frame, then the corresponding magnetic orbitals look like a x^2-y^2 and a xy couple. Then, falling in a sort of orthogonality relationship (null overlap), one may guess that this coupling is prone to ferromagnetic nature, according to the basic paradigms of magneto-structural correlations [85–88]. Also, an almost-orthogonality can be proposed for the contact along the #1–#4 line, having the situation of the first site with one lobe approximately aligned to this axis (behaving then as able for σ-type bonding), while the other metal-based orbital has the lobes roughly perpendicular to the intercentrum direction (as if ready for δ-type overlap). The orbital interaction along the #1–#3 line follows a similar situation. By symmetry, the interaction between the central magnetic orbital and those of the other face undergo a mirroring of mutual overlap relations. Then, the orbital quasi-orthogonality reasons are explaining the ferromagnetic coupling between the central atom and the embedding cluster, obtained from fit and computational simulations.

4. Conclusions

The work deals with the structural chemistry of a series of newly synthesized coordination polynuclear compounds, seen from multiple perspectives, experimental, theoretical and in correlation with the magnetic properties. The common feature of the presented systems is the functionality of pyrazolate bridges as reinforcement of molecular edifices.

The clusters sketched by the Cu(II) centers show remarkable patterns: almost equilateral triangles, regular hexagon and centered trigonal antiprism, for the respective tri-, hexa- and heptanuclears. Remarkably, the last two systems span high space group symmetries, R-3 and R-3c. The opportunity of meeting iconic patterns allows naming the compounds **1**, **2** and **3** as representative for the described distinct topologies. In this way, the methods and outcomes related to their structural analyses show general relevance, extended upon past and future relatives falling in the **1**, **2** or **3** classes.

Whether the actual literature knows many analogs of **1** and several congeners of **2**, compound **3** has only two previously reported relatives, lacking detailed insight. The fact that compound **3** uses the **1**-type subsystems as a starting base affords the prediction that this class can be further grown by systematic synthetic attempts.

The presented compounds occasioned new Kambe-type systematization of energy levels, namely having analytical formulas, linear in terms of existing coupling parameters. Whether the equilateral triangle is a well-known Kambe case, we extended this treatment for two interacting trimers, the model being usable for the effective account of long-range effects in the class of compound **1**. We also found Kambe-like formulas for system **3**. The hexagon from case **2** does not admit Kambe rationalization, but we considered explicit numeric expression for the energy levels, making its treatment transparent. Remarkably, system **2** can be legitimately presented as a case of coordination polynuclear aromaticity. To the best of our knowledge, such developments related with the spin Hamiltonian methodologies were not presented previously.

The excurse is concluded with the simulation of exchange coupling parameters trough BS-DFT calculations (Broken Symmetry). The BS approach consists of numeric experiments aiming at the controlled swap of spin polarization on designated centers. At relatively high nuclearities (as 6 or 7), the BS approach is non-trivial, in the technical respect, demanding proper educated guesses of the orbitals used to initiate the calculations.

The interpretation of BS computational results is somewhat challenging, outlining here clear and simple recipes. The presented systems show, as a distinct signature, the strong antiferromagnetic coupling along the pyrazolate bridges. The spin density maps of various BS position isomers (different symmetry-distinct ways to switch from α to β the site spin, at one or many centers) are certifying the success of the attempted BS configurations. The spin density maps are enabling the view of site-specific magnetic orbitals.

Supplementary Materials: The following are available online at https://www.mdpi.com/2624-8549/3/1/31/s1, Figure S1: X-ray molecular structure of compound **1** with atoms coloring scheme, Figures S2–S4: Packing along *a*, *b* and *c* axes for compound **1**, Figure S5: X-ray molecular structure of compound **2** with atoms coloring scheme, Figures S6–S8: Packing along *a*, *b* and *c* axes for compound **2**, Figure S9: X-ray molecular structure of compound **3** with atoms coloring scheme, Figures S10 and S11: Packing along *a*, *b* and *c* axes for compound **3**, Table S1: Crystal data for complexes **1–3**, Table S2: Selected bond lengths [Å] and angles [°] for **1**, Table S3: Selected bond lengths [Å] and angles [°] for **2**, Table S4: Selected bond lengths [Å] and angles [°] for **3**, Table S5; The count and energies of spin states (including total and intermediate spin quantum numbers) in a copper hexanuclear system, constituted as dimer of trimers, with the topology discussed in the main text, Table S6: Numeric energies in a hexanuclear system with local 1/2 spin centers and hexagonal topology (having unique exchange coupling parameter *J* along the edges), Table S7: The count and energies of spin states in a copper heptanuclear system, constituted as a central atom interacting equivalently with two trimers, Tables S8–S10: The high spin (HS) and broken symmetry (BS) energies calculated for compounds **1–3**.

Author Contributions: Conceptualization, supervision, writing—review and editing, validation, F.C. and M.F.; methodology, software, F.C.; formal analysis, data curation, A.-I.T.; data curation, visualization, investigation, original draft preparation, A.M.T. and M.C.B.; resources, C.M.Z. and L.O.C.; funding acquisition, L.O.C. and M.F. All authors have read and agreed to the published version of the manuscript.

Funding: This research and the APC were partially funded by a grant from the Romanian Ministry of Education and Research, CCCDI-UEFISCDI, project number PN-III-P2-2.1-PED-2019-3009, within PNCDI III and University of Bucharest.

Institutional Review Board Statement: Not applicable.

Conflicts of Interest: The authors declare no conflict of interest.

References

1. Putz, M.V.; Cimpoesu, F.; Ferbinteanu, M. *Structural Chemistry, Principles, Methods, and Case Studies*; Springer: Cham, Switzerland, 2018; pp. 629–633. [CrossRef]
2. Thompson, L.K. Polynuclear coordination complexes – from dinuclear to nonanuclear and beyond. *Coord. Chem. Rev.* **2002**, *233–234*, 193–206. [CrossRef]
3. Halcrow, M.A. Pyrazoles and pyrazolides-flexible synthons in self-assembly. *Dalton Trans.* **2009**, 2059–2073. [CrossRef] [PubMed]
4. Mukherjee, R. Coordination chemistry with pyrazole-based chelating ligands: Molecular structural aspects. *Coord. Chem. Rev.* **2000**, *203*, 151–218. [CrossRef]
5. Solomon, E.I.; Sundaram, U.M.; Machonkin, T.E. Multicopper Oxidases and Oxygenases. *Chem. Rev.* **1996**, *96*, 2563–2606. [CrossRef]
6. Cole, A.P.; Root, D.E.; Mukherjee, P.; Solomon, E.I.; Stack, T.D. A trinuclear intermediate in the copper-mediated reduction of O_2: Four electrons from three coppers. *Science* **1996**, *273*, 1848–1850. [CrossRef]
7. Kaim, W.; Rall, J. Copper—A "Modern" Bioelement. *Angew. Chem. Int. Ed.* **1996**, *35*, 43–60. [CrossRef]
8. Lee, S.K.; George, S.D.; Antholine, W.E.; Hedman, B.; Hodgson, K.O.; Solomon, E.I. Nature of the intermediate formed in the reduction of O_2 to H_2O at the trinuclear copper cluster active site in native laccase. *J. Am. Chem. Soc.* **2002**, *124*, 6180–6193. [CrossRef]
9. Tsukerblat, B.S.; Kuyavskaya, B.Y.; Belinskii, M.I.; Ablov, A.V.; Novotortsev, V.M.; Kalinnikov, V.T. Antisymmetric Exchange in the Trinuclear Clusters of Copper (II). *Theor. Chim. Acta* **1975**, *38*, 131–138. [CrossRef]

10. Ferrer, S.; Lloret, F.; Bertomeu, I.; Alzuet, G.; Borras, J.; Garcya-Granda, S.; Liu-Gonzalez, M.; Haasnoot, J.G. Cyclic Trinuclear and Chain of Cyclic Trinuclear Copper(II) Complexes Containing a Pyramidal Cu$_3$O(H) Core. Crystal Structures and Magnetic Properties of [Cu$_3$(μ3-OH)(aaat)$_3$(H$_2$O)$_3$](NO$_3$)2·H$_2$O [aaat = 3-Acetylamino-5-amino-1,2,4-triazolate] and {[Cu$_3$(μ3-OH)(aat)$_3$(μ3-SO$_4$)]·6H$_2$O}$_n$ [aat = 3-Acetylamino-1,2,4-triazolate]: New Cases of Spin-Frustrated Systems. *Inorg. Chem.* **2002**, *41*, 5821–5830. [CrossRef]

11. Rudra, I.; Wu, Q.; Voorhis, T.V. Predicting exchange coupling constants in frustrated molecular magnets using density functional theory. *Inorg. Chem.* **2007**, *46*, 10539–10548. [CrossRef]

12. Wang, L.; Sun, Y.; Yu, Z.; Qi, Z.; Liu, C. Theoretical investigation on triagonal symmetry copper trimers: Magneto-structural correlation and spin frustration. *J. Phys. Chem. A* **2009**, *113*, 10534–10539. [CrossRef] [PubMed]

13. Yoon, J.; Solomon, E.I. Ground-state electronic and magnetic properties of a μ3-Oxo-bridged trinuclear Cu(II) complex: Correlation to the native intermediate of the multicopper oxidases. *Inorg. Chem.* **2005**, *44*, 8076–8086. [CrossRef] [PubMed]

14. Afrati, T.; Dendrinou-Samara, C.; Raptopoulou, C.; Terzis, A.; Tangoulis, V.; Tsipis, A.; Kessissoglou, D.P. Experimental and theoretical study of the antisymmetric magnetic behavior of copper inverse-9-metallacrown-3 compounds. *Inorg. Chem.* **2008**, *47*, 7545–7555. [CrossRef] [PubMed]

15. Canon-Mancisidor, W.; Spodine, E.; Paredes-Garcia, V.; Venegas-Yazigi, D. Theoretical description of the magnetic properties of μ3-hydroxo bridged trinuclear copper(II) complexes. *J. Mol. Model.* **2013**, *19*, 2835–2844. [CrossRef]

16. Angaroni, M.; Ardizzoia, G.A.; Beringhelli, T.; La Monica, G.; Gatteschi, D.; Masciocchi, N.; Moret, M. Oxidation reaction of [{Cu(Hpz)$_2$Cl}$_2$](Hpz = pyrazole): Synthesis of the trinuclear copper(II) hydroxo complexes [Cu$_3$(OH)(pz)$_3$(Hpz)$_2$Cl$_2$]·solv (solv = H$_2$O or tetrahydrofuran). Formation, magnetic properties, and X-ray crystal structure of [Cu$_3$(OH)(pz)$_3$(py)$_2$Cl$_2$]·py (py = pyridine). *J. Chem. Soc. Dalton Trans.* **1990**, 3305–3309. [CrossRef]

17. Sakai, K.; Yamada, Y.; Tsubomura, T.; Yabuki, M.; Yamaguchi, M. Synthesis, Crystal Structure, and Solution Properties of a Hexacopper (II) Complex with Bridging Hydroxides, Pyrazolates, and Nitrates. *Inorg. Chem.* **1996**, *35*, 542–544. [CrossRef]

18. Angaridis, P.A.; Baran, R.; Boča, R.; Cervantes-Lee, F.; Haase, W.; Mezei, G.; Raptis, R.G.; Werner, R. Synthesis and Structural Characterization of Trinuclear Cu^{II} − Pyrazolato Complexes Containing μ3-OH, μ3-O, and μ3-Cl Ligands. Magnetic Susceptibility Study of [PPN]$_2$[(μ3-O)Cu$_3$(μ-pz)$_3$Cl$_3$]. *Inorg. Chem.* **2002**, *41*, 2219–2228. [CrossRef]

19. Mezei, G.; Baran, P.; Raptis, R.G. Anion Encapsulation by Neutral Supramolecular Assemblies of Cyclic Cu^{II} Complexes: A Series of Five Polymerization Isomers, [{cis-Cu^{II}(μ-OH)(μ-pz)}$_n$], n=6, 8, 9, 12, and 14. *Angew. Chem. Int. Ed.* **2004**, *43*, 574–577. [CrossRef]

20. Fernando, I.R.; Surmann, S.A.; Urech, A.A.; Poulsen, A.M.; Mezei, G. Selective total encapsulation of the sulfate anion by neutral nano-jars. *Chem. Commun.* **2012**, *48*, 6860–6862. [CrossRef]

21. Canon-Mancisidor, W.; Gomez-Garcia, C.J.; Espallargas, G.M.; Vega, A.; Spodine, E.; Venegas-Yazigi, D.; Coronado, E. Structural re-arrangement in two hexanuclear Cu^{II} complexes: From a spin frustrated trigonal prism to a strongly coupled antiferromagnetic soluble ring complex with a porous tubular structure. *Chem. Sci.* **2014**, *5*, 324–332. [CrossRef]

22. Henkelis, J.J.; Jones, L.F.; de Miranda, M.P.; Kilner, C.A.; Halcrow, M.A. Two Heptacopper(II) Disk Complexes with a [Cu$_7$(μ_3-OH)$_4$(μ-OR)$_2$]$^{8+}$ Core. *Inorg. Chem.* **2010**, *49*, 11127–11132. [CrossRef]

23. Mohamed, A.A.; Ricci, S.; Burini, A.; Galassi, R.; Santini, C.; Chiarella, G.M.; Melgarejo, D.Y.; Fackler, J.P., Jr. Halide and Nitrite Recognizing Hexanuclear Metallacycle Copper(II) Pyrazolates. *Inorg.Chem.* **2011**, *50*, 1014–1020. [CrossRef] [PubMed]

24. Li, H.-X.; Ren, Z.-G.; Liu, D.; Chen, Y.; Lang, J.-P.; Cheng, Z.-P.; Zhu, X.-L.; Abrahams, B.F. Single-crystal-to-single-crystal structural transformations of two sandwich-like Cu(II) pyrazolate complexes and their excellent catalytic performances in MMA polymerization. *Chem. Commun.* **2010**, *46*, 8430–8432. [CrossRef] [PubMed]

25. Yakovleva, M.A.; Kushan, E.V.; Boltacheva, N.S.; Filyakova, V.I.; Nefedov, S.E. Deprotonation of pyrazole and its analogs by aqueous copper acetate in the presence of triethylamine. *Russ. J. Inorg. Chem.* **2012**, *57*, 181–192. [CrossRef]

26. Di Nicola, C.; Garau, F.; Gazzano, M.; Guedes da Silva, M.F.C.; Lanza, A.; Monari, M.; Nestola, F.; Pandolfo, L.; Pettinari, C.; Pombeiro, A.J.L. New Coordination Polymers and Porous Supramolecular Metal Organic Network Based on the Trinuclear Triangular Secondary Building Unit [Cu$_3$(μ3-OH)(μ-pz)$_3$]$^{2+}$ and 4,4′-Bypiridine. *Cryst. Growth Des.* **2012**, *12*, 2890–2901. [CrossRef]

27. Di Nicola, C.; Garau, F.; Gazzano, M.; Lanza, A.; Monari, M.; Nestola, F.; Pandolfo, L.; Pettinari, C.; Zorzi, F. Reactions of a Coordination Polymer Based on the Triangular Cluster [Cu$_3$(μ3-OH)(μ-pz)$_3$]$^{2+}$ with Strong Acids. Crystal Structure and Supramolecular Assemblies of New Mono-, Tri-, and Hexanuclear Complexes and Coordination Polymers. *Cryst. Growth Des.* **2010**, *10*, 3120–3131. [CrossRef]

28. Zhou, Q.; Liu, Y.; Wang, R.; Fu, J.; Xu, J.; Lou, J. Synthesis, crystal structure and magnetic properties of a trinuclear Cu(II)-pyrazolate complex containing μ3-OH. *J. Coord. Chem.* **2009**, *62*, 311–318. [CrossRef]

29. Zhou, J.H.; Liu, Z.; Li, Y.Z.; Song, Y.; Chen, X.T.; You, X.Z. Synthesis, structures and magnetic properties of two copper(II) complexes with pyrazole and pivalate ligands. *J. Coord. Chem.* **2006**, *59*, 147–156. [CrossRef]

30. Davydenko, Y.M.; Demeshko, S.; Pavlenko, V.A.; Dechert, S.; Meyer, F.; Fritsky, I.O. Synthesis, Crystal Structure, Spectroscopic and Magnetically Study of Two Copper(II) Complexes with Pyrazole Ligand. *Z. Anorg. Allg. Chem.* **2013**, *639*, 1472–1476. [CrossRef]

31. Bala, S.; Bhattacharya, S.; Goswami, A.; Adhikary, A.; Konar, S.; Mondal, R. Designing Functional Metal–Organic Frameworks by Imparting a Hexanuclear Copper-Based Secondary Building Unit Specific Properties: Structural Correlations with Magnetic and Photocatalytic Activity. *Cryst. Growth Des.* **2014**, *14*, 6391–6398. [CrossRef]

32. Condello, F.; Garau, F.; Lanza, A.; Monari, M.; Nestola, F.; Pandolfo, L.; Pettinari, C. Synthesis and Structural Characterizations of New Coordination Polymers Generated by the Interaction Between the Trinuclear Triangular SBU [Cu$_3$(μ_3-OH)(μ-pz)$_3$]$^{2+}$ and 4,4′-Bipyridine. 3° *Cryst. Growth Des.* **2015**, *15*, 4854–4862. [CrossRef]

33. Morgan, G.T.; Ackerman, I. CLII.—Substitution in the pyrazole series. Halogen derivatives of 3:5-dimethylpyrazole. *J. Chem. Soc. Trans.* **1923**, *123*, 1308–1318. [CrossRef]

34. Hüttel, R.; Schäfer, O.; Jochum, P. Die Jodierung der Pyrazole. *Justus Liebigs Ann. Chem.* **1955**, *593*, 200–207. [CrossRef]

35. Buchner, E.; Fritsch, M. VI. Darstellung und Derivate des freien Pyrazols. *Justus Liebigs Ann. Chem.* **1893**, *273*, 256–266. [CrossRef]

36. Kahn, O. *Molecular Magnetism*; VCH Publishers: New York, NY, USA, 1993.

37. *Crystal Structure 4.2: Crystal Structure Analysis Package, Rigaku Corporation (2000–2015)*; Rigaku Corporation: Tokyo, Japan.

38. Dolomanov, O.V.; Bourhis, L.J.; Gildea, R.J.; Howard, J.A.K.; Puschmann, H. OLEX2: A complete structure solution, refinement and analysis program. *J. Appl. Cryst.* **2009**, *42*, 339–341. [CrossRef]

39. Sheldrick, G.M. Crystal structure refinement with SHELXL. *Acta Crystallogr. Sect. C Struct. Chem.* **2015**, *71*, 3–8. [CrossRef] [PubMed]

40. Macrae, C.F.; Edgington, P.R.; McCabe, P.; Pidcock, E.; Shields, G.P.; Taylor, R.; Towler, M.; Van De Streek, J. Mercury: Visualization and analysis of crystal Structures. *J. Appl. Crystallogr.* **2006**, *39*, 453–457. [CrossRef]

41. *POVRAY, Version 3.5*; Persistence of Vision Raytracer Pty. Ltd.: Williamstown, VIC, Australia, 2002.

42. Koch, W.; Holthausen, M.C. *A Chemist's Guide to Density Functional Theory*; VCH Publisher: Berlin, Germany, 2001.

43. Becke, A.D. Density-functional thermochemistry. III. The role of exact exchange. *J. Chem. Phys.* **1993**, *98*, 5648–5652. [CrossRef]

44. Frisch, M.J.; Pople, J.A.; Binkley, J.S. Self-consistent molecular orbital methods 25. Supplementary functions for Gaussian basis sets. *J. Chem. Phys.* **1984**, *80*, 3265–3269. [CrossRef]

45. Schmidt, M.W.; Baldridge, K.K.; Boatz, J.A.; Elbert, S.T.; Gordon, M.S.; Jensen, J.H.; Koseki, S.; Matsunaga, N.; Nguyen, K.A.; Su, S.; et al. General Atomic and Molecular Electronic Structure System. *J. Comput. Chem.* **1993**, *14*, 1347–1363. [CrossRef]

46. Addison, A.W.; Rao, T.N.; Reedijk, J.; van Rijn, J.; Verschoor, G.C. Synthesis, structure, and spectroscopic properties of copper(II) compounds containing nitrogen–sulphur donor ligands; the crystal and molecular structure of aqua[1,7-bis(*N*-methylbenzimidazol-2′-yl)-2,6-dithiaheptane]copper(II) perchlorate. *J. Chem. Soc. Dalton Trans.* **1984**, *7*, 1349–1356. [CrossRef]

47. Yang, L.; Powell, D.R.; Houser, R.P. Structural variation in copper(I) complexes with pyridylmethylamide ligands: Structural analysis with a new four-coordinate geometry index, τ_4. *Dalton Trans.* **2007**, *9*, 955–964. [CrossRef] [PubMed]

48. Casarin, M.; Corvaja, C.; di Nicola, C.; Falcomer, D.; Franco, L.; Monari, M.; Pandolfo, L.; Pettinari, C.; Piccinelli, F.; Tagliatesta, P. Spontaneous Self-Assembly of an Unsymmetric Trinuclear Triangular Copper(II) Pyrazolate Complex, [Cu$_3$(μ_3-OH)(μ-pz)$_3$(MeCOO)$_2$(Hpz)] (Hpz = Pyrazole). Synthesis, Experimental and Theoretical Characterization, Reactivity, and Catalytic Activity. *Inorg. Chem.* **2004**, *43*, 5865–5876. [CrossRef] [PubMed]

49. Heisenberg, W. Zur theorie des ferromagnetismus. *Z. Phys.* **1928**, *49*, 619–636. [CrossRef]

50. van Vleck, J.H.; Sherman, A. The Quantum Theory of Valence. *Rev. Mod. Phys.* **1935**, *7*, 167–228. [CrossRef]

51. Kambe, K. On the Paramagnetic Susceptibilities of Some Polynuclear Complex Salts. *J. Phys. Soc. Jpn.* **1950**, *5*, 48–51. [CrossRef]

52. Sinn, E. Magnetic Exchange in Polynuclear Metal Complexes. *Coord. Chem. Rev.* **1970**, *5*, 313–347. [CrossRef]

53. Bersuker, I.B. Modern aspects of the Jahn-Teller effect theory and applications to molecular problems. *Chem. Rev.* **2001**, *101*, 1067–1114. [CrossRef]

54. Cage, B.; Cotton, F.A.; Dalal, N.S.; Hillard, E.A.; Rakvin, B.; Ramsey, C.M. Observation of Symmetry Lowering and Electron Localization in the Doublet-States of a Spin-Frustrated Equilateral Triangular Lattice: Cu$_3$(O$_2$C$_{16}$H$_{23}$)·1.2C$_6$H$_{12}$. *J. Am. Chem. Soc.* **2003**, *125*, 5270–5271. [CrossRef]

55. Vreugdenhil, W.; Haasnoot, J.G.; Schoondergang, M.F.J. Reedijk, Spectroscopic and magnetic properties of transition metal(II) triflouromethanesulfonate compounds with bridging asymmetric 3,4-dialkyl substituted 1,2,4-triazole ligands. *J. Inorg. Chim. Acta* **1987**, *130*, 235–242. [CrossRef]

56. Antolini, L.; Fabretti, A.C.; Gatteschi, D.; Giusti, A.; Sessoli, R. Synthesis, crystal and molecular structure, and magnetic properties of bis[tris(.mu.-3,5-diamino-1,2,4-triazole-N1,N2)tris(thiocyanato-N)nickel(II)]nickel(II) hexahydrate. *Inorg. Chem.* **1990**, *29*, 143–145. [CrossRef]

57. Antolini, L.; Fabretti, A.C.; Gatteschi, D.; Giusti, A.; Sessoli, R. Synthesis, crystal and molecular structure, and magnetic properties of bis[(.mu.-3,5-diamino-1,2,4-triazole-N1,N2-bis(.mu.-3,5-diamino-1,2,4-triazolato-N1,N2)triaquacobalt(II))]cobalt(III) trichloride nonahydrate. *Inorg. Chem.* **1991**, *30*, 4858–4860. [CrossRef]

58. Kolnaar, J.J.A.; van Dijk, G.; Kooijman, H.; Spek, A.L.; Ksenofontov, V.G.; Gütlich, P.; Haasnoot, J.G.; Reedijk, J. Synthesis, Structure, Magnetic Behavior, and Mössbauer Spectroscopy of Two New Iron(II) Spin-Transition Compounds with the Ligand 4-Isopropyl-1,2,4-triazole. X-ray Structure of [Fe$_3$(4-isopropyl-1,2,4-triazole)$_6$(H$_2$O)$_6$](tosylate)$_6$·2H$_2$O. *Inorg. Chem.* **1997**, *36*, 2433–2440. [CrossRef] [PubMed]

59. Garcia, Y.; van Koningsbruggen, P.J.; Lapouyade, R.; Fournès, L.; Rabardel, L.; Kahn, O.; Ksenofonov, V.; Levchenko, G.P.; Gütlich, P. Influences of Temperature, Pressure, and Lattice Solvents on the Spin Transition Regime of the Polymeric Compound [Fe(hyetrz)$_3$]A$_2$·3H$_2$O (hyetrz = 4-(2′-hydroxyethyl)-1,2,4-triazole and A$^-$ = 3-nitrophenylsulfonate). *Chem. Mater.* **1998**, *10*, 2426–2433. [CrossRef]

60. Roubeau, O.; Gomez, J.M.A.; Balskus, E.; Kolnaar, J.J.A.; Haasnoot, J.G.; Reedijk, J. Spin-transition behaviour in chains of Fe^{II} bridged by 4-substituted 1,2,4-triazoles carrying alkyl tails. *New J. Chem.* **2001**, *25*, 144–150. [CrossRef]

61. Spielberg, E.T.; Fittipaldi, M.; Geibig, D.; Gatteschi, D.; Plass, W. Electronic and magnetic structure of a triacetylphloroglucinol-bridged C_3-symmetric trinuclear copper complex: Magnetic characterization, ESR spectroscopy, and DFT calculations. *Inorg. Chim. Acta* **2010**, *363*, 4269–4276. [CrossRef]

62. Shaik, S.; Hiberty, P.C. *A Chemist's Guide to Valence Bond. Theory*; John Wiley & Sons: Hoboken, NJ, USA, 2007.

63. Garciabach, M.A.; Blaise, P.; Malrieu, J.P. Dimerization of polyacetylene treated as a spin-Peierls distortion of the Heisenberg Hamiltonian. *Phys. Rev. B Condens. Matter.* **1992**, *46*, 15645–15651. [CrossRef]

64. Mohamed, A.A.; Burini, A.; Galassi, R.; Paglialunga, D.; Galan-Mascaros, J.-R.; Dunbar, K.R.; Fackler, J.P., Jr. Self-assembly of a High-Nuclearity Chloride-Centered Copper(II) Cluster. Structure and Magnetic Properties of [Au(PPh$_3$)$_2$][trans-Cu$_6$(μ-OH)$_6${μ-(3,5-CF$_3$)$_2$pz}$_6$Cl]. *Inorg. Chem.* **2007**, *46*, 2348–2349. [CrossRef]

65. Borrás-Almenar, J.J.; Clemente-Juan, J.M.; Coronado, E.; Tsukerblat, B.S. MAGPACK A Package to Calculate the Energy Levels, Bulk Magnetic Properties, and Inelastic Neutron Scattering Spectra of High Nuclearity Spin Clusters. *J. Comput. Chem.* **2001**, *22*, 985–991. [CrossRef]

66. Rumer, G. Zum theorie der spinvalenz. *Nachr. Ges. Wiss. Göttingen Math. Phys. Kl.* **1932**, *1932*, 337–341.

67. Cimpoesu, F.; Chihaia, V.; Stanica, N.; Hirao, K. The Spin Hamiltonian Effective Approach to the Vibronic Effects. Selected Cases. *Adv. Quant. Chem.* **2003**, *44*, 273–288. [CrossRef]

68. Pauling, L.; Wheland, G.W. The Nature of the Chemical Bond. V. The Quantum-Mechanical Calculation of the Resonance Energy of Benzene and Naphthalene and the Hydrocarbon Free Radicals. *J. Chem. Phys.* **1933**, *1*, 362–374. [CrossRef]

69. Ferbinteanu, M.; Buta, C.; Toader, A.M.; Cimpoesu, F. The spin coupling in the polyaromatic hydrocarbons and carbon-based materials. In *Carbon-Related Materials-in Recognition of Nobel Lectures by Prof. Akira Suzuki in ICCE*; Kaneko, S., Mele, P., Endo, T., Eds.; Springer: Cham, Switzerland, 2017; pp. 327–371. [CrossRef]

70. Toader, A.M.; Buta, C.M.; Frecus, B.; Mischie, A.; Cimpoesu, F. Valence Bond Account of Triangular Polyaromatic Hydrocarbons with Spin: Combining Ab Initio and Phenomenological Approaches. *J. Phys. Chem. C* **2019**, *123*, 6869–6880. [CrossRef]

71. Noodleman, L. Valence bond description of antiferromagnetic coupling in transition metal dimers. *J. Chem. Phys.* **1981**, *74*, 5737–5743. [CrossRef]

72. Noodleman, L.; Davidson, E.R. Ligand spin polarization and antiferromagnetic coupling in transition metal dimers. *Chem. Phys.* **1986**, *109*, 131–143. [CrossRef]

73. Ruiz, E.; Cano, J.; Alvarez, S.; Alemany, P. Broken symmetry approach to calculation of exchange coupling constants for homobinuclear and heterobinuclear transition metal complexes. *J. Comp. Chem.* **1999**, *20*, 1391–1400. [CrossRef]

74. Bencini, A.; Totti, F.; Daul, C.A.; Doclo, K.; Fantucci, P.; Barone, V. Density Functional Calculations of Magnetic Exchange Interactions in Polynuclear Transition Metal Complexes. *Inorg. Chem.* **1997**, *36*, 5022–5030. [CrossRef]

75. Shoji, M.; Koizumi, K.; Kitagawa, Y.; Kawakami, T.; Yamanaka, S.; Okumura, M.; Yamaguchi, K. A general algorithm for calculation of Heisenberg exchange integrals *J*. in multispin systems. *Chem. Phys. Lett.* **2006**, *432*, 343–347. [CrossRef]

76. Ruiz, E. Theoretical study of the exchange coupling in large polynuclear transition metal complexes using DFT methods. In *Principles and Applications of Density Functional Theory in Inorganic Chemistry II. Structure and Bonding*; Springer: Berlin/Heidelberg, Germany, 2004; Volume 113, pp. 71–102. [CrossRef]

77. Atitoaie, A.; Tanasa, R.; Enachescu, C. Size dependent thermal hysteresis in spincrossover nanoparticles reflected within a Monte Carlo based Ising-like model. *J. Magn. Magn. Mater.* **2012**, *324*, 1596–1600. [CrossRef]

78. Varret, F.; Salunke, S.A.; Boukheddaden, K.; Bousseksou, A.; Codjovi, É.; Enachescu, C.; Linares, J. The Ising-like model applied to switchable inorganic solids: Discussion of the static properties. *Comptes Rendus Chim.* **2003**, *6*, 385–393. [CrossRef]

79. Leininger, T.; Stoll, H.; Werner, H.-J.; Savin, A. Combining long-range configuration interaction with short-range density functionals. *Chem. Phys. Lett.* **1997**, *275*, 151–160. [CrossRef]

80. Kohn, W.; Meir, Y.; Makarov, D.E. van der Waals Energies in Density Functional Theory. *Phys. Rev. Lett.* **1998**, *80*, 4153. [CrossRef]

81. Iikura, H.; Tsuneda, T.; Yanai, T.; Hirao, K. A long-range correction scheme for generalized-gradient-approximation exchange functionals. *J. Chem. Phys.* **2001**, *115*, 3540–3544. [CrossRef]

82. Tsuneda, T.; Kamiya, M.; Hirao, K. Regional self-interaction correction of density functional theory. *J. Comput. Chem.* **2003**, *24*, 1592–1598. [CrossRef]

83. Grimme, S. Accurate description of van der Waals complexes by density functional theory including empirical corrections. *J. Comput. Chem.* **2004**, *25*, 1463–1473. [CrossRef] [PubMed]

84. Grimme, S.; Ehrlich, S.; Goerigk, L. Effect of the damping function in dispersion corrected density functional theory. *J. Comput. Chem.* **2011**, *32*, 1456–1465. [CrossRef]

85. Hay, P.J.; Thibeault, J.C.; Hoffmann, R. Orbital interactions in metal dimer complexes. *J. Am. Chem. Soc.* **1975**, *97*, 4884–4899. [CrossRef]

86. Kahn, O.; Galy, J.; Journaux, Y.; Morgenstern-Badarau, I. Synthesis, crystal structure and molecular conformations, and magnetic properties of a copper-vanadyl (CuII-VOII) heterobinuclear complex: Interaction between orthogonal magnetic orbitals. *J. Am. Chem. Soc.* **1982**, *104*, 2165–2176. [CrossRef]
87. Kahn, O. Magnetism of the Heteropolymetallic Systems. In *Theoretical Approaches*; Structure and Bonding; Springer: Berlin/Heidelberg, Germany, 1987; Volume 68. [CrossRef]
88. Ruiz, E.; Alemany, P.; Alvarez, S.; Cano, J. Structural Modeling and Magneto− Structural Correlations for Hydroxo-Bridged Copper (II) Binuclear Complexes. *Inorg. Chem.* **1997**, *36*, 3683–3688. [CrossRef]

Article

Spin-Crossover 2-D Hofmann Frameworks Incorporating an Amide-Functionalized Ligand: *N*-(Pyridin-4-yl)benzamide

Xandria Ong [1], Manan Ahmed [1], Luonan Xu [1], Ashley T. Brennan [1], Carol Hua [2], Katrina A. Zenere [3], Zixi Xie [3], Cameron J. Kepert [3], Benjamin J. Powell [4] and Suzanne M. Neville [1,*]

[1] The School of Chemistry, The University of New South Wales, Sydney 2052, Australia; xandria.ong@unsw.edu.au (X.O.); manan.ahmed@unsw.edu.au (M.A.); luonan.xu@unsw.edu.au (L.X.); ashley.brennan@unsw.edu.au (A.T.B.)

[2] School of Chemistry, The University of Melbourne, Parkville 3010, Australia; carol.hua@unimelb.edu.au

[3] The School of Chemistry, The University of Sydney, Sydney 2006, Australia; katrina.zenere@sydney.edu.au (K.A.Z.); zixi.xie@sydney.edu.au (Z.X.); cameron.kepert@sydney.edu.au (C.J.K.)

[4] School of Mathematics and Physics, The University of Queensland, St. Lucia 4072, Australia; powell@physics.uq.edu.au

* Correspondence: s.neville@unsw.edu.au

Abstract: Two analogous 2-D Hofmann-type frameworks, which incorporate the novel ligand *N*-(pyridin-4-yl)benzamide (benpy) [Fe^{II}(benpy)$_2$M(CN)$_4$]·2H$_2$O (M = Pd (**Pd(benpy)**) and Pt (**Pt(benpy)**)) are reported. The benpy ligand was explored to facilitate spin-crossover (SCO) cooperativity via amide group hydrogen bonding. Structural analyses of the 2-D Hofmann frameworks revealed benpy-guest hydrogen bonding and benpy-benpy aromatic contacts. Both analogues exhibited single-step hysteretic spin-crossover (SCO) transitions, with the metal-cyanide linker (M = Pd or Pt) impacting the SCO spin-state transition temperature and hysteresis loop width (**Pd(benpy)**: $T_{\frac{1}{2}}\downarrow\uparrow$: 201, 218 K, ΔT: 17 K and **Pt(benpy)**: $T_{\frac{1}{2}}\downarrow\uparrow$: 206, 226 K, ΔT: 20 K). The parallel structural and SCO changes over the high-spin to low-spin transition were investigated using variable-temperature, single-crystal, and powder X-ray diffraction, Raman spectroscopy, and differential scanning calorimetry. These studies indicated that the ligand–guest interactions facilitated by the amide group acted to support the cooperative spin-state transitions displayed by these two Hofmann-type frameworks, providing further insight into cooperativity and structure–property relationships.

Keywords: spin crossover; Hofmann framework; hydrogen bonding

Citation: Ong, X.; Ahmed, M.; Xu, L.; Brennan, A.T.; Hua, C.; Zenere, K.A.; Xie, Z.; Kepert, C.J.; Powell, B.J.; Neville, S.M. Spin-Crossover 2-D Hofmann Frameworks Incorporating an Amide-Functionalized Ligand: *N*-(Pyridin-4-yl)benzamide. *Chemistry* **2021**, *3*, 360–372. https://doi.org/10.3390/chemistry3010026

Received: 14 December 2020
Accepted: 20 February 2021
Published: 1 March 2021

Publisher's Note: MDPI stays neutral with regard to jurisdictional claims in published maps and institutional affiliations.

1. Introduction

SCO can occur in d^4 to d^7 transition metals, whereby a reversible transition between low-spin (LS) and high-spin (HS) states occurs with the application of external stimuli (e.g., magnetic field, light irradiation, temperature, pressure) [1–6]. Of these, Fe^{II} complexes are the most widely reported as they exhibit versatile SCO behaviors, including abrupt and hysteretic transitions and multi-stepped pathways [7–10]. Solid-state cooperativity, defined as the effectiveness of the propagation of SCO effects throughout the material, arises in Fe^{II} complexes as there is a strong coupling of spin-state switching to the lattice due the large volume change between HS and LS states (5–12%). [10–13] As cooperativity is dependent on the strength of communication between metal centers, the direct bridging of switching metal ions in extended networks (i.e., coordination polymers (CPs) and metal-organic frameworks (MOFs)) is an effective approach [9,14–22], as well as via supramolecular interactions, such as hydrogen bonds and aromatic contacts [10]. One of the most intensively explored SCO CP families is Hofmann frameworks, which are typically characterized by Fe^{II} sites bridged by square planar [M(CN)$_4$]$^{2-}$ groups (M = Pd, Pt, or Ni) and connected (3-D) or spaced (2-D) into frameworks by *N*-donor aromatic organic ligands [9,14–22]. Effective SCO communication in these materials comes from the metallocyanide groups

163

and can be modulated by the organic ligand and encapsulated guest molecules. Hence, the choice of organic ligand incorporated into Hofmann frameworks has a significant effect on the SCO properties observed, and, thus, substantial focus has been directed at structure-function studies with various ligand functional groups and lengths. For example, in the 2-D Hofmann framework [Fe^{II}(pyridine)$_2$M(CN)$_4$] (M = Pd or Pt) a thermally induced, single-step hysteretic SCO transition was observed ($T_{\frac{1}{2}}\downarrow\uparrow$: 208, 213 K; ΔT = 5 and $T_{\frac{1}{2}}\downarrow\uparrow$: 208, 216 K; ΔT = 8, for the Pd and Pt analogues, respectively) [15]. When pyrazine was used in place of pyridine to form the 3-D Hofmann analogue [Fe^{II}(pyrazine)Pd(CN)$_4$], a higher transition temperature and a wider hysteresis loop ($T_{\frac{1}{2}}\downarrow\uparrow$: 220, 240 K; ΔT = 20 K) was observed [15]. However, framework dimensionality alone is not the only possible influencing factor on SCO cooperativity. For example, pyridine functionalization in the 2-D framework [Fe^{II}(4-phenylpyridine)$_2$Pd(CN)$_4$] resulted in an increased transition temperature compared to the pyridine analogue and an impressive hysteresis loop width of 40 K [23]. Structural studies on this analogue highlight that, alongside the framework connectivity, aromatic contacts play a large role in enhancing the SCO communication. Likewise, the use of the ligand 3-aminopyridine in the 2-D Hofmann framework [Fe(3-aminopyridine)$_2$M(CN)$_4$] (M = Ni, Pd or Pt) emphasizes the significance of hydrogen bonds on cooperativity enhancement. High transition temperatures and wide (>20 K) thermal hysteresis loops were observed for both the Pt and Pd analogues (Pd: $T_{\frac{1}{2}}\downarrow\uparrow$: 169, 206 K; ΔT = 37 K and Pt: $T_{\frac{1}{2}}\downarrow\uparrow$: 183, 213 K; ΔT = 30) [24]. Hence, the concept of targeting increased SCO cooperativity by both incorporating SCO sites into CPs and also including functionalized ligands with hydrogen bonding and/or aromatic interaction capacity serves as the basis for the materials prepared in this report. We utilized the ligand *N*-(pyridin-4-yl)benzamide (benpy; Figure 1) [25] to facilitate SCO cooperativity via both hydrogen bonding from the amide functional group and aromatic contacts via the aromatic rings (Figure 1). Amide-functionalized ligands were successfully applied to a range of SCO framework previously, resulting in a range of interesting structures and SCO features [26–30] based on interaction with the amide group. Here, we report the structure and SCO properties of the 2-D Hofmann frameworks' analogues [Fe^{II}(benpy)$_2$(M(CN)$_4$)] (M = Pd and Pt) with an interest in the impact of amide group contacts on the structure and cooperativity of the spin-state transition.

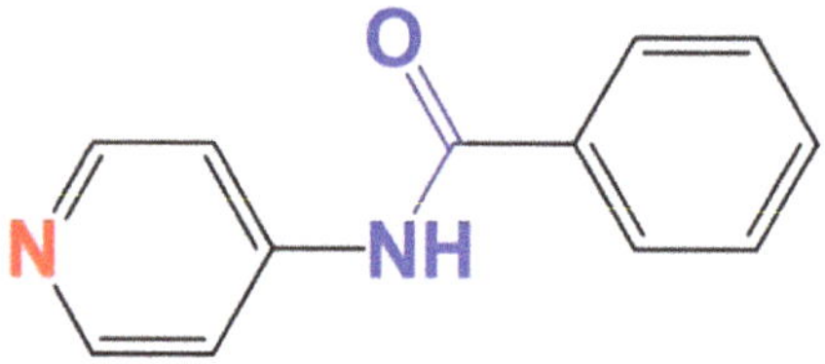

Figure 1. *N*-(pyridin-4-yl)benzamide (benpy) with metal coordination site (red), potential hydrogen bonding (blue), and aromatic (black) interaction sites indicated.

2. Materials and Methods

Reagents were purchased commercially from Sigma-Aldrich or Merck and not further purified before use. Care was taken, handling Fe(ClO$_4$)$_2$·6H$_2$O only in small amounts to avoid any potential explosions.

2.1. Synthesis of N-(pyridin-4-yl)benzamide (benpy)

The 4-aminopyridine (1.69 g, 18.0 mmol) and triethylamine (2.28 g, 22.5 mmol) were dissolved in CH$_2$Cl$_2$ (30 mL). This solution was placed in an ice bath, wherein benzoyl chloride (2.11 g, 15.0 mmol) was added with continuous stirring once the temperature reached 0 °C. The mixture was then allowed to return to room temperature while stirring for 1 h. The translucent, needle-like precipitate was collected via filtration and washed with excess CH$_2$Cl$_2$. The crude product was recrystallized using 1:1 EtOH:H2O (1.63 g, 8.22 mmol, 54.8%). The ^{1}H NMR (d_6-DMSO, 300 MHz): δ = 7.53–7.66 (m, J_{HH} = 15.22 Hz,

3H), 7.78 (d, J_{HH} = 1.46 Hz, 2H), 7.96 (d, J^{HH} = 6.90 Hz, 2H), 8.47 (d, J_{HH} = 6.27 Hz, 2H), and 10.59 (s, 1H). The ^{13}C NMR (d_6-DMSO, 75 MHz): δ = 114.0, 127.9, 128.6, 132.1, 145.9, 150.5, and 166.5. ESI-MS (ESI$^+$, m/z) calculated [M + H]$^+$ for $C_{12}H_{10}N_2O$: 198.08 found 199.08. IR (solid, v/cm^{-1}): 3241.3 (w), 3158.6 (w), 1673.5 (vs), 1586.3 (vs), 1503.1 (vs), 1414.7 (m), 1327.4 (s), 1292.0 (s), 1262.4 (w), 1207.5 (s), 1109.8 (w), 1090.8 (w), 1074.4 (w), 1027.4 (m), 991.1 (m), 932.4 (w), 892.9 (m), 822.6 (vs). Elemental analysis (%) calculated C 72.71, H 5.09, N 14.13 and found C 72.02, H 4.64, N 14.03.

2.2. Synthesis of [Fe^{II}(benpy)$_2$Pd(CN)$_4$]·2H$_2$O (Pd(benpy))

Vial-in-vial slow diffusion methods were used to prepare a bulk crystalline sample. Solid Fe(ClO$_4$)$_2$·6H$_2$O (15.0 mg, 0.059 mmol) was weighed into a small vial and K$_2$Pd(CN)$_4$ (17.0 mg, 0.059 mmol) and benpy (23.4 mg, 0.118 mmol) were weighed into a larger vial. The small vial was placed within the larger vial before 1:1 EtOH:H$_2$O (~20 mL) solvent was added slowly to fill both vials to the top. Yellow crystals with square-plate morphology formed over six weeks. IR (solid, v/cm^{-1}): 3533.4 (m), 2170.8 (s), 1594.5 (vs), 1523.2 (s), 1422.3 (s), 1337.8 (s), 1299.5 (s), 1263.1 (m), 1210.8 (s), 1097.0 (w), 1069.3 (w), 1015.7 (m), 997.7 (w), 929.9 (w), 892.8 (w), 830.4 (s), 707.1 (vs), 682.8 (s). Elemental analysis (air-dried sample) calculated (%) C 50.74 H 3.04 N 16.91 and found (%) C 50.57 H 3.28 N 17.01.

2.3. Synthesis of [Fe^{II}(benpy)$_2$Pt(CN)$_4$]·2H$_2$O (Pt(benpy))

Vial-in-vial slow diffusion methods were used to prepare a bulk crystalline sample. Solid Fe(ClO$_4$)$_2$·6H$_2$O (15.0 mg, 0.059 mmol) was weighed into a small vial and K$_2$Pt(CN)$_4$ (22.3 mg, 0.059 mmol) and benpy (23.4 mg, 0.118 mmol) were weighed into a larger vial. The small vial was placed within the larger vial before 1:1 EtOH:H2O (~20 mL) solvent was added slowly to fill both vials to the top. Yellow crystals with square-plate morphology formed over six weeks (100% yield). IR (solid, v/cm^{-1}): 3544.0 (b), 2168.0 (s), 1594.4 (br), 1524.0 (br), 1422.1 (m), 1337.6 (m), 1299.3 (m), 1262.3 (w), 1210.8 (m), 1096.3 (w), 1069.1 (aw), 1015.5 (w), 967.9 (w), 928.5 (w), 892.3 (w), 830.4 (s), 732.9 (w), 706.5 (s), 666.2 (w). Elemental analysis (air-dried sample) calculated (%) C 44.71 H 2.88 N 14.56 and found (%) C 43.97 H 2.98 N 14.75.

2.4. Thermogravimetric Analysis (TGA)

Crystalline samples were loaded onto a platinum pan and measured on a TA Instruments Discovery TGA Analyzer at a ramp rate of 1 °C min^{-1} (RT–500 °C) under a dry N$_2$ gas environment (20 mL min^{-1}).

2.5. Variable-Temperature, Single-Crystal X-ray Diffraction Analysis (VT-SCXRD)

Data were collected at the Australian Synchrotron macromolecular crystallography (MX1) beamline radiation source [31]. Data were collected using a Dectris Eiger 9M detector and Si<111> monochromated synchrotron radiation (λ = 0.71073 Å). The data were collected using the BlueIce software [32] and were processed using the XDS software [33]. All structures were solved using SHELXT [34] and refined using SHELXL [35] within the OLEX2 graphical user interface [36]. Hydrogen atoms were placed at idealized positions and refined using the riding model. Details of the crystallographic data collection and refinement parameters are summarized in Table 1 and selected parameters are presented in Table S2. The data have been deposited in the Cambridge Crystallographic Data Centre (CCDC) and are freely available at CCDC: 2049372-2049375.

Table 1. Representative structural details for **Pd(benpy)** and **Pt(benpy)** in the LS and HS states.

	$Fe^{II}(benpy)_2Pd(CN)_4\cdot 2H_2O$		$Fe^{II}(benpy)_2Pt(CN)_4\cdot 2H_2O$	
Spin state	LS	HS	LS	HS
Temperature (K)	100	250	100	250
Average Fe–N≡C	1.9976	2.1411	1.926	2.125
Average Fe–N(py)	1.93775	2.2042	1.997	2.216
Average Fe–N (Å)	1.9577	2.1621	1.950	2.155
O·water (Å)	2.767(2)	2.846(2)	2.730(2)	2.805(2)
water·N(H) (Å)	2.852(4)	2.904(2)	2.895(2)	2.915(3)
Σ (Fe) (°)	13.20	8.160	12.82	12.40
Ligand torsion (°)	24.3	35.5	13.8	40.3

2.6. Variable-Temperature, Powder X-ray Diffraction Analysis (VT-PXRD)

Data were collected at the Powder Diffraction beamline BL-10 (20.0 keV, 0.589062 Å) at the Australian Synchrotron [37]. A polycrystalline sample was loaded into 0.7-mm-diameter borosilicate glass capillary and mounted onto the powder diffraction beamline. The capillaries were rotated at ~1 Hz during data collection to assist in powder averaging. The beamline was set up with a nominal wavelength of 1.0 Å; the wavelength was determined accurately using NIST SRM660b LaB_6 standard. Temperature-dependent data were then collected continuously over the range 175–250 K. Data were collected using a Mythen microstrip detector [38] from 1.5° to 75° in 2θ. To cover the gaps between detector modules, two data sets, each of 60 s in duration, were collected with the detector set 5° apart and they were then merged to give a single data set using PDViper [39]. A slit size of 2 mm was used to ensure that the fraction of the capillary illuminated by the X-ray beam was the same as the isothermal zone on the cryostream. Le Bail analysis and peak fitting were performed using the TOPAS software package [40].

2.7. Variable-Temperature Magnetic Susceptibility

Variable-temperature magnetic susceptibility data were obtained using a Quantum Design VersaLab magnetometer with a Vibrating Sample Magnetometer (VSM) accessory attached. Measurements were taken continuously over the range 150–300 K under a 0.3 T magnetic field with sweep rates of 0.5, 1, 2, and 4 K min^{-1}. The crystalline samples were loaded into polypropylene holders (Formolene 4100 N) and clipped into a brass half-tube for analysis.

2.8. Differential Scanning Calorimetry (DSC)

Measurements were taken at 10 K min^{-1} (294–190 K) using a Netzsch DSC 204 F1 Phoenix with a liquid nitrogen cooling cryostat. The crystals were placed into a 40-μL aluminum crucible and sealed. Instrument calibrations were made using sapphire standards with the sample holder kept dry and under a constant, dry N_2 flow (20 mL min^{-1}). The Netzsch Proteus Analysis 2019 software was used to process the data.

2.9. Calculation of the Magnetic Susceptibility

The magnetic susceptibility was calculated from the Slichter–Drickamer model [41,42], which predicts that the HS fraction, γ_{HS}, is given by

$$\gamma_{HS}(T) = \frac{1}{1 + \exp\left[\frac{\Delta H + \Gamma(1 - 2\gamma_{HS}(T))}{RT} - \frac{\Delta S}{R}\right]}, \tag{1}$$

where ΔH and ΔS are, respectively, the enthalpy and entropy differences between the HS and LS states, R is the gas constant, and Γ parameterizes the interactions between metal centers. We neglected magnetic interactions between metals in the HS state, whence one finds that the magnetic susceptibility is given by

$$\chi(T) = [1 - \gamma_{HS}(T)]\chi_{LS} + \gamma_{HS}\chi_{HS}, \tag{2}$$

where χ_{LS} (χ_{HS}) is the susceptibility in the LS (HS) state. For both **Pd(benpy)** and **Pd(benpy)** we obtained the best fits on constraining $\chi_{LS} = 0$ and $\chi_{HS}T = 3.6$ cm^3 K mol^{-1}. No closed form solution was known for the transcendental Equation (1), so a numerical solution was obtained self-consistently for a range of parameters (ΔH, ΔS, and Γ), with the best fit taken as the parameters with the lowest root-mean-square error relative to the measured susceptibility. We swept through parameters three times, each time narrowing the range but moving through finer steps, to obtain accurate solutions in a reasonable time.

The Slichter–Drickamer model is equivalent to a mean field treatment of the Ising–Wajnflasz–Pick model [42,43]. Like all mean field theories, it is more accurate in higher dimensions. So long as the interlayer elastic interactions are not too small (compared to $k_B T$), **Pd(benpy)** and **Pd(benpy)** are three-dimensional materials and mean field theory should be reasonable. The anisotropy between the intra- and interlayer elastic interactions is only captured on average but does not affect the applicability of the theory. If the interlayer elastic interactions are small compared to $k_B T$, then **Pd(benpy)** and **Pd(benpy)** are two-dimensional materials. As we did not discuss critical behavior below a mean field, approximation is reasonable in either case.

2.10. Variable-Temperature Raman Spectroscopy

Raman spectra were collected on a Renishaw inVia-Qontor upright microscope. The excitation wavelength of 785 nm (20 mW at 10% laser power, L ×50 objective) was used to acquire the Raman spectra in the range 150–2300 cm^{-1} with a 10-s exposure time. The samples were loaded into 0.7-mm-diameter glass capillaries and mounted on a Linkam-FTIR 600 variable temperature stage. Liquid nitrogen was continuously flowed onto the Linkam stage with a temperature programmer and cooling pump connected. Data were collected over the range 170–250 K at 10-K steps.

3. Results and Discussion

3.1. Synthesis and Characterization

The two materials **Pd(benpy)** and **Pt(benpy)** were successfully synthesized as bulk crystalline solids via slow diffusion methods. Their purity was confirmed via elemental analysis (CHN), IR spectroscopy, Le Bail analysis of the PXRD data (Figures S4 and S5) and TGA measurements. The TGA data (Figure S1) showed an initial mass loss below ~50 °C (**Pd(benpy)**: 4.90% and **Pt(benpy)**: 4.33%) corresponding to a theoretical loss of two solvent water molecules per Fe^{II} site (calculated for **Pd(benpy)**: 4.58% and **Pt(benpy)**: 4.07%). There was no further mass loss until ~300 °C, when framework decomposition occurred.

3.2. Spin-Crossover Properties

3.2.1. Variable-Temperature Magnetic Susceptibility

A color change from yellow to red was observed when the crystals of **Pd(benpy)** and **Pt(benpy)** were cooled with liquid N$_2$, indicative of a HS to LS transition (Figure S2). Variable-temperature magnetic susceptibility measurements (300–150 K) revealed hysteretic, single-step SCO transitions (Figure 2a,b). The $\chi_M T$ values remained constant at ~3.50 cm^3 Kmol^{-1} from 300 to 196 K for **Pd(benpy)** and 300–203 K for **Pt(benpy)**, corresponding to HS Fe^{II} sites ($S = 2$) [10]. Below this, a rapid decrease in $\chi_M T$ values occurred, to ~0.1 cm^{3}Kmol^{-1}, indicative of a complete HS to LS ($S = 0$) transition of the Fe^{II} sites. The scan rate dependence of the SCO transition was explored (rates: 0.5, 1, 2, and 4 K min^{-1}; Figure 2a,b. Extrapolation of the scan rate versus transition temperature (Figure S3) revealed zero-scan rate transition temperatures of $T_{\frac{1}{2}}\downarrow\uparrow$: 201, 218 K ($\Delta T$: 17 K) for **Pd(benpy)** and $T_{\frac{1}{2}}\downarrow\uparrow$: 206, 226 K ($\Delta T$: 20 K) for **Pt(benpy)**. The higher transition temperature observed in the Pt analogue compared to the Pd analogue was consistent with other Hofmann-type frameworks [9,14–22]. The abrupt transition character and approximate transition temperatures and hysteresis loop width were consistent with other 2-D Hofmann frameworks [44].

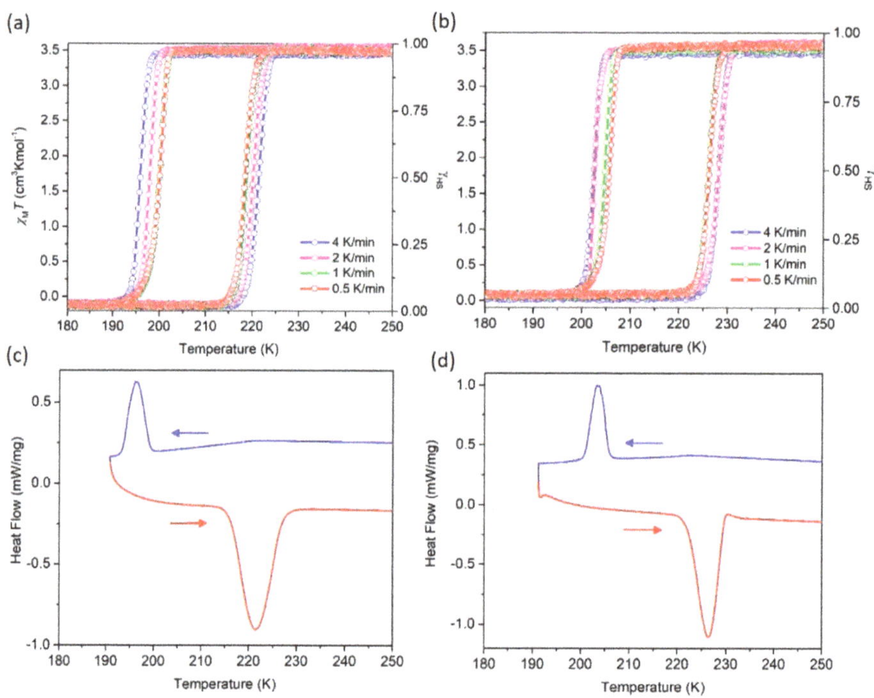

Figure 2. $\chi_M T$ (and HS fraction) versus temperature for (**a**) **Pd(benpy)** and (**b**) **Pt(benpy)**, variable scan rate data are shown. Differential scanning calorimetry data for (**c**) **Pd(benpy)** and (**d**) **Pt(benpy)**.

3.2.2. Differential Scanning Calorimetry

DSC measurements (Figure 2c,d) show an endothermic peak on cooling and an exothermic peak on heating for both **Pd(benpy)** and **Pt(benpy)**. The peak maxima, **Pd(benpy)**: 196, 222 K and **Pt(benpy)**: 204, 226 K, were consistent with the $T_{\frac{1}{2}}$ values recorded by magnetic susceptibility (The DSC data were collected at 10 Kmin^{-1}, thus a slight difference was anticipated compared to the magnetic susceptibility data). The thermodynamic parameters from these data were **Pd(benpy)**: $\Delta H_{avg} = 15.7 \pm 2$ kJ mol^{-1}, $\Delta S_{avg} = 75.3 \pm 5$ J K^{-1} mol^{-1} and **Pt(benpy)**: $\Delta H_{avg} = 17.9 \pm 2$ kJ mol^{-1} and $\Delta S_{avg} = 83.3 \pm 4$ J K^{-1} mol^{-1}, obtained from an average of the cooling and heating peaks. Enthalpy and entropy values of 10–20 kJ mol^{-1} and 35–90 J K^{-1} mol^{-1} values are typical for Fe^{II} SCO [45], in particular, Hofmann frameworks [9,14–22].

3.2.3. Magnetic Susceptibility Calculation

The spin transition curves were calculated from the Slichter–Drickamer model (Figures S4 and S5) [41,42]. The parameters extracted from the fits for **Pd(benpy)** were $\Delta H = 14.5$ kJ mol^{-1}, $\Delta S = 68.9$ J K^{-1} mol^{-1}, and $\Gamma = 5.38$ kJ K^{-1} mol^{-1} and for **Pt(benpy)** were $\Delta H = 20.8$ kJ mol^{-1}, $\Delta S = 96.1$ J K^{-1} mol^{-1}, and $\Gamma = 5.90$ kJ mol^{-1}. The ΔH and ΔS values agreed well with the DSC data.

There were three contributions to the entropy difference between the HS and LS states, $\Delta S = \Delta S_{spin} + \Delta S_{orb} + \Delta S_{vib}$. For Fe^{II} the spin contribution was $\Delta S_{spin} = R \ln 5 \simeq 13.4$ J K^{-1} mol^{-1} and the orbital contribution was $\Delta S_{orb} = R \ln 3 \simeq 9.1$ J K^{-1} mol^{-1}, thus the vibrational contribution was $\Delta S_{vib} = 46.4\ \text{J K}^{-1}\ \text{mol}^{-1}$ for **Pd(benpy)** and $\Delta S_{vib} = 73.6\ \text{J K}^{-1}\ \text{mol}^{-1}$ **Pt(benpy)**, based on the fits to the Slichter–Drickamer model. (Using the calorimetry values of ΔS yielded $\Delta S_{vib} = 52.8\ \text{J K}^{-1}\ \text{mol}^{-1}$ for **Pd(benpy)** and $\Delta S_{vib} = 60.8\ \text{J K}^{-1}$ mol^{-1} **Pt(benpy)**).

It can be shown [43] that the Slichter–Drickamer model is equivalent to the mean-field solution of the long-range Ising–Wajnflasz–Pick model and, thence, that

$$\Gamma = 2 \sum_{n,m} J_{n,m}, \tag{3}$$

where $J_{n,m}$ is the coupling between metal centers n unit cells apart in the crystallographic a-direction and m unit cells apart in the crystallographic b-direction. Thus, Γ measures the sum of the microscopic interactions, $J_{n,m}$, which, in turn, can be related to the underlying elastic interactions in the material [43]. Thus, the large positive values of Γ (37% of ΔH in **Pd(benpy)** and 28% of ΔH in **Pt(benpy)**) showed that the sum of the elastic interactions in both materials was significant and ferroelastic. Overall, the modeling showed that the spin-state transitions in both **Pd(benpy)** and **Pt(benpy)** were strongly cooperative.

3.3. Variable-Temperature Structural Analysis
3.3.1. Variable-Temperature Powder X-ray Diffraction

Data were collected to monitor the structure variation over the SCO transition (250–175 K). Figure 3a and Figure S8a show the evolution of a selected Bragg peak (2θ = 6.5–6.8°, *hkl*: 011) for **Pd(benpy)** and **Pt(benpy)**, respectively, over the HS to LS to HS transition. These data are also presented as peak position versus temperature (Figures S8b and S9), showing SCO character and transition temperatures consistent with magnetic susceptibility data. All of the data over the temperature range 250–175–250 K were fitted by Le Bail refinement. These data: (1) confirmed the bulk purity of the samples (Figures S6 and S7), (2) confirmed the symmetry and space group, and (3) allowed the temperature dependence of the unit cell parameters to be assessed. Figure 3b and Figure S11d show the unit cell volume evolution for **Pd(benpy)** and **Pt(benpy)**, respectively, confirming the abrupt single-step volumetric change over the reversible spin-state transition and an overall volume change of ~5.5%, consistent with a complete HS to LS transition. The evolution of the unit cell lengths for **Pd(benpy)** and **Pt(benpy)** are shown in Figures S10 and S11, respectively. The transition temperatures from VT-PXRD data were ca. 5 K lower than that observed by magnetic susceptibility. This is not uncommon due to the differences in experimental conditions. Here, the ramp rate for data collection was much faster for VT-PXRD than magnetic susceptibility.

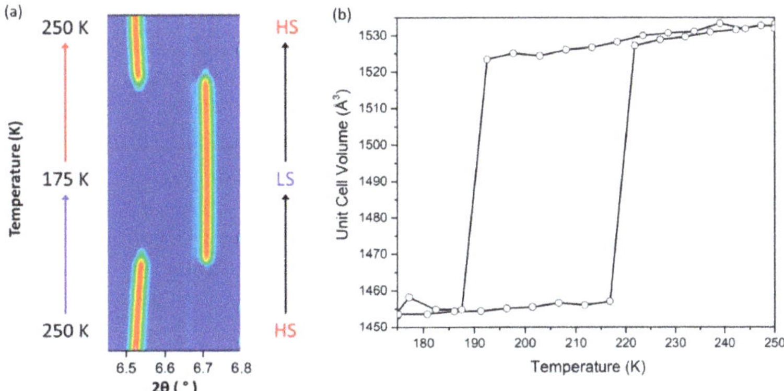

Figure 3. VT-PXRD data for **Pd(benpy)** showing (**a**) the evolution of a single Bragg peak (011) versus temperature (250–175–250 K) and (**b**) the temperature dependence of the unit cell volume extracted from Le Bail analysis of the individual PXRD patterns.

3.3.2. Variable-Temperature Raman Spectroscopy

VT-Raman spectra were collected (150–2300 cm^{-1}; 250–150 K) to monitor vibrational transitions that occur with spin-state change. Figure 4a and Figure S12 show representative HS and LS patterns for **Pd(benpy)** and **Pt(benpy)**, respectively. The largest frequency shifts between spin states were the bands corresponding to the Fe–N and C≡N stretching, in agreement with the literature on Hofmann frameworks [44]. The characteristic Fe–N stretching peaks (<250 cm^{-1}) and C≡N bands (~2200 cm^{-1}) were upshifted from the HS to the LS state (Figure 4b and Figures S13 and S14). The C≡N bands also increased in intensity over the HS to LS transition (Figure 4b and Figure S14b, **Pd(benpy)** and **Pt(benpy)**, respectively).

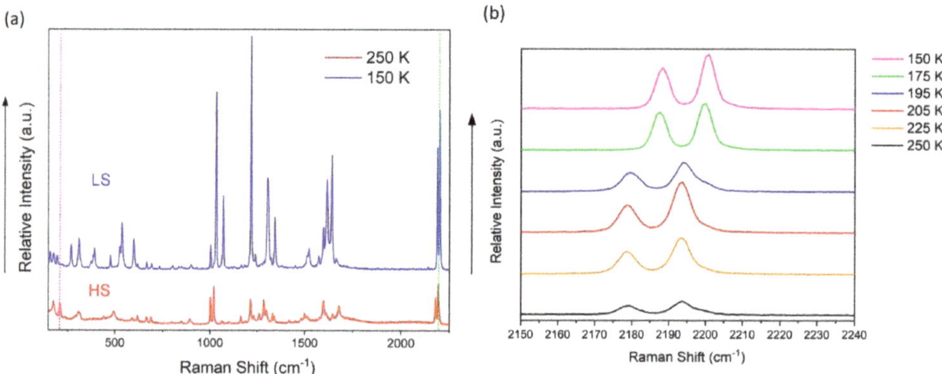

Figure 4. VT-Raman data for **Pd(benpy)** showing (**a**) representative HS (red) and LS (blue) spectra; the Fe–N and C≡N stretching regions are indicated by pink and green dotted lines, respectively. (**b**) The temperature dependence of the C≡N vibration.

3.3.3. Variable-Temperature, Single-Crystal X-ray Diffraction
Overall Structural Description

Powder diffraction analyses at 250 K confirmed that **Pd(benpy)** crystallizes in the primitive monoclinic space group: $P2_1/n$ and **Pt(benpy)** in the monoclinic C-centered space group $C2/m$ (Figures S6 and S7). SCXRD data collected on each agreed with this symmetry and space group assignment (Table 1). The asymmetric unit (ASU; Figure S15) of **Pd(benpy)** contained one Fe^{II} site, a half of a [Pd(CN)$_4$)]$^{2-}$ group, one benpy ligand, and one water molecule. The ASU (Figure S16) of **Pt(benpy)** contained one Fe^{II} site, a quarter of a [Pd(CN)$_4$)]$^{2-}$ group, one benpy ligand, and one water molecule. The main distinction between the two is that the benpy ligands showed a long-range, ordered, torsional twist around the amide bond in **Pd(benpy)** (Figure 5a). In **Pt(benpy)**, the benzyl ring of the ligand showed two-fold disorder (Figure 5b). Hence, the same ligand torsional ligand was observed but with short-range order, thus accounting for the space group difference between the analogues. The water molecule in **Pt(benpy)** was also disordered over two positions across the mirror plane.

Aside from the difference in symmetry, the overall formula for both was [Fe^{II}(benpy)$_2$ M(CN)$_4$]·2H$_2$O, where M = Pd or Pt, and the Hofmann framework structures were similar. For each, the Fe^{II} sites were arranged in a distorted octahedral environment with an overall [FeN$_6$] coordination, with the equatorial sites occupied by [M(CN)$_4$]$^{2-}$ groups, bound via the N atoms, and the axial sites occupied by benpy ligands, coordinated via a pyridyl N atom. The bridging of each Fe^{II} site by four [M(CN)$_4$]$^{2-}$ units, which, in turn, acted as four coordinating linkages, resulted in the formation of a 2-D layered-grid structure of the type [Fe^{II}(M(CN)$_4$)], as shown in Figure 6. The benpy ligands extended above and below the [Fe^{II}(M(CN)$_4$)] layers (Figure 6b,c). Neighboring benpy ligands along the *a*-direction were oriented in the same direction (Figure S17a), i.e., within each ligand the oxygen and amide groups pointed in opposite directions and were oriented in the same direction compared to adjacent ligands. The *trans*-benpy ligands, i.e., those coordinated to the same Fe^{II} site, were related by inversion symmetry and, thus, the amide functional groups were facing in opposite directions (Figure S17b). Additionally, the [Fe^{II}(M(CN)$_4$)] layers undulated along the *ab*-plane, as illustrated in Figure 6c.

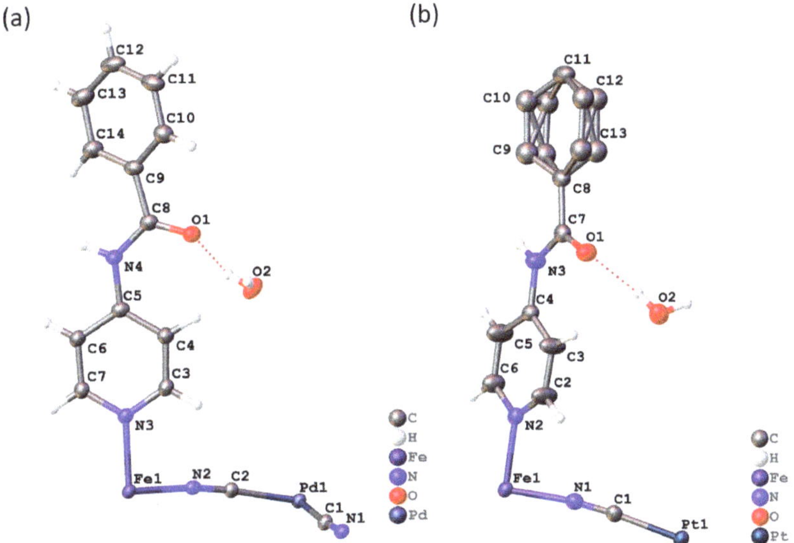

Figure 5. SCXRD thermal ellipsoid representation (50%) of the ASU of (**a**) **Pd(benpy)** and (**b**) **Pt(benpy)** at 100 K. Two-fold disorder in **Pt(benpy)** is shown.

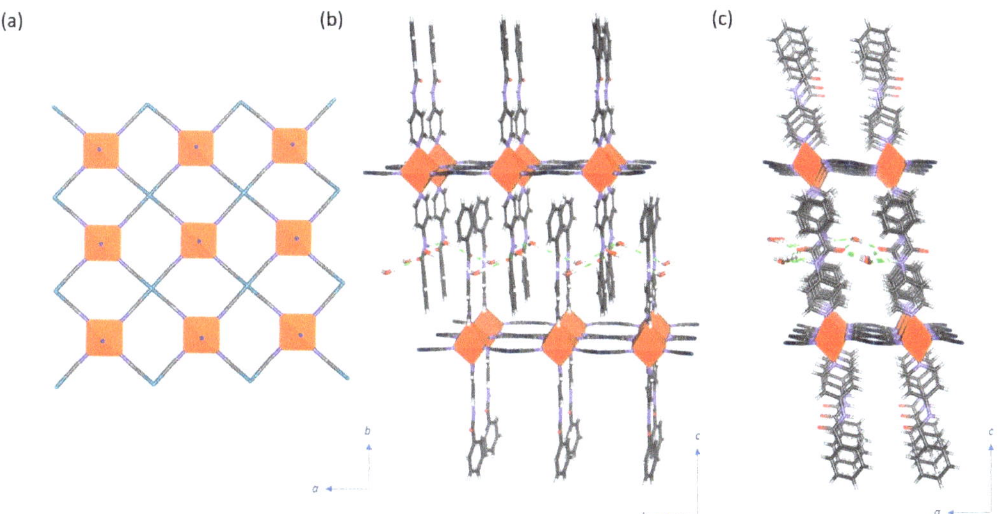

Figure 6. SCXRD structural representation of **Pd(benpy)** showing (**a**) the 2-D, grid-like Hofmann layers of the type [Fe^{II}(Pd(CN)$_4$)], (**b**) the interdigitation of benpy ligands on adjacent 2-D layers, and (**c**) the head-to-tail overlap of interdigitated benpy ligand. The water molecules are located in the interlayer spacing and form hydrogen-bonding interactions with the benpy ligands (green).

Neighboring 2-D [Fe^{II}(M(CN)$_4$)(benpy)$_2$] layers were stacked along the *c*-axis, with efficient packing provided by ligand interdigitation from neighboring layers (Figure 6b,c). The ligands of adjacent layers stacked in a head-to-tail fashion (Figure 6c) with near overlap. The nearly eclipsed overlap of the interdigitated ligands and close packing along the *b*-axis allowed for weak offset π-stacking, evidenced by the centroid-centroid distances between benpy ligands of ~3.5 Å (Figure 7). Two water molecules per Fe^{II} site were located in the interlayer spacing. These water molecules were engaged in a range of host–guest hydrogen-bonding interactions with the amide group of the benpy ligands (Figure 7). Similar arrays of host–guest interactions involving amide functional groups have been observed in other SCO frameworks [26–30].

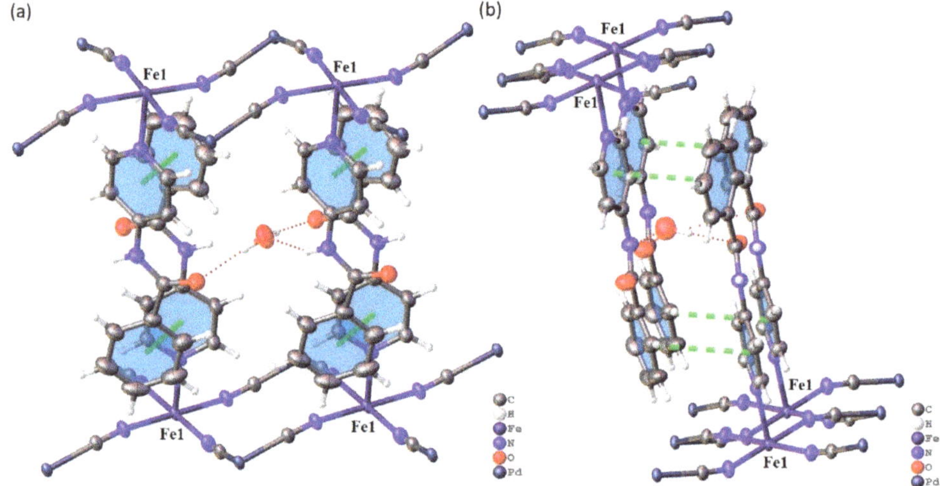

Figure 7. SCXRD structural representation of **Pd(benpy)** showing the position of the water molecules in the interlayer spacing, (**a**) the hydrogen-bonding interactions with the benpy ligands, and (**b**) the aromatic contacts.

Variable-Temperature Data

VT-SCXRD data were collected at 10-K intervals over the range of 250–100 K (HS to LS, cooling) for **Pd(benpy)** and 100–250 K (LS to HS, warming) for **Pt(benpy)**. Plots of the unit cell parameters (*a*-, *b*-, and *c*-axes) and unit cell volume versus temperature (Figures S18 and S19) matched the transition temperature anticipated from magnetic susceptibility. An overall 5.0% volume change was observed for both analogues over the HS to LS transition, consistent with a complete spin-state conversion. Notably, the evolution of the unit cell parameters was different for **Pd(benpy)** for **Pt(benpy)**. In particular, both the interlayer spacing (i.e., the *c*-axis) and the β angle increased over the HS to LS transition for **Pd(benpy)** but decreased for **Pt(benpy)**. The reason for this difference was not entirely clear but may have arisen from the relative structural support provided by the long- versus short-range order of the ligand torsion and water molecules for **Pd(benpy)** for **Pt(benpy)**, respectively.

Detailed structural analyses for **Pd(benpy)** for **Pt(benpy)** were conducted at 250 and 100 K to assess the structure changes over the HS to LS transition. Relevant parameters are summarized in Table S2. The overall variation to average Fe-N bond length for both was ~0.2 Å, which was consistent with a complete spin-state transition [10]. The octahedral distortion parameter (Σ), defined as the sum of the deviation from 90° of the 12 *cis*-N-Fe-N angles, is also a useful indicator of spin-state change. Typically, it is more regular in the LS state compared to the HS state [10]. However, for **Pd(benpy)** the degree of distortion was seen to increase over the HS to LS transition (HS = 8.16° and LS = 13.20°) and for **Pt(benpy)** there was no substantial change over the SCO transition. This is often an indicator of elastic frustration [46–50], which arises from a competition between interactions and the volumetric change over the spin-state transition. In most cases, this results in multi-stepped SCO transitions, but here the cooperative one-step transition suggests that there was a degree of elastic frustration. However, it was not great enough to overcome the preferred energetics of the bulk material to attain spin-state homogeneity. Other minor changes to the structures included a contraction in host–guest hydrogen bonds and ligand–ligand aromatic contacts, as well as a more planar Hofmann layer in the LS state, as evidenced by change in Fe–N≡C angles toward 180°. Furthermore, the benpy torsional twist for both analogues decreased substantially with HS to LS transition. A notable distinction between **Pd(benpy)** and **Pt(benpy)** was that the guest-accessible void space was larger in **Pt(benpy)**, but for both analogues the void volume decreased with HS to LS transition (**Pd(benpy)**: 5% (73 Å^3) to 3% (37 Å^3) and **Pt(benpy)**: 13% (195 Å^3) to 6% (94 Å^3)). This distinction may underlie the difference in long-range and short-range ordering of the ligand torsional twist in **Pd(benpy)** and **Pt(benpy)**. Indeed, on closer inspection, the ligand–water contacts were slightly longer in **Pt(benpy)** compared to **Pd(benpy)** in the HS state (O1·water: 2.137(2) and 2.006(2) Å for Pt and Pd, respectively). These contacts may play a role in the ligand ordering.

4. Conclusions

Two isostructural 2-D Hofmann-like framework materials, [Fe^{II}(benpy)$_2$M(CN)$_4$]·2H$_2$O (M = Pd, Pd), were prepared with a novel amide-functionalized ligand (benpy). Variation of M (Pd or Pt) resulted in subtle changes to structure and SCO properties, which we comprehensively assessed by variable temperature magnetic susceptibility, DSC, PXRD, Raman spectroscopy, and SCXRD analyses, and validated by calculation of the magnetic susceptibility data. The main change to the structure between the Pd and Pt analogues was the long- or short-range ordering of a torsional twist in the benpy ligand around the amide group. Structural analyses suggested this difference arose due to the relative steric bulk of Pd versus Pt and, hence, the packing efficiency and contacts that can be achieved. Both the Pd and Pt analogues showed abrupt and hysteretic one-step SCO transitions, with the Pt transition occurring at a slightly higher temperature. This difference was ascribed to a subtle variation HS stability arising from a combination of the greater electron density and size of Pt compared to Pd. Comparison of the HS and LS structures for both analogues revealed a degree of elastic frustration due to the increase in distortion over the HS to LS transition, opposed to the normal decrease, in particular, for the Pd analogue. Although, rather than a multi-step SCO transition, which can arise in such cases, a cooperative one-step character prevailed, indicating that the frustration effects were not large enough to overcome the preferential energetics of spin-state homogeneity. Consistent with this theory, we found that the net interactions were ferroelastic and that the transition was strongly cooperative. Overall, the inclusion of an amide-functionalized ligand into these SCO Hofmann frameworks successfully led to a range of through-bond (i.e., coordination bridges) and through-space (i.e., supramolecular contacts) pathways. Rather than functioning antagonistically, these functioned collectively to produce cooperative spin-state switching properties. This study supports the role of various structural modes of elastic coupling propagation in SCO materials.

Supplementary Materials: Additional synthetics, structural, and characterization data are available online at https://www.mdpi.com/2624-8549/3/1/26/s1.

Author Contributions: Conceptualization, M.A. and S.M.N.; investigation, X.O., M.A., L.X., A.T.B., C.H., K.A.Z. and Z.X.; calculations, B.J.P.; writing, X.O., C.J.K., B.J.P. and S.M.N. All authors have read and agreed to the published version of the manuscript.

Funding: This research was funded by The Australian Research Council (ARC; DP200100305).

Informed Consent Statement: Not applicable.

Data Availability Statement: Crystallographic data are freely available at the Cambridge Crystallographic Data Centre under 2049372-2049375.

Acknowledgments: We thank the ARC for Fellowships and Discovery Project funding. Access and use of the facilities of the Australian Synchrotron was supported by ANSTO. This research was facilitated by access to Sydney Analytical, a core research facility at the University of Sydney.

Conflicts of Interest: The authors declare no conflict of interest.

References

1. Kahn, O.; Kröber, J.; Jay, C. Spin Transition Molecular Materials for displays and data recording. *Adv. Mater.* **1992**, *4*, 718–728. [CrossRef]
2. Létard, J.-F.; Guionneau, P.; Goux-Capes, L. Towards Spin Crossover Applications. *Top. Curr. Chem.* **2004**, *235*, 221–249.
3. Gütlich, P.; Goodwin, H.A. Spin Crossover—An Overall Perspective. *Top. Curr. Chem.* **2004**, *233*, 1–47.
4. Gütlich, P.; Gaspar, A.B.; Garcia, Y. Spin state switching in iron coordination compounds. *Beilstein J. Org. Chem.* **2013**, *9*, 342–391. [CrossRef]
5. Halcrow, M.A. *Spin-Crossover Materials: Properties and Applications*; John Wiley & Sons: London, UK, 2013.
6. Kumar, K.S.; Ruben, M. Emerging trends in spin crossover (SCO) based functional materials and devices. *Coord. Chem. Rev.* **2017**, *346*, 176–205. [CrossRef]
7. Gütlich, P.; Hauser, A.; Spiering, H. Thermal and optical switching of Iron(II) Complexes. *Angew. Chem. Int. Ed.* **1994**, *33*, 2024. [CrossRef]
8. Real, J.A.; Gaspar, A.B.; Niel, V.; Muñoz, M.C. Communication between iron(II) building blocks in cooperative spin transi-tion phenomena. *Coord. Chem. Rev.* **2003**, *236*, 121–141. [CrossRef]
9. Real, J.A.; Gaspar, A.B.; Muñoz, M.C. Thermal, pressure and light switchable spin-crossover materials. *Dalton Trans.* **2005**, 2062–2079. [CrossRef]
10. Collet, E.; Guionneau, P. Structural analysis of spin-crossover materials: From molecules to materials. *C. R. Chim.* **2018**, *21*, 1133–1151. [CrossRef]
11. Spiering, H. Elastic Interaction in Spin-Crossover Compounds. *Top. Curr. Chem.* **2006**, 171–195. [CrossRef]
12. Nishino, M.; Boukheddaden, K.; Konishi, Y.; Miyashita, S. Simple Two-Dimensional Model for the Elastic Origin of Coopera-tivity among Spin States of Spin-Crossover Complexes. *Phys. Rev. Lett.* **2007**, *98*, 247203. [CrossRef]
13. Nishino, M.; Boukheddaden, K.; Miyashita, S. Molecular dynamics study of thermal expansion and compression in spin-crossover solids using a microscopic model of elastic interactions. *Phys. Rev. B* **2009**, *79*, 012409. [CrossRef]
14. Kitazawa, T.; Gomi, Y.; Takahashi, M.; Takeda, M.; Enomoto, M.; Miyazaki, A. Spin-crossover behaviour of the coordination polymer FeII(C5H5N)2NiII(CN)4. *J. Mater. Chem.* **1996**, *6*, 119. [CrossRef]
15. Niel, V.; Martinez-Agudo, J.M.; Munoz, M.C.; Gaspar, A.B.; Real, J.A. Cooperative spin crossover behavior in cya-nide-bridged Fe(II)−M(II) bimetallic 3D Hofmann-like networks (M= Ni, Pd, and Pt). *Inorg. Chem.* **2001**, *40*, 3838–3839. [CrossRef]
16. Muñoz, M.C.; Real, J.A. Thermo-, piezo-, photo- and chemo-switchable spin crossover iron(II)-metallocyanate based coordination polymers. *Coord. Chem. Rev.* **2011**, *255*, 2068–2093. [CrossRef]
17. Ni, Z.-P.; Liu, J.-L.; Hoque, N.; Liu, W.; Li, J.-Y.; Chen, Y.-C.; Tong, M.-L. Recent advances in guest effects on spin-crossover behavior in Hofmann-type metal-organic frameworks. *Coord. Chem. Rev.* **2017**, *335*, 28–43. [CrossRef]
18. Ohtani, R.; Hayami, S. Guest-Dependent Spin-Transition Behavior of Porous Coordination Polymers. *Chem. Eur. J.* **2017**, *23*, 2236–2248. [CrossRef] [PubMed]
19. Ohba, M.; Yoneda, K.; Agustí, G.; Muñoz, M.C.; Gaspar, A.B.; Real, J.A.; Yamasaki, M.; Ando, H.; Nakao, Y.; Sakaki, S.; et al. Bidirectional Chemo-Switching of Spin State in a Microporous Framework. *Angew. Chem. Int. Ed.* **2009**, *48*, 4767–4771. [CrossRef] [PubMed]
20. Southon, P.D.; Liu, L.; Fellows, E.A.; Price, D.J.; Halder, G.J.; Chapman, K.W.; Moubaraki, B.; Murray, K.S.; Létard, J.-F.; Kepert, C.J. Dynamic Interplay between Spin-Crossover and Host−Guest Function in a Nanoporous Metal−Organic Framework Material. *J. Am. Chem. Soc.* **2009**, *131*, 10998–11009. [CrossRef] [PubMed]
21. Klein, Y.M.; Sciortino, N.F.; Ragon, F.; Housecroft, C.E.; Kepert, C.J.; Neville, S.M. Spin crossover intermediate plateau stabilization in a flexible 2-D Hofmann-type coordination polymer. *Chem. Commun.* **2014**, *50*, 3838–3840. [CrossRef]
22. Liu, W.; Peng, Y.-Y.; Wu, S.-G.; Chen, Y.-C.; Hoque, N.; Ni, Z.-P.; Chen, X.-M.; Tong, M. Guest-Switchable Multi-Step Spin Transitions in an Amine-Functionalized Metal-Organic Framework. *Angew. Chem. Int. Ed.* **2017**, *56*, 14982–14986. [CrossRef]
23. Seredyuk, M.; Gaspar, A.B.; Ksenofontov, V.; Verdaguer, M.; Villain, F.; Gütlich, P. Thermal- and Light-Induced Spin Crossover in Novel 2D Fe(II) Metalorganic Frameworks {Fe(4-PhPy)2[MII(CN)x]y}·sH2O: Spectroscopic, Structural, and Mag-netic Study. *Inorg. Chem.* **2009**, *48*, 6130–6141. [CrossRef] [PubMed]
24. Liu, W.; Wang, L.; Su, Y.-J.; Chen, Y.-C.; Tucek, J.; Zboril, R.; Ni, Z.-P.; Tong, M.-L. Hysteretic Spin Crossover in Two-Dimensional (2D) Hofmann-Type Coordination Polymers. *Inorg. Chem.* **2015**, *54*, 8711–8716. [CrossRef] [PubMed]
25. Noveron, J.C.; Lah, M.S.; Del Sesto, R.E.; Arif, A.M.; Miller, J.S.; Stang, P.J. Engineering the Structure and Magnetic Properties of Crystalline Solids via the Metal-Directed Self-Assembly of a Versatile Molecular Building Unit. *J. Am. Chem. Soc.* **2002**, *124*, 6613–6625. [CrossRef] [PubMed]

26. Valverde-Muñoz, F.J.; Bartual-Murgui, C.; Piñeiro-López, L.; Muñoz, M.C.; Real, J.A. Influence of host-guest and host-host interactions on the spin-crossover of 3D Hofmann-type clathrates {FeII(pina)[MI(CN)2]2}·xMeOH (MI = Ag, Au). *Inorg. Chem.* **2019**, *58*, 10038–10046. [CrossRef]

27. Lan, W.; Valverde-Muñoz, F.J.; Dou, Y.; Hao, Y.; Muñoz, M.C.; Zhou, Z.; Liu, H.; Liu, Q.; Real, J.A.; Zhang, D. A thermal- and light-induced switchable one-dimensional rare loo-like spin crossover coordination polymer. *Dalton Trans.* **2019**, *48*, 17014. [CrossRef]

28. Hao, X.; Dou, T.; Cao, L.; Yang, L.; Liu, H.; Li, D.; Zhang, D.; Zhou, Z. Tuning of crystallization method and ligand confor-mation to give a mononuclear compound or two-dimensional SCO coordination polymer based on a new semi-rigid V-shaped bis-pyridyl bis-amide ligand. *Acta Cryst.* **2020**, *C76*, 412–418.

29. Mondal, D.J.; Roy, S.; Yadav, J.; Zeller, M.; Konar, S. Solvent-induced reversible spin-crossover in a 3D Hofmann-type coor-dination polymer and unusual enhancement of the lattice cooperativity at the desolvated state. *Inorg. Chem.* **2020**, *59*, 13024–13028. [CrossRef] [PubMed]

30. Ahmed, M.; Xie, Z.; Thoonen, S.; Hua, C.; Kepert, C.J.; Price, J.R.; Neville, S.M. A new spin crossover FeII coordination environment in a two-fold interpenetrated 3-D Hofmann-type framework material. *Chem. Commun.* **2021**, *57*, 85–88. [CrossRef]

31. Cowieson, N.P.; Arago, D.; Clift, M.; Ericsson, D.J.; Gee, C.; Harrop, S.J.; Mudie, N.; Panjikar, S.; Price, J.R.; Ribol-di-Tunnicliffe, A.; et al. MX1: A bending-magnet crystallography beamline serving both chemical and macromolecular crystallographic communities at the Australian Synchrotron. *J. Synch. Rad.* **2015**, *22*, 187–190. [CrossRef]

32. McPhillips, T.M.; McPhillips, S.E.; Chiu, H.-J.; Cohen, A.E.; Deacon, A.M.; Ellis, P.J.; Garman, E.; Gonzalez, A.; Sauter, N.K.; Phiza-ckerley, R.P. Blu-Ice and the Distributed Control System: Software for data acqui-sition and instrument control at macromolecular crystallography beamlines. *J. Synch. Rad.* **2002**, *9*, 401–406. [CrossRef] [PubMed]

33. Kabsch, W. XDS. *Acta Cryst. D* **2010**, *66*, 125–132. [CrossRef]

34. Sheldrick, G.M. SHELXT–Integrated space-group and crystal-structure determination. *Acta Cryst. Sect. A* **2015**, *71*, 3–8. [CrossRef] [PubMed]

35. Sheldrick, G.M. Crystal structure refinement with SHELXL. *Acta Cryst. Sect. C* **2015**, *71*, 3–8. [CrossRef] [PubMed]

36. Dolomanov, O.V.; Bourhis, L.J.; Gildea, R.J.; Howard, J.A.K.; Puschmann, H. OLEX2: A complete structure solution, refinement and analysis program. *J. Appl. Crystallogr.* **2009**, *42*, 339–341. [CrossRef]

37. Wallwork, K.S.; Kennedy, B.J.; Wang, D.C. The High Resolution Powder Diffraction Beamline for the Australian Synchrotron. In *Fourth Huntsville Gamma-Ray Burst Symposium*; AIP Publishing: New York, NY, USA, 2007; Volume 879, pp. 879–882.

38. Bergamaschi, A.; Cervellino, A.; DiNapoli, R.; Gozzo, F.; Henrich, B.; Johnson, I.; Kraft, P.; Mozzanica, A.; Schmitt, B.; Shi, X. The MYTHEN detector for X-ray powder diffraction experiments at the Swiss Light Source. *J. Synchrotron Radiat.* **2010**, *17*, 653–668. [CrossRef] [PubMed]

39. Jong, L.; Ruben, G.; Spear, K. *PDViPeR*; Synchrotron Light Source Australia Pty Ltd.: Melbourne, Australia, 2014.

40. Coelho, A. *TOPAS Version 4.1*; Bruker AXS GmbH: Karlsruhe, Germany, 2007.

41. Wajnflasz, J.; Pick, R. Transitions «Low spin»-«High spin» dans les complexes de Fe^{2+}. *J. Phys.* **1971**, *32*, 90–91. [CrossRef]

42. Pavlik, J.; Boča, R. Established Static Models of Spin Crossover. *Eur. J. Inorg. Chem.* **2013**, *5–6*, 697–709. [CrossRef]

43. Ruzzi, G.; Cruddas, J.; McKenzie, R.H.; Powell, B.J. Equivalence of elastic and Ising models for spin crossover materials. *arXiv* **2020**, arXiv:2008.08738.

44. Ragon, F.; Yaksi, K.; Sciortino, N.F.; Chastanet, G.; Létard, J.-F.; D'Alessandro, D.M.; Kepert, C.J.; Neville, S.M. Thermal Spin Crossover Behaviour of Two-Dimensional Hofmann-Type Coordination Polymers Incorporating Photoactive Ligands. *Aust. J. Chem.* **2014**, *67*, 1563–1573. [CrossRef]

45. Roubeau, O.; Castro, M.; Burriel, R.; Haasnoot, J.G.; Reedijk, J. Calorimetric Investigation of Triazole-Bridged Fe(II) Spin-Crossover One-Dimensional Materials: Measuring the Cooperativity. *J. Phys. Chem. B* **2011**, *115*, 3003–3012. [CrossRef] [PubMed]

46. Molnár, G.; Mikolasek, M.; Ridier, K.; Fahs, A.; Nicolazzi, W.; Bousseksou, A. Molecular Spin Crossover Materials: Review of the Lattice Dynamical Properties. *Annalen Physik* **2019**, *531*, 1900076. [CrossRef]

47. Kulmaczewski, R.; Olguín, J.; Kitchen, J.A.; Feltham, H.L.C.; Jameson, G.N.L.; Tallon, J.L.; Brooker, S. Remarkable Scan Rate Dependence for a Highly Constrained Dinuclear Iron(II) Spin Crossover Complex with a Wide Thermal Hysteresis Loop. *J. Am. Chem. Soc.* **2014**, *136*, 878–881. [CrossRef] [PubMed]

48. Paez-Espejo, M.; Sy, M.; Boukheddaden, K. Elastic Frustration Causing Two-Step and Multistep Transitions in Spin-Crossover Solids: Emergence of Complex Antiferroelastic Structures. *J. Am. Chem. Soc.* **2016**, *138*, 3202–3210. [CrossRef]

49. Cruddas, J.; Powell, B.J. Structure-property relationships and the mechanism of multistep transitions in spin crossover mate-rials and frameworks. *Inorg. Chem. Front.* **2020**, *7*, 4424–4437. [CrossRef]

50. Cruddas, J.; Powell, B.J. Spin-State Ice in Elastically Frustrated Spin-Crossover Materials. *J. Am. Chem. Soc.* **2019**, *141*, 19790–19799. [CrossRef] [PubMed]

chemistry

Communication

Electrochemical Switching of First-Generation Donor-Acceptor Stenhouse Adducts (DASAs): An Alternative Stimulus for Triene Cyclisation

Nicholas D. Shepherd [1], Harrison S. Moore [1], Jonathon E. Beves [2] and Deanna M. D'Alessandro [1],*

[1]   School of Chemistry, The University of Sydney, Sydney, NSW 2006, Australia; nshe0025@uni.sydney.edu.au (N.D.S.); hmoo4552@uni.sydney.edu.au (H.S.M.)

[2]   School of Chemistry, The University of New South Wales, Sydney, NSW 2052, Australia; j.beves@unsw.edu.au

*   Correspondence: deanna.dalessandro@sydney.edu.au

Abstract: Donor-acceptor Stenhouse adducts (DASAs) are a photo-switch class that undergoes triene cyclisation in response to visible light. Herein, electrochemical oxidation is demonstrated as an effective alternative stimulus for the triene cyclisation commonly associated with photo-switching.

Keywords: photoswitching; electrochemistry; stimuli responsive

Citation: Shepherd, N.D.; Moore, H.S.; Beves, J.E.; D'Alessandro, D.M. Electrochemical Switching of First-Generation Donor-Acceptor Stenhouse Adducts (DASAs): An Alternative Stimulus for Triene Cyclisation. *Chemistry* **2021**, *3*, 728–733. https://doi.org/10.3390/chemistry3030051

Academic Editor: Catherine Housecroft

Received: 23 February 2021
Accepted: 28 June 2021
Published: 7 July 2021

Organic photo-switches have gained significant attention due to the contrasting properties of their initial and metastable states [1]. Accessing these states via controlled light irradiation allows efficient modulation of their properties which are relevant to applications including chemical sensing [2,3], drug delivery [2,3], heterogenous catalysis [4], molecular machines [5,6], and data storage [7].

Donor-acceptor Stenhouse adducts (DASAs) [8–12] are a class of photo-switches that are visible light-responsive, thus minimising photochemical degradation [13]. DASAs are comprised of donor (provided by the nucleophile used during synthesis) and acceptor (provided by a carbon acid, usually Meldrum's or 1,3-dimethyl barbituric acids) components linked by a central triene chain (Figure 1c–h) [8–10]. Typically, excitation with white light results in triene chain's cyclisation to yield a cyclopentenone zwitterionic species (Figure 2d) [8–10]. Further appeal for this photo-switch class is based on the relatively simple two-step synthetic procedure [10]. Syntheses detailed in the literature involve the preparation of an activated furan precursor through Knoevenagel condensation between a carbon acid and furan [8–10]. The respective DASA is completed through nucleophilic addition of the respective donor component [8–10]. In addition to the aforementioned applications [14–16], DASAs have also been used in preparing orthogonal photo-switches [17] and photolithography [18,19].

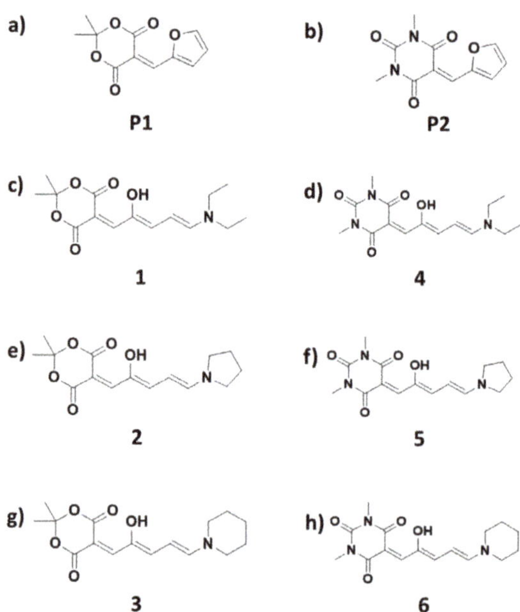

Figure 1. First-generation DASA compounds were prepared from precursors **P1** and **P2** (**a**,**b**, respectively). DASAs subject to voltammetry and UV-vis SEC analyses are designated as **1–6** (**c–h**).

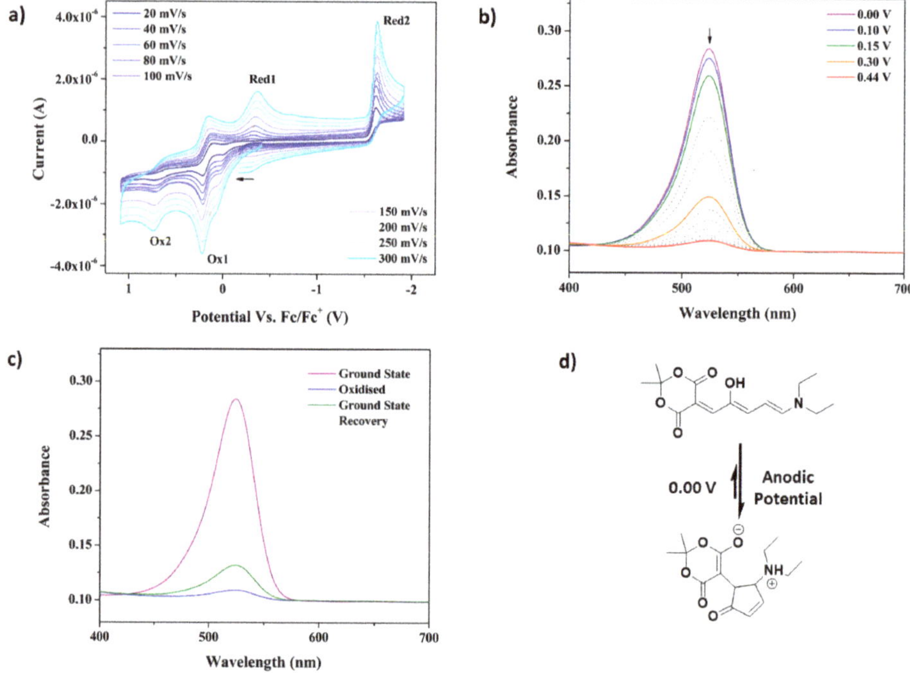

Figure 2. CV recorded at 20–300 mV/s scan rates (**a**); UV-vis SEC spectra from 400–700 nm (14,250–25,750 cm^{-1}) (**b**); UV-vis spectra following SEC indicating recovery of the linear isomer in the same range (arrow shows the direction of change) (**c**); and proposed quasi-reversible cyclisation in response to an anodic redox potential (**d**) for compound **1**. Electrochemical experiments were conducted in 0.1 M TBAPF$_6$ in MeCN.

While most DASA studies have focused on photo-switching behaviour, acidic and basic conditions have been shown to stabilise the metastable state [15,20]. Furthermore, other organic photo-switch classes have responded to alternative stimuli such as heating [21,22], pH [21,22], and redox potential [21–26]. In particular, select azobenzene [25,26], chromene [23,24,27,28], and diarylethene [29,30] photo-switches can undergo their photo-switching behaviour in response to an applied redox potential.

Herein, a series of first-generation DASA compounds (denoted **1–6** in Figure 1) were subjected to voltammetric and UV-vis spectroelectrochemical (UV-vis SEC) studies to elucidate their redox activity and establish the potential for electrochemical switching. These inquiries were expected to demonstrate a link with photo-controlled DASA cyclisation. DASAs included structures prepared from P1 and P2 precursors, as shown in Figure 1.

DASA compounds and their precursors were prepared in accordance with procedures from the literature [8,9]. Successful syntheses were confirmed using ^{1}H NMR (ESI, Figures S1–S8), ESI-MS (ESI, Figures S9–S16), and ATR-FTIR (ESI, Figures S17–S24) spectroscopies. UV-vis spectra (ESI, Figures S25–S30) in MeCN indicated a transition in the visible region for 1–6. This signal was consistent with the S_0–S_1 transition associated with the respective DASAs' triene chain [8–10]. λ_{max} was considerably higher for compounds prepared from 1,3-dimethyl barbituric acid instead of Meldrum's acid (**1**: 524 nm, **2**: 526 nm, and **3**: 523 nm, compared to **4**: 549 nm, **5**: 552 nm, and **6**: 548 nm). As expected, the peak intensity decreased following irradiation with white LEDs (4 Watts) as a result of expected photo-switching behaviour [8–10]. Compound 2 was the only DASA that exhibited near quantitative conversion in MeCN, based on the S_0–S_1 transition intensity reducing by ca. 98%. Other DASA species underwent incomplete photo-switching in MeCN (ca. 68–87%) based on the partial disappearance of the S_0–S_1 band. Spectral changes following photo-irradiation in MeCN were irreversible.

Cyclic and square wave voltammetry (CV and SQW, respectively) for **1–6** (Figure 2a and ESI, Figures S31–S36) indicated two quasi-reversible oxidation processes of interest. These were designated as Ox1 and Ox2 with average $E_{1/2}$ values of ca. 0.117 and 0.680 V (ESI, Table S1), respectively. Oxidation processes were consistent with the 4π-eletrocyclisation step where the alcohol functional group was oxidised as part of the photo-switching mechanism [31]. It is expected that the electrolyte would stabilise the zwitterionic state as observed in redox active spiropyran functional groups [24]. $E_{1/2}$ values for Ox1 and Ox2 were higher in DASAs **1–3** compared to **4–6**. This was attributed to the carbon acid (either Meldrum's or 1,3-dimethylbarbituric acid), where a reduced electron withdrawing effect in **4–6** promoted electron transfer [8,10,32]. The exceptions were compounds **3** and **6**, where the latter had a higher $E_{1/2}$ compared to the former. Additionally, $E_{1/2}$ values for Ox1 varied based on the amine derivative used. Potentials increased for DASAs derived from piperidine (**3**: 0.118, **6**: 0.0355 V), pyrrolidine (**2**: 0.166 V, **5**: 0.0840 V), and Et$_2$NH (**1**: 0.190 V, **4**: 0.111 V) nucleophiles. The electron-donating effects of the donor component of **1–6** was expected to reduce $E_{1/2}$ based on documented nucleophile effects on DASA photo-switching wavelengths [8,10,32]. Scan rate dependence plots for Ox1 and Ox2 indicated that these processes were affected by diffusion (ESI, Figures S39–S44). Red2 in **1–6** was consistent with redox processes associated with **P1** and **P2** precursors (ESI, Figures S37–S38 and Table S1) and were not attributed to the triene cyclisation.

UV-vis SEC measurements linked electrochemical oxidation processes to the triene cyclisation. Applying an anodic potential resulted in a decrease in the S_0–S_1 band intensity in all DASA compounds, with representative data for **1** provided in Figure 2b (see also ESI, Figures S45a–S49a). The trends in $E_{1/2}$ values calculated for **1–6** from voltammetric experiments were consistent with the potentials required for electrochemical oxidation during SEC experiments (ESI, Table S2).

The spectral behaviour noted during UV-vis SEC experiments indicated complete conversion to the cyclised state following application of a potential, compared to typically incomplete switching under photo-irradiation. The reduction in S_0–S_1 band intensity in response to an anodic potential also indicated that it was a more effective stimulus than photo-irradiation. Most DASA compounds studied herein showed a more substantial decrease in band intensity during SEC experiments. Photo-switching was likely, in part, hindered by the DASA concentration, as outlined by Lui et al. [33] for DASAs in chloroform and toluene solvents. This would be compounded by MeCN being a polar solvent that has inhibited switching behaviour in past studies [11]. In this respect, an anodic potential was able to overcome these impediments, as both switching experiments in the current work were conducted at approximately the same concentration. Additionally, the charge carrier effect of the electrolyte likely further enhances the efficiency of DASA cyclisation via electrochemical oxidation. The main exception was compound **6**, where photo-irradiation (ca. 87% conversion) was more effective than electrochemical oxidation (ca. 77% conversion) at inducing cyclisation. Additionally, for compound **2**, both stimuli resulted in nearly quantitative yield of the cyclised state. It is likely that this is in part attributable to the electron donating nature of the respective nucleophile [34], although this will require further investigation for confirmation.

Removing the potential resulted in a partial return of the S_0–S_1 band (Figure 2c and ESI, Figure S45b–Figure S49b), in keeping with quasi-reversible behaviour observed in CV data. The reappearance of the S_0-S_1 band also suggests that the Red1 peak detected in CV data for **1–6** (Figure 1a and ESI, Figures S31–S36) was associated with a reduction in the linear isomer. Note the discrepancy between voltammetry and SEC voltages is due to the Fc/Fc$^+$ internal reference in CV experiments. Recovery ranged from 4–47% recovery (ESI, Table S2) across the series, with 4 and 3 representing the lower and upper limits, respectively. It was further noted that the nucleophile appeared to affect recovery, where Et$_2$NH < pyrrolidine < piperidine (ESI, Table S2). Trends in recovery were also attributed to the electron withdrawing effect of the carbon acid as noted previously [8,10,32]. It was expected that the electron withdrawing behaviour of the acceptor component would adversely affect reversibility. Additionally, the diminished fatigue resistance compared to that observed in photo-switching studies [9] is also attributable to the harsh nature associated with an electrochemical potential.

The SEC data confirmed that application of an anodic potential resulted in the conversion of the linear isomer to the cyclised zwitterion (Figure 2d). Similar behaviour has also been observed in spiropyran-based materials where the electrolyte stabilises the zwitterionic merocyanine species [24]. The immediate decrease in S_0–S_1 band intensity confirmed that the Ox1 processes in **1–6** were associated with the 4π-electrocyclisation step, precluding a multistep process as observed during photo-switching [31,35]. The initial conversion demonstrated an alternative stimulus for DASA switching. The poor recovery of the linear configuration following electrochemical oxidation indicates exposure to this stimulus would impede long-term switching function during practical applications. Therefore, compounds and materials with improved fatigue resistance are necessary to exploit this switching behaviour in a functional context.

In summary, selected first-generation DASAs exhibited the cyclisation traditionally associated with their photo-switching behaviour in response to an anodic potential. Furthermore, electrochemical oxidation was demonstrated to be more effective than photo-irradiation for cyclising most of the DASAs studied in the current work. Despite this, the alternative stimulus was only partially effective, with poor recovery of the linear triene isomer. This work points to the potential for electrochemical switching of DASA-incorporated systems of interest for electrochemical sensing, electrocatalysis, and electrochemically driven molecular machines. These studies are currently underway in our laboratory.

Supplementary Materials: The following are available online at https://www.mdpi.com/article/10.3390/chemistry3030051/s1, Figures S1–S8: NMR data, Figures S9–S16: ESI-MS data, Figures S17–S24: ATR-FTIR data, Figures S25–S30: UV-Vis spectra from 400–700 nm, Figures S31–S38: CV data, Figures S39-S44: Scan rate dependence data, Figures S45–S49: Spectroelectrochemical data, Tables S1 and S2: Redox processes for DASAs and precursors.

Author Contributions: N.D.S. completed the experimental work and compiled the manuscript. H.S.M. and J.E.B. provided collaborative support with the measurement and interpretation of data. D.M.D. supervised the overall project. All authors have read and agreed to the published version of the manuscript.

Funding: Australian Research Council for funding (FT170100283, FT170100094) as well as the University of Sydney (USyd) for support.

Data Availability Statement: Data available on request from the corresponding author.

Acknowledgments: The authors would like to thank the Australian Research Council for funding (FT170100283, FT170100094) as well as the University of Sydney (USyd) for support. N.D.S. gratefully acknowledges the School of Chemistry, USyd for support provided by the 'Postgraduate Scholarship for Photoactive Metal-Organic Frameworks'.

Conflicts of Interest: The authors declare no conflict of interest.

References

1. Natali, M.; Giordani, S. Molecular Switches as Photocontrollable "Smart" Receptors. *Chem. Soc. Rev.* **2012**, *41*, 4010–4029. [CrossRef]
2. Tylkowski, B.; Trojanowska, A.; Marturano, V.; Nowak, M.; Marciniak, L.; Giamberini, M.; Ambrogi, V.; Cerruti, P. Power of Light—Functional Complexes Based on Azobenzene Molecules. *Coord. Chem. Rev.* **2017**, *351*, 205–217. [CrossRef]
3. Avella-Oliver, M.; Morais, S.; Puchades, R.; Maquieira, Á. Towards Photochromic and Thermochromic Biosensing. *TrAC-Trends Anal. Chem.* **2016**, *79*, 37–45. [CrossRef]
4. Gong, L.L.; Yao, W.T.; Liu, Z.Q.; Zheng, A.M.; Li, J.Q.; Feng, X.F.; Ma, L.F.; Yan, C.S.; Luo, M.B.; Luo, F. Photoswitching Storage of Guest Molecules in Metal-Organic Framework for Photoswitchable Catalysis: Exceptional Product, Ultrahigh Photocontrol, and Photomodulated Size Selectivity. *J. Mater. Chem. A* **2017**, *5*, 7961–7967. [CrossRef]
5. Terao, F.; Morimoto, M.; Irie, M. Light-Driven Molecular-Crystal Actuators: Rapid and Reversible Bending of Rodlike Mixed Crystals of Diarylethene Derivatives. *Angew. Chemie-Int. Ed.* **2012**, *51*, 901–904. [CrossRef]
6. Feringa, B.L. In Control of Motion: From Molecular Switches to Molecular Motors. *Acc. Chem. Res.* **2001**, *34*, 504–513. [CrossRef]
7. Andréasson, J.; Pischel, U. Storage and Processing of Information Using Molecules: The All-Photonic Approach with Simple and Multi-Photochromic Switches. *Isr. J. Chem.* **2013**, *53*, 236–246. [CrossRef]
8. Helmy, S.; Oh, S.; Leibfarth, F.A.; Hawker, C.J.; Read De Alaniz, J. Design and Synthesis of Donor-Acceptor Stenhouse Adducts: A Visible Light Photoswitch Derived from Furfural. *J. Org. Chem.* **2014**, *79*, 11316–11329. [CrossRef] [PubMed]
9. Helmy, S.; Leibfarth, F.; Oh, S.; Poelma, J.; Hawker, C.J.; Read De Alaniz, J. Photoswitching Using Visible Light: A New Class of Organic Photochromic Molecules. *J. Am. Chem. Soc.* **2014**, *136*, 8169–8172. [CrossRef]
10. Lerch, M.M.; Szymański, W.; Feringa, B.L. The (Photo)Chemistry of Stenhouse Photoswitches: Guiding Principles and System Design. *Chem. Soc. Rev.* **2018**, *47*, 1910–1937. [CrossRef] [PubMed]
11. Mallo, N.; Foley, E.D.; Iranmanesh, H.; Kennedy, A.D.W.; Luis, E.T.; Ho, J.; Harper, J.B.; Beves, J.E. Structure-Function Relationships of Donor-Acceptor Stenhouse Adduct Photochromic Switches. *Chem. Sci.* **2018**, *9*, 8242–8252. [CrossRef]
12. Mallo, N.; Tron, A.; Andréasson, J.; Harper, J.B.; Jacob, L.S.D.; McClenaghan, N.D.; Jonusauskas, G.; Beves, J.E. Hydrogen-Bonding Donor-Acceptor Stenhouse Adducts. *ChemPhotoChem* **2020**, *4*, 407–412. [CrossRef]
13. Bléger, D.; Hecht, S. Visible-Light-Activated Molecular Switches. *Angew. Chemie-Int. Ed.* **2015**, *54*, 11338–11349. [CrossRef]
14. Poelma, S.O.; Oh, S.S.; Helmy, S.; Knight, A.S.; Burnett, G.L.; Soh, H.T.; Hawker, C.J.; Read De Alaniz, J. Controlled Drug Release to Cancer Cells from Modular One-Photon Visible Light-Responsive Micellar System. *Chem. Commun.* **2016**, *52*, 10525–10528. [CrossRef] [PubMed]
15. Zhong, D.; Cao, Z.; Wu, B.; Zhang, Q.; Wang, G. Polymer Dots of DASA-Functionalized Polyethyleneimine: Synthesis, Visible Light/PH Responsiveness, and Their Applications as Chemosensors. *Sensors Actuators, B Chem.* **2018**, *254*, 385–392. [CrossRef]
16. Balamurugan, A.; Lee, H. A Visible Light Responsive On–Off Polymeric Photoswitch for the Colorimetric Detection of Nerve Agent Mimics in Solution and in the Vapor Phase. *Macromolecules* **2016**, *49*, 2568–2574. [CrossRef]
17. Lerch, M.M.; Hansen, M.J.; Velema, W.A.; Szymanski, W.; Feringa, B.L. Orthogonal Photoswitching in a Multifunctional Molecular System. *Nat. Commun.* **2016**, *7*, 1–10. [CrossRef] [PubMed]
18. Sinawang, G.; Wu, B.; Wang, J.; Li, S.; He, Y. Polystyrene Based Visible Light Responsive Polymer with Donor–Acceptor Stenhouse Adduct Pendants. *Macromol. Chem. Phys.* **2016**, *217*, 2409–2414. [CrossRef]
19. Singh, S.; Friedel, K.; Himmerlich, M.; Lei, Y.; Schlingloff, G.; Schober, A. Spatiotemporal Photopatterning on Polycarbonate Surface through Visible Light Responsive Polymer Bound DASA Compounds. *ACS Macro Lett.* **2015**, *4*, 1273–1277. [CrossRef]
20. Safar, P.; Povazanec, F.; Pronayova, N.; Baran, P.; Kickelbick, G.; Kosizek, J.; Breza, M. No Title. *Collect. Czech. Chem. Commun.* **2000**, *65*, 1911–1938.
21. Klajn, R. Spiropyran-Based Dynamic Materials. *Chem. Soc. Rev.* **2014**, *43*, 148–184. [CrossRef] [PubMed]
22. Lukyanov, B.S.; Lukyanova, M.B. Spiropyrans: Synthesis, Properties, and Application. (Review)*. *Chem. Heterocycl. Compd.* **2005**, *41*, 281–311. [CrossRef]

23. Saad, A.; Oms, O.; Marrot, J.; Dolbecq, A.; Hakouk, K.; El Bekkachi, H.; Jobic, S.; Deniard, P.; Dessapt, R.; Garrot, D.; et al. Design and Optical Investigations of a Spironaphthoxazine/Polyoxometalate/ Spiropyran Triad. *J. Mater. Chem. C* **2014**, *2*, 4748–4758. [CrossRef]

24. Wagner, K.; Robert, B.; Zanoni, M.; Gambhir, S.; Dennany, L.; Breukers, R.; Higgins, M.; Wagner, P.; Diamond, D.; Wallace, G.G.; et al. A Multiswitchable Poly(Terthiophene) Bearing a Spiropyran Functionality: Understanding Photo- and Electrochemical Control. *J. Am. Chem. Soc.* **2011**, *133*, 5453–5462. [CrossRef]

25. Goulet-Hanssens, A.; Utecht, M.; Mutruc, D.; Titov, E.; Schwarz, J.; Grubert, L.; Bléger, D.; Saalfrank, P.; Hecht, S. Electrocatalytic Z → E Isomerization of Azobenzenes. *J. Am. Chem. Soc.* **2017**, *139*, 335–341. [CrossRef]

26. Goulet-Hanssens, A.; Rietze, C.; Titov, E.; Abdullahu, L.; Grubert, L.; Saalfrank, P.; Hecht, S. Hole Catalysis as a General Mechanism for Efficient and Wavelength-Independent Z → E Azobenzene Isomerization. *Chem* **2018**, *4*, 1740–1755. [CrossRef]

27. Li, Z.; Zhou, Y.; Peng, L.; Yan, D.; Wei, M. A Switchable Electrochromism and Electrochemiluminescence Bifunctional Sensor Based on the Electro-Triggered Isomerization of Spiropyran/Layered Double Hydroxides. *Chem. Commun.* **2017**, *53*, 8862–8865. [CrossRef]

28. Garling, T.; Tong, Y.; Darwish, T.A.; Wolf, M.; Kramer Campen, R. The Influence of Surface Potential on the Optical Switching of Spiropyran Self Assembled Monolayers. *J. Phys. Condens. Matter* **2017**, *29*, 414002–414010. [CrossRef]

29. Areephong, J.; Logtenberg, H.; Browne, W.R.; Feringa, B.L. Symmetric Six-Fold Arrays of Photo- and Electrochromic Dithienylethene Switches. *Org. Lett.* **2010**, *12*, 2132–2135. [CrossRef]

30. Peters, A.; Branda, N.R. Electrochromism in Photochromic Dithienylcyclopentenes. *J. Am. Chem. Soc.* **2003**, *125*, 3404–3405. [CrossRef]

31. Lerch, M.M.; Wezenberg, S.J.; Szymanski, W.; Feringa, B.L. Unraveling the Photoswitching Mechanism in Donor-Acceptor Stenhouse Adducts. *J. Am. Chem. Soc.* **2016**, *138*, 6344–6347. [CrossRef]

32. Hemmer, J.R.; Poelma, S.O.; Treat, N.; Page, Z.A.; Dolinski, N.D.; Diaz, Y.J.; Tomlinson, W.; Clark, K.D.; Hooper, J.P.; Hawker, C.; et al. Tunable Visible and Near Infrared Photoswitches. *J. Am. Chem. Soc.* **2016**, *138*, 13960–13966. [CrossRef]

33. Lui, B.F.; Tierce, N.T.; Tong, F.; Sroda, M.M.; Lu, H.; Read de Alaniz, J.; Bardeen, C.J. Unusual Concentration Dependence of the Photoisomerization Reaction in Donor–Acceptor Stenhouse Adducts. *Photochem. Photobiol. Sci.* **2019**, *18*, 1587–1595. [CrossRef] [PubMed]

34. Liu, H.; Pu, S.; Liu, G.; Chen, B. Photochromism of Asymmetrical Diarylethenes with a Pyrimidine Unit: Synthesis and Substituent Effects. *Dye. Pigment.* **2014**, *102*, 159–168. [CrossRef]

35. Bull, J.N.; Carrascosa, E.; Mallo, N.; Scholz, M.S.; Da Silva, G.; Beves, J.E.; Bieske, E.J. Photoswitching an Isolated Donor-Acceptor Stenhouse Adduct. *J. Phys. Chem. Lett.* **2018**, *9*, 665–671. [CrossRef] [PubMed]

MDPI

St. Alban-Anlage 66

4052 Basel

Switzerland

Tel. +41 61 683 77 34

Fax +41 61 302 89 18

www.mdpi.com

Chemistry Editorial Office

E-mail: chemistry@mdpi.com

www.mdpi.com/journal/chemistry